Preface For the *Old School Calculus* Series

If we turn to the problems to which the calculus owes its origin, we find that not merely, not even primarily, geometry, but every other branch of mathematical physics—astronomy, mechanics, hydrodynamics, elasticity, gravitation, and later electricity and magnetism—in its fundamental concepts and basal laws contributed to its development and that the new science became the direct product of these influences.

— William Osgood

The birth of calculus in the 18th century may constitute the single most significant intellectual advance in the history of mankind since the development of classical geometry. Indeed, with the possible exceptions of fire, agriculture, spoken languages, writing and the counting numbers, it may be the single most important creation of homo sapiens in terms of the sheer magnitude of its consequences.

The quote by the old Harvard University master above-who knew quite a bit about both the content and teaching of calculus- underscores both the importance of the subject and its deep connection to the study of the empirical world. Before calculus, the most basic calculations of the properties of natural phenomena-such as the area under curves, volumes under surfaces and the velocity of moving objects- required significant ingenuity utilizing exhaustively detailed geometric and algebraic computations. For example, Galileo attempted to compute the area under the curve generated by a rolling circle in 1599 by an elaborate procedure of tracing and cutting out the curve and its' generating circle in sheet metal and weighing them to produce proportional areas. Interestingly, it gave erroneous results. The problem was officially solved nearly 40 years later by Galileo's students Torricelli and Viviana, using the tangent methods of analytic geometry that anteceded the differential calculus. (It should be noted that the correct area had been produced prior in unpublished works by several others.)

Of course, using the methods of calculus, this area can be very easily calculated now directly from the formula for the cycloid. Indeed, without the calculus, so many ideas in both the physical and social sciences we take as obvious today-such as work, acceleration, compound interest and temperature-were incapable of being defined precisely, let alone determined with any accuracy. For that reason alone, the importance of at least an intuitive understanding of calculus by most educated people is of paramount importance.

A common comment by mathematicians is that every standard introductory calculus textbook is a version of the 18th century calculus textbooks by Leanord Euler, *Institutiones calculi differentialis* and *Institutionum calculi integralis* . While this statement is ludicrous if taken literally, it's intent is quite truthful, namely that the basic substance of a calculus class for a general audience hasn't changed dramatically in topic selection or presentation since the great German master wrote his texts over 2 centuries ago. A basic calculus course has traditionally focused on the applications of the subject to geometry and physical/social sciences rather than the rigorous theory, which is now usually reserved for later courses in analysis. Since the firm foundation of calculus is only 120 years old, only the teaching of the applied aspects of the subject have existed throughout its entire existence.

After World War II with the birth of the Space Program, there was a Renaissance in America of mathematical and scientific training at universities and high schools, leading to a dramatic upgrading of undergraduate programs in mathematics and the hard sciences. The college calculus courses in those days were introductions to modern mathematics for beginners, brimming with proofs of important theorems, careful definitions and inequality computations as well as numerous applications to the sciences, primarily mechanics. This was tragically short lived. Over the last 40 years, the training of students at these levels has degraded gradually but steadily, particularly in elementary calculus courses.

Today's calculus courses are-at most universities, particularly American ones- a fading shadow of what they once were. Once, not so long ago, calculus was a true mathematics course-aimed at serious students who, while not necessarily going on to careers in mathematics, nevertheless, were eager to pit themselves against this significantly difficult course and learn some important skills they would use the rest of their lives. Indeed, many colleges in the mid-20th century onwards imposed a requirement of at least a semester of calculus upon pre-med and business students to act as a sieve to limit the number of applicants to the professional schools.

(Alright, this is not a "fact" in the sense I can provide a ton of evidence that this was the intent of the requirement. But this has been testified to by many in academia, far too many to quote here. I suggest going to your favorite search engine and begin finding testimonies.)

My point is that the calculus courses of yesteryear, unlike today, required actual mathematical skills that most of today's students do not possess. When these courses were standard calculus classes, average entering freshmen at most American universities had solid backgrounds in Euclidean geometry, algebra and trigonometry from high school and they were expected to be able to use this background with minimal review. In fact, not only do they not have such skills entering from high school, but they are actively discouraged from developing in their formative years, when it would be easiest for them to acquire them.

The big advantage of the rigorous calculus texts of the past was that it was almost impossible for such students to con their way to a good grade-the fact that rigorous mathematics was an **essential** part of the structure of the course ensured they actually had to learn something to do reasonably well. And the course acted to ensure that students with impure motives who didn't even try **didn't** get good grades.

I remember as a premed sitting around with a number of students taking calculus using Stewart and the discussion of the exam was like they were talking about a football game and how they were going to "beat" the exam. They came up with codes, mnemonics,word games-not a single theorem or concept or proof. I made the idiotic mistake of asking if anyone actually learned the material and the whole table erupted with laughter. The President of the Student Medical Association smiled at me like The Grinch.

"Winning is about APPEARING to know what you're doing,not actually doing it. Don't worry- you can always work taking out the trash in my office on 5th avenue."

Our society rewards this kind of behavior. Why? Because letting these monsters use Stewart and get their A's without learning anything is good for business, that's why. The university gets to pack the classes with 200 **paying** students by making this a required course, the students get their A's which the college can use to improve its' ranking standing so that administrators get promoted for making so much money and helping public relations and off they go to Ivy League medical schools thinking urea is made in the kidney-and worse, not giving a shit.

And 5 years later they're killing and crippling patients left and right and being acquitted at malpractice trials because the only one in the room who's a better liar then they are is the son of a bitch defending them.

Today, pre-med, pre-law and business students encourage each other to take calculus as an "easy A"-and *not* because they're mathematical geniuses. It's a consequence of the purely pragmatic, plug and chug nature of most university calculus courses-which of course lends itself easily to having programmable calculators and IPhones doing all the work. Worse, not only do students not have to do any real thinking in these courses, they're *encouraged* not to because the final grade is all that matters to both students and universities.

In other words, if you're teaching basic calculus given in the mathematics department of a university and your students leave the course thinking it was an "easy A"-*you're doing it wrong*.

Many mathematicians and scientists in other fields of a mathematical bent often mourn this state of affairs. They often complain about the guts of calculus being removed from present day textbooks in order to make the book as easy and accessible as possible for the maximum number of students, many of whom don't give a damn about the theory of calculus. They perceive most students that come into their calculus courses as not only poorly prepared, but completely indifferent to why calculus is important or why it works. Poor preparation is an inconvenience and a headache for both teacher and student, but if the student cares and is willing to work hard, this is not an insurmountable difficulty. But if the students see calculus as an annoying and meaningless hurdle they have to scam their way through before they can go out into the Real World and foreclose on old people's homes for fun-well, there's no book or course in the world that's going to make teaching those students anything but a chore.

I have great sympathy for the pessimism of most practicing mathematicians in regards to the subject. I've discussed my perceptions of the current state of mathematical academia in the mission statement of my company at the main website for our books. What I said there, I think, bears paraphrasing here to encapsulate the current status of calculus courses and textbooks and provide context for the myriad of calculus textbooks I'm preparing for publication under this series banner in addition to the one you're holding in your hand.

To me, there are 2 major factors ruining calculus today for students and professors alike at most universities: The first is the perceived academic psychology of students taking today's calculus courses and the second is the financial incentivizing of the resulting structure. These factors are by no means independent of each other-indeed, focusing on these 2 factors vastly oversimplifies the situation. But since these at the 2 factors that are most determinate in the direct planning of calculus courses each semester by faculty and staff, we'll focus on these. Since they are closely intertwined, we need to analyze them together.

We begin with a quote from one of my favorite sources:

The two biggest obstacles to the success of the Moore method (or, for that matter, of teaching of any kind) are students who don't want to be there and students who want to be somewhere else. The two are not the same thing; let me explain. By students who don't want to be there, I refer to required courses. If a student comes to me and asks my help to learn something that I already know, I am overjoyed. I am sure I can teach it to him, whatever it is, and I expect the process to be pleasant for both of us. If, however, he comes to me and says "I don't really want to know this stuff, but it's required that I get a C in it before I can go out and make a lot of money", then I'm unhappy………... I dream of the ideal university, full of students who are full of intellectual curiosity. The subset of those among them who take a mathematics course do so because they want to know mathematics. They may be future doctors or chemists or executives in a shirt factory, but, for whatever reason, they want to find out what this mathematics stuff is about, and they come to me free willing and ask me to teach it to them. Oh, joy!-Paul Halmos, **I Want To Be A Mathematician**

Ok, there's a lot to unpack here that reflects directly on the teaching of calculus today compared with those of yesteryear that this book represents. To Halmos then, there are essentially 3 kinds of students in any university mathematics class:

- # 1) Geeks who absolutely live for understanding mathematics and its related subjects.
- # 2) Students in other disciplines who, while not mathematics or physical science students, have the same curiosity and passion about learning in general. These students are willing to work hard and try and learn something new, especially if you can convince them it'll be helpful in whatever field they chose.
- # 3) Sociopathic scam-artists who break into the professors' office to get the final exam the night before, program the entire textbook in code into their calculators and know a hundred other ways to cheat. .

Clearly, in the quote above, Halmos is referring to students of types (2) and (3). "Students that don't want to be there" is type (2) and "students that would rather be somewhere else" are of type (3).

Granted, I'm not describing these students in anywhere near as polite a manner as Halmos is. But having been on the receiving end of such students' mocking and watching them be rewarded so heavily for their evil-well, you can forgive me for being somewhat tactless.

I would argue that the American academic system over the last 20 or so years has virtually eliminated students of type (2) and heavily encouraged students of type (3).Calculus courses today are geared specifically on these assumptions in order to maximize the profit generated by making these courses required.

It would take an entire book-which I hope to write someday-to fully explain this statement and all its implications. But Halmos almost certainly hit the nail on the head about the fundamental difference between students of type (2) and type (3): Students of type (2) are true students and have that most critical requirements for the mastery of any academic discipline: *curiosity*. They want to learn. They enjoy learning, even if it's in an area unrelated to their chosen pursuit and they aren't initially passionate about.

Students of type (3) aren't real students. They aren't really curious about anything. To them, learning anything is for suckers. To them, this is a game they need to play to win at any cost to get to their real goals: ***Money and temporal power.***

Because America is a society which doesn't encourage honesty or values-it encourages people to devour each other-such students are rewarded for cheating any way they can. They mockingly laugh at students that actually study and try and learn things to get high marks while they get straight A's while never understanding a thing. And sure enough, most get into Ivy League medical, dental, law or business schools and eventually kill someone through either cruelty, incompetence or indifference before calling their own 1000 dollar an hour lawyers

And most universities-especially those at the bottom of the academic food chain-couldn't be happier to hand those kids A's and look the other way about cheating because it's great for their bottom line. They get to say they have kids with great GPAs and get to double tuition accordingly.

Let's face it-**that's** the "academic" reason most universities buy non-mathematical 4 pound calculus books with new editions every year. It's so the premeds, accounting students, actuaries, pre-law and all the rest of the master cheaters that form the vast majority of bodies filling the enormous lecture halls of the average 200 student registration-per-semester of the required calculus course can program the solutions of all their exams into their programmable calculators.

And this is where the financial factor of student debt, which is a lifetime burden now for most American students considering going to higher education that aren't wealthy-has its impact differently on each of the student types above.

Because of the college debt deliberately shackling an entire generation, most students who aren't wealthy simply won't be able to stay in college no matter how much they want to learn if they can't succeed. In the academic paths to most lucrative careers, truly superior grades are required to even have a chance at entering these fields. Of most non-wealthy students have to work nearly full time in addition to going to school to be able to live-and that's the best case scenario, assuming they don't have any additional ongoing personal problems. Therefore, when put in an impossible position to compete for their life's dreams, they simply either dispose of their ethics and join group (3) or drop out. And those that manage to stay can only do so by taking the absolute minimum of coursework they can afford to borrow enough to support. Most financial support today for such students is loan-based-actual free federal and state student financial aid has been slashed in most states to the point its' insignificant.

In addition, in the case of calculus courses, the impact of this limiting debt and its resulting pressures has been magnified by the ludicrous prices of calculus textbooks. While nearly all standard American college textbooks have reached astronomical prices-the 400 dollar textbook is now a reality as of this writing in summer 2018-calculus textbooks have become notorious in this regard. This is not only for their substantial price-a complete new 8th edition of the very commonly used *Calculus* by James Stewart is now running over 300 USD at Amazon, with a used set running a still-painful 168 USD-these books are constantly "updated" with "new" editions where insignificant changes are routinely made and accompanied by a significant bump in price.

Again, a complete discussion of the mechanics of the textbook monopoly in America and how its' exacerbating the student debt crisis would take us too far afield here. But in the particular circumstance of calculus, the steadily climbing prices of relatively indistinguishable new editions is impossible to justify beyond simple greed. The situation was very well summed up by an anonymous calculus professor at UCLA in 2015:

The subject of calculus did not change much in the last hundred years….[T]here are no reasons why the textbooks have to be updated every 5 years or even more frequently." (For the record, the sixth edition of Stewart's "Calculus" was published in 2007, and the seventh in 2012.)
….New illustrations are sometimes added, exercises are shuffled and so on, but these do not substantially affect teaching/learning. Textbook publishers produce new editions solely as a means to sell more books and make more profit.(https://www.collegian.psu.edu/opinion/columnists/article_5d3d65b4-9b8d-11e4-8abd-5f63ba6c5556.html *)*

Given the enormous financial and academic pressures on non-wealthy students in today's sink or swim American university environment, it's not hard to understand why the selective pressures at work favor only the extreme student types of (1) and (3). Group (1) students will always be there and sadly, they'll always be in the minority. These students will always register for the hardest classes and buy, beg for and borrow mathematics books to help them master as much as time allows. Unless they suffer a personal tragedy that literally makes it impossible for them to perform at 100 percent, they'll find a way to persevere. Equally sadly, group (3) students will always be there as well and they'll always find a way to game the system. And again, because this is America, they'll be rewarded for it because both they and the people that rely on them profit from their amoral behavior. Its' why while they probably (hopefully?) won't be the majority of students taking calculus, they'll always be far more numerous than the math/science geeks.

The disturbing trend is how the current system actively discourages bright, curious students who are not mathematics or hard science majors-basically Halmos' type (2) students, "students who don't want to be there'-and who would benefit enormously from a meaty introduction to the calculus. It would not only teach them an enormous number of basic skills that are applicable in any number of careers, it would teach them the value of thinking mathematically. Indeed, with the right textbook and/or teacher, it might inspire some of them who never considered it before to pursue a career in mathematics or the hard sciences. For many students in today's universities, calculus is their first and only real exposure to mathematics. Since this exposure most of the curious non-science students gets is not only mediocre, pragmatic and uninspiring, in content, but extremely stressful, both financially and in terms of the student's precious time. In the end, most students see this course as a boring, mechanical, expensive, unpleasant and time consuming chore they resent being forced by their department to muddle through and leave without the slightest clue what the hell they wasted all that time and effort for. Why would any of these students ever take another mathematics course again after that unless it was at academic knifepoint? And if that's the only impression they ever get about mathematics, why would any of them ever see it in a positive light?

****

The Old School Calculus Series will attempt to rectify this situation by making available inexpensively a number of full calculus textbooks and supplementary sources which it is hoped will inspire both teachers and students to take calculus seriously again. The *Old School* series banner doesn't refer so much to the fact many of these books are republished out of print texts from decades ago as the fact all books in this series will attempt to recapture the spirit, comprehensiveness and rigor of these books of yesteryear to inspire a new generation of students. As with many of Blue Collar Scholar initial publications, the first few will be classic out of print texts brought into the 21st century with new original supplementary material. Eventually, we hope to publish original works for all mathematical subjects and calculus will be foremost among them.

These books will range considerably in academic difficulty level. From run of the mill, "plug and chug" non-rigorous textbooks for average beginners, which we hope will appeal to talented high school and weak freshman college students to accessible but somewhat-higher-than-today's-standard-level calculus textbooks, which we hope will raise the bar on commonly used calculus texts and put "the guts" back into ordinary undergraduate calculus classes which to completely rigorous, no-nonsense honors calculus texts for strong, advanced students beginning study to eventually enter top graduate programs in both pure and applied mathematics as well as the physical sciences. Most of all, we hope the inexpensive prices of these books will encourage self-study in this most critical and important of sciences, fostering a rebirth of general interest in both mathematics and the hard sciences in students of all social and financial backgrounds.

Karo Maestro

Founder, Editor, Blue Collar Scholar (LLC pending)

New York City

Summer 2018

CALCULUS
Differential and Integral
THROUGH PROBLEMS AND SOLUTIONS

by
G. M. Petersen
R. F. Graesser

New Prefaces And References by Karo Maestro

The Old School Calculus Series

Blue Collar Scholar/Createspace Publishing

About the Book

1. Representative problems with step by step solutions help the student to avoid the pitfalls of the subject.

2. Ample discussions and examples guide the student in discovering the correct method for solving problems.

3. Helps the student first to obtain a grasp of the ideas and an appreciation of the material involved; and then to return a second time for the more rigorous details.

4. Emphasizes the need to memorize as few formulas as possible by reasoning out the correct attack in any situation.

Copyright, 1956
by LITTLEFIELD, ADAMS & CO.

Reprinted 1958, 1959, 1961, 1963

Library of Congress Catalog Card No. 56-33

All New Material @2018 Karo Maestro

Printed in the United States of America

PREFACE

It is recorded that the impatient youth who became Alexander the Great complained to his tutor about the time he had to spend on his geometry lessons and asked if there were no short cuts to knowledge. His tutor made that memorable reply, "There is no Royal Road to learning." The authors of this book do not lay claim to having presented a short cut to calculus. Far from it, they believe that a mastery of the subject can be gained only by long hours of concentration and study. The purpose of this book is twofold. The authors hope, in the first place, to help the student avoid some of the pitfalls of the subject, by showing him representative problems on each topic properly worked out. These problems and their discussion enable the student to see the correct way of solving problems and to avoid the useless toil of working with some incorrect procedure. The student is not absolved from working out some problems on his own. Second, the authors have often appealed to the reader's intuition in theoretical discussions. They have done this in the belief that it is better that the student first obtain a grasp of the ideas and an appreciation of the material involved, to return a second time for the rigorous details.

Calculus need not be made easy; it is easy already. However, it is a subject which cannot be mastered by sheer memory. On the contrary, the student must accustom himself to memorizing as few formulas as possible and to reasoning out the correct attack in any situation. Many authorities contend that it is not natural for man to reason and this is often cited as a foundation of the contention that calculus is difficult. In the light of this, perhaps it is meet that we close with the following:

> Happy is the man that findeth wisdom and the man that getteth understanding. For the merchandise of it is better than the merchandise of silver, and the gain thereof than fine gold.   PROV. 3:13,14.

Preface To The 2018 Edition

'Obvious' is the most dangerous word in mathematics.- E. T. Bell

Do not worry about your difficulties in Mathematics. I can assure you mine are still greater. - Albert Einstein

All science requires mathematics. The knowledge of mathematical things is almost innate in us. This is the easiest of sciences, a fact which is obvious in that no one's brain rejects it; for laymen and people who are utterly illiterate know how to count and reckon. -Roger Bacon

Sometimes, the most important properties of something are in the negative-it's the properties something *doesn't* possess. For example, we can't use the property of flight to categorize what a bird is because that would exclude flightless birds like the ostrich. Therefore, the zoological criteria we use is far more sophisticated and difficult to express simply-but it doesn't exclude the ostrich.

Let me begin by saying what this book *isn't*.

It's *not* a calculus textbook.

It's *not* a workbook of exercises for added practice in calculus.

It's not really even a study guide to calculus for harried students in the usual sense of the word- although it certainly *can* be used as one if one is working diligently with a calculus text to master calculus simultaneously and is merely using it as a supplement.

"Ok, well then what is it and why should I care enough to buy one?"

It's a book of solved problems showing part of the guts of calculus that today's textbooks go out of their way to avoid.

Huh?

Ok, let me explain in some detail.

As I rambled on at some length in the introduction to this series, for a number of reasons, the current basic calculus course is a mere shadow of what it was in previous generations. One of the main symptoms of this decline has been the essential automation of these courses where students no longer have to do any real thinking. This is because the course is built around exact answer problems in closed form, where calculators and computers can easily spit out answers. While a lot of mathematicians decry the loss of rigorous proofs in such courses, such as calculating limits precisely with inequalities, that's really just the most obvious loss. A more subtle loss has been the downplaying of the roles of algebra and elementary geometry that were so essential to pre-university training and played such critical roles not only in those calculus courses, but physics and engineering as well. Being able to set up and grind out these kinds of computations forced students to not only understand calculus, but the preliminary skills they learned in high school far better.

Here's an example from Ginsberg's excellent collection (1)-which by the way, I highly recommend:

1039. By Torricelli's law, the velocity of water flowing out from a hole in a water container is $k\sqrt{2gh}$, where k is an empirical coefficient, g is the acceleration due to gravity, and h is the height of the water surface above the hole. Find the time needed to empty a water-filled cylinder of height H and cross section A after a hole of area a is made in its base.

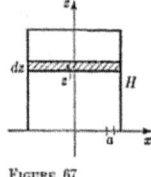
FIGURE 67

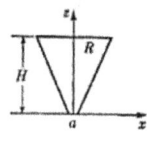
FIGURE 68

Solution. See Figure 67. Suppose the surface of the water is at height z. Then the amount of water flowing out from the cylinder in time dt is $av\,dt = ak\sqrt{2gz}\,dt$. The same amount can also be computed by $-A\,dz$ (negative because dz is negative and the amount of water positive). We have thus

$$-A\,dz = \sqrt{2gz}ka\,dt$$

or

$$dt = -\frac{A}{ka\sqrt{2g}} \cdot \frac{dz}{\sqrt{z}}.$$

Between $t = 0$ and $t = T$, z changes from H to 0 (T is the time of emptying the cylinder). Consequently

$$\int_0^T dt = \int_H^0 -\frac{A}{ka\sqrt{2g}} \cdot \frac{dz}{\sqrt{z}},$$

$$T = -\frac{A}{ka\sqrt{2g}} 2\sqrt{z}\Big|_H^0 = \frac{A}{ka}\sqrt{\frac{2H}{g}}.$$

The key to problem solving is one needs to understand exactly what's being asked and understanding the setup well enough to be able to extract the desired information. Here, it was critical to formulate the correct geometry for all the variables to set up the integral's limits. This deductive use of analytic geometry used to be critical for success in calculus. Today's calculus no longer relies on such logical intuition.

Another important skill that sadly has been largely lost in today's calculus courses is the ability to do a lengthy computation correctly using algebra skills. For example, consider the following differentiation:

$$f(x) = e^{9x}/(-8x^5 - 4x)$$

$$f'(x) = \frac{(9e^{9x}) * (-8x^5 - 4x) - (e^{9x} * -40x^4 - 4)}{(-8x^5 - 4x)^2}$$

$$\frac{(9e^{9x}) * (-8x^5 - 4x) - e^{9x} * (-40x^4 - 4)}{(-8x^5 - 4x)^2}$$

$$\frac{(9e^{9x}) * (-2x^5 - x) - e^{9x} * (-10x^4 - 1)}{(4x^5 + 2x)^2}$$

$$\frac{e^{9x}(-18x^5 + 10x^4 - 9x + 1)}{4x^2(2x^4 + 1)^2}$$

Today of course it's easy to do this computation and far lengthier ones using a computer algebra system such as MATHEMATICA or MAPLE-or if one has some programming skills, to write a short algorithm in some high level language. But serious mathematics students and teachers understand the validity of the old quote-usually attributed to George Polya, but no one seems to be certain if he was the author-*mathematics is not a spectator sport*. Can you truly understand a computation like this if you do not attempt it yourself? How many calculus students today could do one on command?

To which one might reasonably ask, "But if that's true, then what's the point of republishing a collection of solved problems for us to read? Isn't that just a whole book of mathematical spectating?" Yes-which is why neither I nor the original authors make any claim of it being a textbook in calculus that one could actually *learn* calculus by reading.

But reading through long, tedious examples like this-coupled with enough practice-could certainly assist in students *understanding* such problems and eventually mastering them.

The republication of this wonderful collection by Petersen and Grasser, among other things, looks to expose today's students to the solutions of these kinds of problems to improve these skills. It also provides an excellent snapshot of what the kinds of calculus courses we mourn the loss of consisted of and that the OSC series is looking to resurrect.

Petersen and Grasser, as we've said, isn't really a study guide to calculus, it's not complete enough for that. It doesn't have exercises for a student to work out nor detailed discussions of the topics each problem solution represents. Its' purpose is to give a rapid overview of the subject through detailed solutions of problems representative of all the major topics that were typically covered in such a course when it was published.

As I said earlier in the preface, any level of calculus cannot be mastered by strictly studying a bunch of solved problems, no matter how representative of the syllabus they are. This is even more true of the more mathematical, more rigorous calculus courses the *OSC Series* is trying to revive as the basis for standard course work. But a collection of solved problems such as the G&P can indeed make a very valuable supplement for such a course, particularly if the textbook or the professor's lectures are devoid of such examples.

Such collections were far more common in those bygone years since, in general, many university calculus courses were far more sparse in examples then today's are. Indeed, before the 1930's, it was traditional for many mathematical textbooks to refer to the exercises as examples because that was the original function of included exercises: *for students to construct examples in!* (Textbooks based on Cambridge University courses in the UK, such as G.H.Hardy's famous introduction to analysis, *A Course In Pure Mathematics*, were notorious for this.) Therefore, it's understandable there'd be a real educational need for collections like this.

In today's calculus textbooks and courses, of course, this is no longer a problem. Most modern calculus texts are loaded with solved examples. In fact, many come with complete solutions manuals.

So in many ways, from the standpoint of today's calculus, the book is a relic.

But you won't find lengthy, conceptually deep solved examples like the ones contained therein in today's calculus sources. Oh, if you search hard enough online, you'll find websites that have collected problems of this nature such as *Brilliant*- but you'll only find a handful of such sites and even fewer published resources.

One of the main purposes of the OSC series is to republish relics we believe represent the apex of calculus education in the United States, in order to encourage its' resurrection. The hope of republishing this collection-along with the many actual calculus textbooks in the OSC series at the low prices we'll be offering them-will begin the encouragement of this resurrection in modern calculus classrooms and self-study.

I hope this collection is as much a joy and a help to future generations as it has been to past ones.

Karo Maestro

New York City

Summer 2018

References

    1) Ginsberg, A. , *Calculus: Problems And Solutions*, Dover Books, 1995

0.1. Prerequisites for the study of calculus. In what follows it is assumed that the student has already studied elementary geometry, plane trigonometry, college algebra, plane and solid analytic geometry. Because the student may not have readily at hand his textbooks in these subjects, we list in subsequent articles some of the important results from them.

0.2. System of references. In this book articles are indicated by the number of the chapter followed by a period and the number of the article within the chapter. Thus the second article of Chapter 1 is indicated by 1.2. Formulas and equations that are numbered are indicated by the number of the chapter, a period, the number of the article, a period, and then the number of the formula or equation within the article. Thus 3.4.6 refers to the sixth formula of the fourth article of Chapter 3. Problems are numbered like formulas except that **boldface** is used. Thus **3.4.6** refers to the sixth problem of the fourth article of Chapter 3.

0.3. Results from elementary geometry. In the following formulas r denotes radius, h altitude, l slant height, b length of base, B area of base, θ central angle expressed in radians.

Triangle. Area $= \frac{1}{2}bh$

Trapezoid. Area $= \frac{1}{2}h(b_1 + b_2)$

Circle. Area $= \pi r^2$; arc $= r\theta$; circumference $= 2\pi r$; area of a section $= \frac{1}{2}r^2\theta$; area of a segment $= \frac{1}{2}r^2(\theta - \sin\theta)$

Pyramid. Volume $= \frac{1}{3}Bh$

Right circular cone. Lateral area $= \pi r l$; volume $= \frac{1}{3}\pi r^2 h$

Sphere. Area $= 4\pi r^2$; volume $= \frac{4}{3}\pi r^3$; volume of a spherical segment $= \frac{1}{3}\pi h^2(3r - h)$

0.4. Results from algebra

Exponents. If p and q denote positive integers, and m and n are rational numbers, then

$$a^{p/q} = \sqrt[q]{a^p}; \quad a^{-n} = \frac{1}{a^n}; \quad a^0 = 1; \quad a^n a^m = a^{n+m};$$

$$\frac{a^n}{a^m} = a^{n-m} = \frac{1}{a^{m-n}}; \quad (abc)^n = a^n b^n c^n;$$

$$\left(\frac{a}{b}\right)^n = \frac{a^n}{b^n}; \quad (a^n)^m = a^{nm}.$$

Logarithms. $\log N \cdot M = \log N + \log M$;

$$\log \frac{N}{M} = \log N - \log M; \quad \log N^m = m \log N;$$

$$\log \sqrt[q]{N^p} = \frac{p}{q} \log N.$$

These formulas apply to logarithms to any base; $\log N$ usually means logarithm of N to the base 10; $\log_b N$ means the logarithm of N to the base b; $\ln N$ means the logarithm of N to the base e, where $e = 2.7182818 \ldots$.

$$\log_b 1 = 0; \quad \log_a N = \frac{\log_b N}{\log_b a} = \log_b N \cdot \log_a b;$$

$$\log_a b = \frac{1}{\log_b a}.$$

Quadratic equations. If $ax^2 + bx + c = 0$, where $a \neq 0$, then $x = \dfrac{-b \pm \sqrt{b^2 - 4ac}}{2a}$. If a, b, and c are real numbers, the roots are real and distinct if $b^2 - 4ac > 0$; they are equal if $b^2 - 4ac = 0$; they are imaginary if $b^2 - 4ac < 0$.

Factorials. $n! = \underline{/n} = 1 \cdot 2 \cdot 3 \ldots n$

Combinations. The number of combinations of n things taken r at a time, or $_nC_r = \dfrac{n!}{r!(n-r)!} = {_nC_{n-r}}$.

Binomial expansion. $(a + x)^n = a^n + na^{n-1}x + \dfrac{n(n-1)}{2!}$
$\cdot a^{n-2}x^2 + \ldots + \dfrac{n!}{r!(n-r)!} a^{n-r}x^r + \ldots + nax^{n-1} + x^n$,
where n is a positive integer.

Progressions. If a denotes the first term, l the last term or the nth term, S_n the sum of n terms, S_∞ the limit of the sum of n terms as n becomes infinite, d the common difference, and r the common ratio, then

Arithmetic progressions. $l = a + (n - 1)d$;

$$S_n = \frac{n}{2}(a + l) = \frac{n}{2}[2a + (n - 1)d]$$

Geometric progressions. $l = ar^{n-1}$; $S_n = \dfrac{rl - a}{r - 1} = a\dfrac{r^n - 1}{r - 1}$

if $r \neq 1$; $S_\infty = \dfrac{a}{1 - r}$ if $|r| < 1$

0.5. Results from plane trigonometry

$$180° = \pi \text{ radians}$$

Trigonometric Functions of Common Angles

Angle	0	30°	45°	60°	90°
Sine	0	$\tfrac{1}{2}$	$\tfrac{1}{2}\sqrt{2}$	$\tfrac{1}{2}\sqrt{3}$	1
Cosine	1	$\tfrac{1}{2}\sqrt{3}$	$\tfrac{1}{2}\sqrt{2}$	$\tfrac{1}{2}$	0
Tangent	0	$\tfrac{1}{3}\sqrt{3}$	1	$\sqrt{3}$	$\pm\infty$

Any function of $n \cdot 180° \pm$ an acute angle is $\pm$ the same function of the acute angle, the sign being determined by the quadrant in which the original angle falls.

Identities. $\cos^2 x + \sin^2 x = 1$; $\tan^2 x + 1 = \sec^2 x$;

$\cot^2 x + 1 = \csc^2 x$; $\tan x = \dfrac{\sin x}{\cos x}$; $\cot x = \dfrac{1}{\tan x}$;

$$\csc x = \dfrac{1}{\sin x}; \qquad \sec x = \dfrac{1}{\cos x};$$

$$\sin(x \pm y) = \sin x \cdot \cos y \pm \cos x \cdot \sin y;$$
$$\cos(x \pm y) = \cos x \cdot \cos y \mp \sin x \cdot \sin y;$$
$$\tan(x \pm y) = \dfrac{\tan x \pm \tan y}{1 \mp \tan x \cdot \tan y};$$
$$\sin 2x = 2 \sin x \cdot \cos x;$$
$$\cos 2x = \cos^2 x - \sin^2 x = 2\cos^2 x - 1 = 1 - 2\sin^2 x;$$
$$\tan 2x = \dfrac{2 \tan x}{1 - \tan^2 x}; \qquad \sin\dfrac{x}{2} = \pm\sqrt{\dfrac{1 - \cos x}{2}};$$
$$\cos\dfrac{x}{2} = \pm\sqrt{\dfrac{1 + \cos x}{2}};$$

$$\tan \frac{x}{2} = \pm \sqrt{\frac{1-\cos x}{1+\cos x}} = \frac{\sin x}{1+\cos x} = \frac{1-\cos x}{\sin x};$$

$$\sin x + \sin y = 2 \sin \frac{x+y}{2} \cos \frac{x-y}{2};$$

$$\sin x - \sin y = 2 \cos \frac{x+y}{2} \sin \frac{x-y}{2};$$

$$\cos x + \cos y = 2 \cos \frac{x+y}{2} \cos \frac{x-y}{2};$$

$$\cos x - \cos y = -2 \sin \frac{x+y}{2} \sin \frac{x-y}{2};$$

$$2 \sin A \cdot \cos B = \sin(A+B) + \sin(A-B);$$
$$2 \cos A \cdot \cos B = \cos(A-B) + \cos(A+B);$$
$$2 \sin A \cdot \sin B = \cos(A-B) - \cos(A+B).$$

Triangles. If A, B, C are the angles of a plane triangle, and a, b, c are, respectively, the opposite sides, and $s = \frac{1}{2}(a+b+c)$, then

$$A + B + C = 180°; \qquad \frac{a}{\sin A} = \frac{b}{\sin B} = \frac{c}{\sin C};$$

$$a^2 = b^2 + c^2 - 2bc \cdot \cos A; \qquad \text{area} = \sqrt{s(s-a)(s-b)(s-c)}$$
$$= \tfrac{1}{2} ab \cdot \sin C.$$

0.6. Results from plane analytic geometry

Distance $P_1(x_1, y_1)$ to $P_2(x_2, y_2) = \sqrt{(x_1 - x_2)^2 + (y_1 - y_2)^2}$

Slope of the line $P_1 P_2 = m = \dfrac{y_1 - y_2}{x_1 - x_2}$. For $P(x, y)$ so that

$$\frac{P_1 P}{P P_2} = \frac{r_1}{r_2}; \qquad x = \frac{r_1 x_2 + r_2 x_1}{r_1 + r_2}; \qquad y = \frac{r_1 y_2 + r_2 y_1}{r_1 + r_2}.$$

For θ, the *angle between two lines* with slopes m_1, m_2,

$$\tan \theta = \frac{m_1 - m_2}{1 + m_1 \cdot m_2}.$$

For *parallel lines*, $m_1 = m_2$, and for *perpendicular lines*,

$$m_1 = -\frac{1}{m_2}.$$

Equations of a straight line $y - y_1 = m(x - x_1)$;

$y = mx + b$; $\quad y - y_1 = \dfrac{y_1 - y_2}{x_1 - x_2}(x - x_1)$; $\quad \dfrac{x}{a} + \dfrac{y}{b} = 1$;

$$x \cos \omega + y \sin \omega = P$$

Distance $P_1(x_1, y_1)$ from $Ax + By + C = 0$ is $\dfrac{Ax_1 + By_1 + C}{\pm\sqrt{A^2 + B^2}}$.

If $P(x, y)$ is also $P(r, \theta)$ in polar coordinates, then

$x = r \cos \theta$; $\quad y = r \sin \theta$; $\quad x^2 + y^2 = r^2$; $\quad \theta = \arctan \dfrac{y}{x}$.

Conic sections

Circle with center at (h, k). $\quad (x - h)^2 + (y - k)^2 = r^2$.

Parabolas with vertex at (h, k). $\quad (y - k)^2 = \pm 4P(x - h)$;

$$(x - h)^2 = \pm 4P(y - k).$$

Ellipse with center at (h, k). $\quad \dfrac{(x - h)^2}{a^2} + \dfrac{(y - k)^2}{b^2} = 1$.

Hyperbolas with centers at (h, k). $\quad \dfrac{(x - h)^2}{a^2} - \dfrac{(y - k)^2}{b^2} = \pm 1$,

where the *asymptotes* are $\quad \dfrac{(x - h)^2}{a^2} - \dfrac{(y - k)^2}{b^2} = 0$.

Equilateral hyperbola $xy = c$, where the *asymptotes* are $xy = 0$. The conic $ax^2 + bxy + cy^2 + dx + ey + f = 0$ is an hyperbola, parabola, ellipse according as $b^2 - 4ac$ is greater than zero, equal to zero, or less than zero. (Here a circle is considered to be a special case of an ellipse.)

0.7. Results from solid analytic geometry

Distance $\quad P(x_1, y_1, z_1)$ to $P_2(x_2, y_2, z_2)$

$$= \sqrt{(x_1 - x_2)^2 + (y_1 - y_2)^2 + (z_1 - z_2)^2}.$$

Direction numbers of the line P_1P_2:

$$(x_1 - x_2), (y_1 - y_2), (z_1 - z_2).$$

For $P(x, y, z)$ so that $\dfrac{P_1P}{PP_2} = \dfrac{r_1}{r_2}$. $\quad x = \dfrac{r_1 x_2 + r_2 x_1}{r_1 + r_2}$;

$$y = \dfrac{r_1 y_2 + r_2 y_1}{r_1 + r_2}; \quad z = \dfrac{r_1 z_2 + r_2 z_1}{r_1 + r_2}.$$

For θ the *angle between two lines* with the direction numbers a_1, b_1, c_1 and a_2, b_2, c_2.

$$\cos\theta = \frac{a_1 a_2 + b_1 b_2 + c_1 c_2}{\sqrt{a_1^2 + b_1^2 + c_1^2}\sqrt{a_2^2 + b_2^2 + c_2^2}}$$

Parallel lines $\quad \dfrac{a_1}{a_2} = \dfrac{b_1}{b_2} = \dfrac{c_1}{c_2}.$

Perpendicular lines. $\quad a_1 a_2 + b_1 b_2 + c_1 c_2 = 0.$

Equation of the plane through $P_1(x_1, y_1, z_1)$ perpendicular to a line with the direction numbers a, b, c is

$$a(x - x_1) + b(y - y_1) + c(z - z_1) = 0$$

Distance of $P_1(x_1, y_1, z_1)$ from the plane $Ax + By + Cz + D = 0$ is

$$\frac{Ax_1 + By_1 + Cz_1 + D}{\pm\sqrt{A^2 + B^2 + C^2}}$$

Equations of the line through $P_1(x_1, y_1, z_1)$ with the direction numbers

$$a, b, c \quad \text{are} \quad \frac{x - x_1}{a} = \frac{y - y_1}{b} = \frac{z - z_1}{c}.$$

If $P(x, y, z)$ is also $P(r, \theta, z)$ in *cylindrical coordinates*, then $x = r\cos\theta; y = r\sin\theta; z = z; x^2 + y^2 = r^2; \theta = \arctan\dfrac{y}{x}.$

If $P(x, y, z)$ is also $P(r, \theta, \phi)$ in *spherical coordinates*, then

$$x = r\sin\phi\cos\theta; \quad y = r\sin\phi\sin\theta; \quad z = r\cos\phi.$$

$x^2 + y^2 + z^2 = r^2; \theta = \arctan\dfrac{y}{x}; \phi = \arccos\dfrac{z}{\sqrt{x^2 + y^2 + z^2}}$

Quadric Surfaces

Sphere with center at (h, k, l):

$$(x - h)^2 + (y - k)^2 + (z - l)^2 = r^2$$

Ellipsoid with center at $(0, 0, 0)$: $\quad \dfrac{x^2}{a^2} + \dfrac{y^2}{b^2} + \dfrac{z^2}{c^2} = 1$

Hyperboloid of one sheet $\quad\dfrac{x^2}{a^2}+\dfrac{y^2}{b^2}-\dfrac{z^2}{c^2}=1$

Hyperboloid of two sheets $\quad\dfrac{x^2}{a^2}-\dfrac{y^2}{b^2}-\dfrac{z^2}{c^2}=1$

Hyperbolic paraboloid (saddle surface) $\quad\dfrac{x^2}{a^2}-\dfrac{y^2}{b^2}=cz$

Elliptic paraboloid $\quad\dfrac{x^2}{a^2}+\dfrac{y^2}{b^2}=cz$

CHAPTER 1

FUNCTIONS, LIMITS, AND CONTINUITY

1.1. Functions and the notation for functions. Given tw variables, if for each value of the first within a given range, o or more values of the second are determined, the second is sa to be a function of the first. If y is a function of x, we may d note this by the symbols $y = f(x)$.

1.1.1. If $f(x) = x^2$, find: (a) $f(y)$; (b) $f(z)$; (c) $f(4$ (d) $f(0)$; (e) $f(-1)$

Solution. (a) $f(y) = y^2$; (b) $f(z) = z^2$; (c) $f(4)$ $4^2 = 16$; (d) $f(0) = 0^2 = 0$; (e) $f(-1) = (-1)^2 = 1$

1.1.2. If $H(t) = \dfrac{t-1}{t+1}$, find: (a) $H(t^2)$; (b) $H^2(t)$;

(c) $H(2z)$; (d) $H\left(\dfrac{2}{t}\right)$; (e) $H(t+2)$

Solution. (a) $H(t^2) = \dfrac{t^2-1}{t^2+1}$; (b) $H^2(t) = \left(\dfrac{t-1}{t+1}\right)^2$

(c) $H(2z) = \dfrac{2z-1}{2z+1}$; (d) $H\left(\dfrac{2}{t}\right) = \dfrac{\dfrac{2}{t}-1}{\dfrac{2}{t}+1} = \dfrac{2-t}{2+t}$;

(e) $H(t+2) = \dfrac{t+2-1}{t+2+1} = \dfrac{t+1}{t+3}$

1.1.3. If $G(u) = \dfrac{u}{u+1}$, find: (a) $G\left(\dfrac{1}{u}\right)$; (b) $\dfrac{1}{G(u)}$;

(c) $G[G(u)]$; (d) $G\left[\dfrac{1}{G(u)}\right]$; (e) $G(u + \Delta u) - G(u)$

Solution. (a) $G\left(\dfrac{1}{u}\right) = \dfrac{\dfrac{1}{u}}{\dfrac{1}{u}+1} = \dfrac{1}{1+u}$; (b) $\dfrac{1}{G(u)}$

$= \dfrac{1}{\dfrac{u}{u+1}} = \dfrac{u+1}{u}$; (c) $G[G(u)] = \dfrac{G(u)}{G(u)+1} = \dfrac{\dfrac{u}{u+1}}{\dfrac{u}{u+1}+}$

$$= \frac{u}{2u+1}; \quad \text{(d)} \ G\left[\frac{1}{G(u)}\right] = \frac{\frac{1}{G(u)}}{\frac{1}{G(u)}+1} = \frac{1}{1+G(u)}$$

$$= \frac{1}{1+\frac{u}{u+1}} = \frac{u+1}{2u+1}; \quad \text{(e)} \ G(u+\Delta u) - G(u)$$

$$= \frac{u+\Delta u}{u+\Delta u+1} - \frac{u}{u+1}$$

$$= \frac{u^2 + u\Delta u + u + \Delta u - u^2 - u\Delta u - u}{(u+\Delta u+1)(u+1)}$$

$$= \frac{\Delta u}{(u+\Delta u+1)(u+1)}$$

1.1.4. If $f(x) = 1 + \frac{1}{x}$, show that $f(xy) + f(x) + f(y)$
$= 2 + f(x) \cdot f(y)$

Solution. $f(xy) + f(x) + f(y) = 1 + \frac{1}{xy} + 1 + \frac{1}{x} + 1 + \frac{1}{y}$

$$= 2 + \left(1 + \frac{1}{x}\right)\left(1 + \frac{1}{y}\right) = 2 + f(x)f(y)$$

1.2. Limits. If the numerical value of the difference between a function $f(x)$ and a constant A becomes and remains less than any arbitrarily chosen, preassigned, positive number, however small it may be chosen, whenever the variable x is chosen sufficiently close to, but not equal to, some value a, then $f(x)$ is said to approach A as a limit as x approaches a. This situation is expressed in symbols by $\lim_{x \to a} f(x) = A$.

Also a variable x may approach a only through values larger than a, which situation is expressed symbolically as $x \to a^+$. If the limit is found as a is approached only through values smaller than itself, we write $x \to a^-$. These two ways of approaching a limit may yield different results. For example, $\lim_{x \to 0^-} \frac{|x|}{x} =$
$\lim_{t \to 0^-} (-1) = -1$, while $\lim_{x \to 0^+} \frac{|x|}{x} = \lim_{x \to 0^+} 1 = 1$. This example

also shows that the function need not be defined at a point in order to have a limit at that point; $\frac{|x|}{x}$ has no meaning for $x = 0$ since division by zero is undefined. All methods of approach must give the same limit, or otherwise no limit is defined. For example, if x approaches zero through the set of points $\frac{1}{\pi}, \frac{1}{2\pi}, \frac{1}{3\pi} \ldots$ it would seem that $\lim_{x \to 0} \sin \frac{1}{x} = 0$. On the other hand, if x approaches zero through the set $\frac{2}{\pi}, \frac{2}{5\pi}, \frac{2}{9\pi} \ldots$ it would seem $\lim_{x \to 0} \sin \frac{1}{x} = 1$. Since these two methods of approach yield different results, we say that $\lim_{x \to 0} \sin \frac{1}{x}$ does not exist.

<u>Rule for limits.</u> *The limit of the sum, difference, product, or quotient of functions that approach limits is the sum, difference, product, or quotient of their limits if the separate limits all exist and that of the divisor is not zero.*

If the numerator and denominator of a fraction both approach zero for the same values of x, the limit of their quotient may or may not exist at that point. Such a fraction is said to have the indeterminate form $\frac{0}{0}$. This $\frac{0}{0}$ does not mean that zero is divided by zero, since division by the number zero is undefined. If the numerator does not approach zero as the denominator approaches zero, the fraction will become infinitely large, or increase without end. We would then say that no limit exists, or that the fraction becomes infinite, and we would represent the situation by ∞. Thus $\lim_{x \to 1^+} \frac{x + 3}{x - 1} = \infty$. We may conceive of any variable, say x, as becoming indefinitely large or increasing without end. This we indicate by $x \to \infty$. Or x may remain negative while its numerical value increases indefinitely; this we indicate by $x \to -\infty$. Besides the indeterminate form $\frac{0}{0}$, a fraction may assume the indeterminate form $\frac{\infty}{\infty}$, which means that

oth the numerator and denominator are becoming indefinitely
rge.

It should be evident that

$$\lim_{x \to a} x = a$$

$$\lim_{x \to \infty} c = c, \quad \text{where } c \text{ is a constant}$$

$$\lim_{x \to \infty} \frac{c}{x} = 0, \quad \text{where } c \text{ is a constant}$$

1.2.1. $\lim_{x \to a} x^n$, where n is a positive integer, $= \lim_{x \to a} x \cdot x \cdot x \ldots x$
$= \lim_{x \to a} x \cdot \lim_{x \to a} x \ldots$ to n factors $= a \cdot a \cdot a \ldots$ to n factors $= a^n$

1.2.2. $\lim_{x \to 2}(3x^3 - x^2 + 5) = \lim_{x \to 2} 3x^3 - \lim_{x \to 2} x^2 + \lim_{x \to 2} 5$
$= 3 \lim_{x \to 2} x^3 - \lim_{x \to 2} x^2 + \lim_{x \to 2} 5 = 3 \cdot 2^3 - 2^2 + 5 = 25$ by **1.2.1**.

1.2.3. $\lim_{x \to a} f(x)$, where $f(x)$ is a polynomial, $= f(a)$ by a procedure similar to that used in **1.2.2**.

1.2.4. $\lim_{x \to a} \frac{f(x)}{F(x)}$, where $f(x)$ and $F(x)$ are polynomials,

$$= \frac{\lim_{x \to a} f(x)}{\lim_{x \to a} F(x)} = \frac{f(a)}{F(a)} \text{ by } \mathbf{1.2.3}, \text{ provided that } F(a) \neq 0.$$

1.2.5. $\lim_{x \to \infty} \frac{x^2 + 2x - 3}{2x^3 + 3x^2 + x + 3} = \lim_{x \to \infty} \frac{1 + \dfrac{2}{x} - \dfrac{3}{x^2}}{2x + 3 + \dfrac{1}{x} + \dfrac{3}{x^2}}$

$$= \frac{\lim_{x \to \infty}\left(1 + \dfrac{2}{x} - \dfrac{3}{x^2}\right)}{\lim_{x \to \infty}\left(2x + 3 + \dfrac{1}{x} + \dfrac{3}{x^2}\right)}$$

$$= \frac{\lim_{x \to \infty} 1 + \lim_{x \to \infty} \dfrac{2}{x} - \lim_{x \to \infty} \dfrac{3}{x^2}}{\lim_{x \to \infty} 2x + \lim_{x \to \infty} 3 + \lim_{x \to \infty} \dfrac{1}{x} + \lim_{x \to \infty} \dfrac{3}{x^2}} = \frac{1}{\lim_{x \to \infty} 2x + 3} = 0$$

1.2.6. $\lim_{x \to \infty} \frac{f(x)}{F(x)}$, where $f(x)$ and $F(x)$ are polynomials, and

the degree of $F(x)$ is greater than that of $f(x)$; $\lim\limits_{x\to\infty}\dfrac{f(x)}{F(x)} =$
by a procedure similar to that used in **1.2.5**.

1.2.7. $\lim\limits_{x\to\infty}\dfrac{3x^4 + 2x^2 - x + 4}{2x^2 - 3x - 1} = \lim\limits_{x\to\infty}\dfrac{3x^2 + 2 - \dfrac{1}{x} + \dfrac{4}{x^2}}{2 - \dfrac{3}{x} - \dfrac{1}{x^2}}$

$= \dfrac{\lim\limits_{x\to\infty}\left(3x^2 + 2 - \dfrac{1}{x} + \dfrac{4}{x^2}\right)}{\lim\limits_{x\to\infty}\left(2 - \dfrac{3}{x} - \dfrac{1}{x^2}\right)} = \dfrac{\lim\limits_{x\to\infty} 3x^2 + 2}{2} =$

1.2.8. $\lim\limits_{x\to\infty}\dfrac{f(x)}{F(x)}$, where $f(x)$ and $F(x)$ are polynomials, a the degree of $f(x)$ is greater than that of $F(x)$, is shown to ∞ by a procedure similar to that used in **1.2.7**.

1.2.9. $\lim\limits_{x\to\infty}\dfrac{3x^2 - 2x + 1}{2x^2 + 5x + 3} = \lim\limits_{x\to\infty}\dfrac{3 - \dfrac{2}{x} + \dfrac{1}{x^2}}{2 + \dfrac{5}{x} + \dfrac{3}{x^2}}$

$= \dfrac{\lim\limits_{x\to\infty}\left(3 - \dfrac{2}{x} + \dfrac{1}{x^2}\right)}{\lim\limits_{x\to\infty}\left(2 + \dfrac{5}{x} + \dfrac{3}{x^2}\right)} =$

1.2.10. $\lim\limits_{x\to\infty}\dfrac{a_0 x^n + a_1 x^{n-1} + \cdots + a_{n-1} x + a_n}{b_0 x^n + b_1 x^{n-1} + \cdots + b_{n-1} x + b_n}$

$= \lim\limits_{x\to\infty}\dfrac{a_0 + \dfrac{a_1}{x} + \cdots + \dfrac{a_{n-1}}{x^{n-1}} + \dfrac{a_n}{x^n}}{b_0 + \dfrac{b_1}{x} + \cdots + \dfrac{b_{n-1}}{x^{n-1}} + \dfrac{b_n}{x^n}} =$

where n is a positive integer.

The student should notice that **1.2.11** to **1.2.15** are all of th indeterminate form $\dfrac{0}{0}$.

1.2.11. $\lim\limits_{x\to a}\dfrac{x^2 - a^2}{x - a}$. If $x \neq a$, then $\dfrac{x^2 - a^2}{x - a} = x + a$. Hence

$\lim\limits_{x\to a}\dfrac{x^2 - a^2}{x - a} = \lim\limits_{x\to a}(x + a) = a + a = 2a$

because we conceive that if two variables (or functions) are equal and approach limits, the limits must be equal. Notice, however, that the limits are the same even though the values of the functions at the point $x = a$ are different.

1.2.12. $\lim\limits_{x \to 0} \dfrac{x^3 + 2x^2 + 5x}{x^2 + x} = \lim\limits_{x \to 0} \dfrac{x^2 + 2x + 5}{x + 1}$

$= \dfrac{\lim\limits_{x \to 0}(x^2 + 2x + 5)}{\lim\limits_{x \to 0}(x + 1)} = \dfrac{\lim\limits_{x \to 0} x^2 + \lim\limits_{x \to 0} 2x + \lim\limits_{x \to 0} 5}{\lim\limits_{x \to 0} x + \lim\limits_{x \to 0} 1} = 5$

1.2.13. $\lim\limits_{x \to 0} \dfrac{x}{\sqrt{x + 1} - 1}$

$= \lim\limits_{x \to 0} \dfrac{x}{\sqrt{x + 1} - 1} \cdot \dfrac{\sqrt{x + 1} + 1}{\sqrt{x + 1} + 1}$

$= \lim\limits_{x \to 0} \dfrac{x(\sqrt{x + 1} + 1)}{(x + 1) - 1}$

$= \lim\limits_{x \to 0} (\sqrt{x + 1} + 1) = 2$

1.2.14. $\lim\limits_{x \to 0} \dfrac{\tan x}{\sin x} = \lim\limits_{x \to 0} \dfrac{1}{\cos x} = 1$

1.2.15. $\lim\limits_{\theta \to 0} \dfrac{\sin \theta}{\theta}$ This is a special limit which will be useful to us later. Its evaluation requires a special procedure, one of which is as follows. In the figure we clearly have the inequality, area of triangle OAB < area of sector $OADB$ < area of triangle OCE. Let $\angle COD = \angle DOE = \theta$, which is measured in radians. Let $OD = r$. Then expressing the areas in our inequality in terms of r and θ:

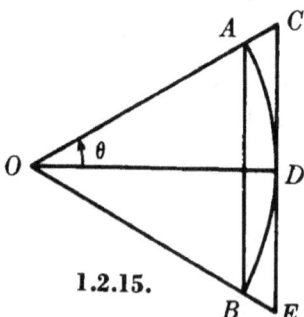

1.2.15.

$$r^2 \sin\theta \cos\theta < r^2\theta < r^2 \tan\theta, \quad \text{or}$$

$$\frac{1}{\sin\theta \cos\theta} > \frac{1}{\theta} > \frac{1}{\tan\theta}. \qquad \frac{1}{\cos\theta} > \frac{\sin\theta}{\theta} > \cos\theta$$

Now as θ approaches zero, both the first and the last members of this inequality approach one. Hence $\lim_{\theta \to 0} \frac{\sin\theta}{\theta} = 1$.

1.3. Continuity. A single-valued function $f(x)$ which is defined in the range $c \leqslant x \leqslant d$ is continuous at a point a of the range if

$$\lim_{x \to a} f(x) = f(a) \qquad (1.3.1)$$

A change in an independent variable x is often indicated by Δx (called an increment of x). Using increments, 1.3.1 may be written

$$\lim_{\Delta x \to 0} f(a + \Delta x) = f(a)$$

$$\lim_{\Delta x \to 0} \Delta y = \lim_{\Delta x \to 0} [f(a + \Delta x) - f(a)] = 0 \qquad (1.3.2)$$

where $f(a + \Delta x) - f(a)$ is indicated by Δy since it represents a change in the function $y = f(x)$.

Notice that for continuity at $x = a$, three conditions are necessary:

$f(x)$ must be defined at $x = a$, i.e., $f(a)$ must exist (1.3.3)

$\lim_{x \to a} f(x)$ must exist (1.3.4)

$\lim_{x \to a} f(x)$ must equal $f(a)$ (1.3.5)

A function is *discontinuous* at a point where it is not continuous. A function is *continuous in an interval* if it is continuous at every point in the interval. Notice that any polynomial is everywhere continuous by virtue of **1.2.3**.

1.3.1. Is $y = \dfrac{x^2 - 4}{x - 2}$ continuous at $x = 2$?

Solution. At $x = 2$, $y = \dfrac{0}{0}$; hence y is undefined, and 1.3.3 is not fulfilled. However, if we define y thus: $y = \dfrac{x^2 - 4}{x - 2}$ if

$x \neq 2$, and $y = 4$ if $x = 2$, then y becomes continuous at $x = 2$. See 1.2.11. Hence the discontinuity at $x = 2$ in the function $y = \dfrac{x^2 - 4}{x - 2}$ is said to be removable. The graph of $y = \dfrac{x^2 - 4}{x - 2}$ is the graph of $y = x + 2$ with the point (2,4) missing.

1.3.2. Is $y = \dfrac{x^2 + 2x + 7}{x - 1}$ continuous at $x = 1$?

Solution. If $x = 1$, $y = \infty$; hence y is undefined at $x = 1$, and 1.3.3 is not satisfied, so that the function is discontinuous at $x = 1$.

1.3.3. Is $y = \dfrac{|x|}{x}$ continuous at $x = 0$?

Solution. As we have seen in 1.2, $\lim\limits_{x \to 0^+} \dfrac{|x|}{x} = 1$, and $\lim\limits_{x \to 0^-} \dfrac{|x|}{x} = -1$. Hence $\lim\limits_{x \to 0} \dfrac{|x|}{x}$ does not exist, and 1.3.4 is not satisfied. Also, at $x = 0$, $y = \dfrac{|0|}{0}$, so that y is not defined, and 1.3.3 is not satisfied.

1.3.4. If $y = \dfrac{x^2 - 1}{x - 1}$ when $x \neq 1$, and $y = 1$ when $x = 1$, is y continuous at $x = 1$?

Solution. $\lim\limits_{x \to 1} \dfrac{x^2 - 1}{x - 1} = \lim\limits_{x \to 1} (x + 1) = 2$. But $y = 1$ when $x = 1$; hence 1.3.5 is not satisfied.

1.3.5. Is $y = \tan x$ continuous at $x = \dfrac{\pi}{2}$?

Solution. Since $\lim\limits_{x \to \pi^+/2} \tan x = \infty$, and $\lim\limits_{x \to \pi^-/2} \tan x = -\infty$, and $\tan \dfrac{\pi}{2}$ is undefined, neither 1.3.3 nor 1.3.4 is satisfied.

1.3.6. Is $y = \sin \dfrac{1}{x}$ continuous at $x = 0$?

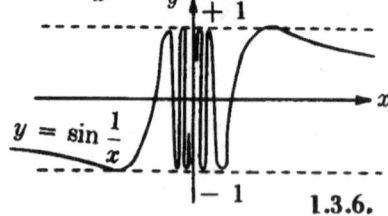

1.3.6.

Solution. Since y is undefined at $x = 0$, y is discontinuous at $x = 0$.

1.3.7. If $y = x \sin \dfrac{1}{x}$ when $x \neq 0$, and $y = 0$ when $x = 0$,

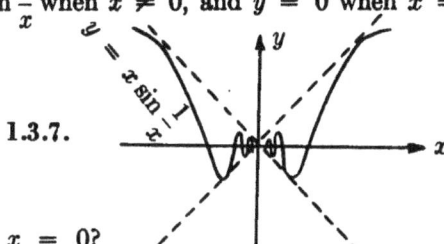

1.3.7.

is y continuous at $x = 0$?

Solution. We have always $\left|\sin \dfrac{1}{x}\right| \leq 1$. Hence $\lim\limits_{x \to 0} x \sin \dfrac{1}{x} = 0$, and y is continuous at $x = 0$.

1.3.8. Is $y = \dfrac{1}{1 - 2^{1/x}}$ continuous at $x = 0$?

Solution. $\lim\limits_{x \to 0^+} \dfrac{1}{1 - 2^{1/x}} = 0$; $\lim\limits_{x \to 0^-} \dfrac{1}{1 - 2^{1/x}} = 1$. Hence 1.3.4 is not satisfied, and y is discontinuous at $x = 0$.

1.3.9. If $y = x^2$ when $x \neq 2$, and $y = 10$ when $x = 2$, is y continuous at $x = 2$?

Solution. Since $\lim\limits_{x \to 2} y = \lim\limits_{x \to 2} x^2 = 4$, and y is defined as 10 when $x = 2$, 1.3.5 is not satisfied, and y is discontinuous at $x = 2$.

1.3.10. The United States domestic postal rate for first class mail is 3 cents per ounce or fraction thereof. Express the cost $f(x)$ of sending first class mail as a function of x, the number of ounces sent. Discuss the continuity of $f(x)$. Then $f(x) = 3$ if $0 < x \leq 1$; $f(x) = 6$ if $1 < x \leq 2$; $f(x) = 9$ if $2 < x \leq 3$; and so on.

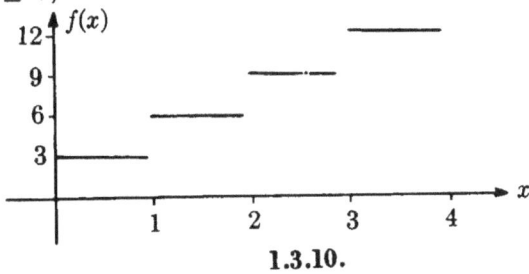

1.3.10.

Solution. $\lim_{x \to 1^-} f(x) = 3$; $\lim_{x \to 1^+} f(x) = 6$. Hence $\lim_{x \to 1} f(x)$ does not exist even though $f(1) = 3$. Therefore 1.3.4 is not satisfied. A similar discussion applies to $x = 2, 3, 4, \ldots$. $f(x)$ is continuous for all nonintegral values of x.

CHAPTER 2

THE DIFFERENCE QUOTIENT AND
THE DERIVATIVE

2.1. Definition of the derivative. The derivative $\dfrac{dy}{dx}$ of a function of x is defined as the limit of a difference quotient $\dfrac{\Delta y}{\Delta x}$, i.e.,

$$\frac{dy}{dx} = \lim_{\Delta x \to 0} \frac{\Delta y}{\Delta x} = \lim_{\Delta x \to 0} \frac{f(x + \Delta x) - f(x)}{\Delta x}$$

The difference quotient represents the average rate of change of y with respect to x. The limit is the instantaneous rate of change of y with respect to x.

This derivative $\dfrac{dy}{dx}$ is also represented by y' and $f'(x)$. A function that has a derivative is said to be *differentiable*. Since the denominator of the fraction $\dfrac{\Delta y}{\Delta x}$ approaches zero, the numerator must also approach zero if the fraction as a whole is to have a limit. But, by the discussion of continuity in 1.3, y is a continuous function of x if Δy approaches zero as Δx approaches zero. (See 1.3.2.) Therefore, in order to be differentiable, a

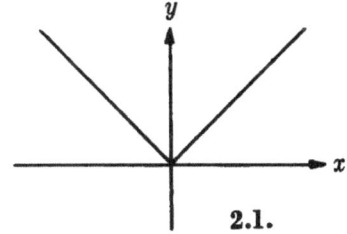

2.1.

function must be continuous. However, a function may be continuous without being differentiable. For example, $y = |x|$, whose graph is shown, is equivalent to $y = x$ if $x \geq 0$, and it is equivalent to $y = -x$ if $x \leq 0$. To find its derivative at $x = 0$, we have, for $\Delta x > 0$, $\Delta y = \Delta x$, $\frac{\Delta y}{\Delta x} = 1$, and $\lim_{\Delta x \to 0^+} \frac{\Delta y}{\Delta x} = 1$. For $\Delta x < 0$, $\Delta y = -\Delta x$, and $\lim_{\Delta x \to 0^-} \frac{\Delta y}{\Delta x} = -1$. Hence $\lim_{\Delta x \to 0} \frac{\Delta y}{\Delta x}$ does not exist even though $\Delta y \to 0$ as $\Delta x \to 0$. Since $\Delta y \to 0$ as $\Delta x \to 0$, $y = |x|$ is continuous at $x = 0$, a fact that is obvious from the graph.

2.2. Differentiation by the four-step rule.

2.2.1. Find $\frac{dy}{dx}$ if $y = x^2 + 3x + 2$.

Solution. $y + \Delta y = (x + \Delta x)^2 + 3(x + \Delta x) + 2$

Step 1 $\qquad y + \Delta y = x^2 + 2x\,\Delta x + (\Delta x)^2$
$\qquad\qquad\qquad\qquad + 3x + 3\,\Delta x + 2$
$\qquad\qquad y = x^2 + 3x + 2$

Step 2 $\qquad\qquad \Delta y = 2x\,\Delta x + (\Delta x)^2 + 3\,\Delta x$

Step 3 $\qquad\qquad \frac{\Delta y}{\Delta x} = 2x + \Delta x + 3$

$\qquad\qquad\lim_{\Delta x \to 0} \frac{\Delta y}{\Delta x} = \lim_{\Delta x \to 0} (2x + \Delta x + 3)$

Step 4 $\qquad\qquad \frac{dy}{dx} = 2x + 3$

The foregoing illustrates the *four-step rule* for differentiation, the four steps being numbered.

Step 1. Give the argument of the function an increment.
Step 2. Compute the corresponding increment in the function.
Step 3. Find the difference-quotient.
Step 4. Take the limit of the difference-quotient as the increment of the argument approaches zero.

2.2.2. Find $\frac{du}{dt}$ if $u = \frac{t}{1+t}$.

Solution. Applying the four-step rule:

Step 1
$$u + \Delta u = \frac{t + \Delta t}{1 + t + \Delta t}$$

$$u = \frac{t}{1+t}$$

$$\Delta u = \frac{t + \Delta t}{1 + t + \Delta t} - \frac{t}{1+t} = \frac{(t+\Delta t)(1+t) - t(1+t+\Delta t)}{(1+t)(1+t+\Delta t)}$$

Step 2
$$\Delta u = \frac{\Delta t}{(1+t)(1+t+\Delta t)}$$

Step 3
$$\frac{\Delta u}{\Delta t} = \frac{1}{(1+t)(1+t+\Delta t)}$$

Step 4
$$\frac{du}{dt} = \lim_{\Delta t \to 0} \frac{\Delta u}{\Delta t} = \frac{1}{(1+t)^2}$$

For each of the following functions, find its derivative with respect to its argument.

2.2.3. $y = \sqrt{x}$
Solution. $y + \Delta y = \sqrt{x + \Delta x}$; $\Delta y = \sqrt{x + \Delta x} - \sqrt{x}$;

$$\frac{\Delta y}{\Delta x} = \frac{\sqrt{x+\Delta x} - \sqrt{x}}{\Delta x} = \frac{\sqrt{x+\Delta x} - \sqrt{x}}{\Delta x} \cdot \frac{\sqrt{x+\Delta x} + \sqrt{x}}{\sqrt{x+\Delta x} + \sqrt{x}}$$

$$= \frac{1}{\sqrt{x+\Delta x} + \sqrt{x}}; \quad \frac{dy}{dx} = \frac{1}{2\sqrt{x}}$$

2.2.4. $y = \frac{2}{x}$

Solution. $y + \Delta y = \frac{2}{x + \Delta x}$; $\Delta y = \frac{2}{x+\Delta x} - \frac{2}{x}$

$$= \frac{-2\Delta x}{(x+\Delta x)x}; \quad \frac{\Delta y}{\Delta x} = \frac{-2}{(x+\Delta x)x}; \quad y' = \frac{-2}{x^2}$$

2.2.5. $y = \frac{3x+1}{2x-5}$

Solution. $y + \Delta y = \frac{3(x+\Delta x) + 1}{2(x+\Delta x) - 5};$

$$= \frac{3(x + \Delta x) + 1}{2(x + \Delta x) - 5} - \frac{3x + 1}{2x - 5} = \frac{-17\Delta x}{(2x + 2\Delta x - 5)(2x - 5)};$$

$$= \frac{-17}{(2x - 5)^2}$$

2.2.6. Find the instantaneous rate of change of the volume a sphere with respect to the radius at the instant that the radius equals 5 inches.

Solution. $V = \frac{4}{3}\pi r^3$; $V + \Delta V = \frac{4}{3}\pi(r + \Delta r)^3$
$= \frac{4}{3}\pi(r^3 + 3r^2 \Delta r + 3r \overline{\Delta r}^2 + \overline{\Delta r}^3);$
$\Delta V = \frac{4}{3}\pi(3r^2 \Delta r + 3r \overline{\Delta r}^2 + \overline{\Delta r}^3);$
$\frac{\Delta V}{\Delta r} = \frac{4}{3}\pi(3r^2 + 3r \Delta r + \overline{\Delta r}^2);$
$\frac{dV}{dr} = 4\pi r^2 = 100 \pi$ cubic inches per inch.

3. Formulas for differentiation. In the following formulas, u and v stand for differentiable functions of x.

$$\frac{d}{dx} K = 0 \qquad (2.3.1)$$

i.e., the derivative of a constant is zero.

$$\frac{d}{dx} x^n = nx^{n-1} \qquad (2.3.2)$$

$$\frac{d}{dx} Ku = K \frac{du}{dx} \qquad (2.3.3)$$

$$\frac{d}{dx} (u \pm v) = \frac{du}{dx} \pm \frac{dv}{dx} \qquad (2.3.4)$$

$$\frac{d}{dx} (uv) = u \frac{dv}{dx} + v \frac{du}{dx} \qquad (2.3.5)$$

$$\frac{d}{dx} (u^n) = nu^{n-1} \frac{du}{dx} \qquad (2.3.6)$$

$$\frac{d}{dx} \left(\frac{u}{v}\right) = \frac{v \frac{du}{dx} - u \frac{dv}{dx}}{v^2} \qquad (2.3.7)$$

$$\frac{dy}{dx} = \frac{1}{\frac{dx}{dy}} \qquad (2.3.8)$$

If y is a function of u and u is a function of x,

$$\frac{dy}{dx} = \frac{dy}{du} \cdot \frac{du}{dx} \qquad (2.3.$$

Formula 2.3.6 is a special case of 2.3.9, for if $y = u^n$, th $\frac{dy}{du} = nu^{n-1}$, and $\frac{dy}{dx} = nu^{n-1}\frac{du}{dx}$.

2.3.1. $\dfrac{d}{dx}(x^2 + 3x + 2) = \dfrac{d}{dx}x^2 + \dfrac{d}{dx}3x + \dfrac{d}{dx}2 = 2x +$

2.3.2. $\dfrac{d}{dx}(3x^{3/2} + 2x^{-2} + 5x^4 + 9) = \tfrac{9}{2}x^{1/2} - 4x^{-3} + 20x^3$

2.3.3. $\dfrac{d}{dx}(3x^2 - 2x + 5)(x^2 + 1)$

$= (3x^2 - 2x + 5)\dfrac{d}{dx}(x^2 + 1) + (x^2 + 1)\dfrac{d}{dx}(3x^2 - 2x + $

$= (3x^2 - 2x + 5)2x + (x^2 + 1)(6x - 2)$

$= 12x^3 - 6x^2 + 16x - 2$

2.3.4. $y = (x^2 - 3x + 2)^{14}$

$\dfrac{dy}{dx} = 14(x^2 - 3x + 2)^{13}\dfrac{d}{dx}(x^2 - 3x + 2)$

$= 14(x^2 - 3x + 2)^{13}(2x - 3)$

2.3.5. $y = (x^3 + 1)^{-2/3}$

$y' = -\tfrac{2}{3}(x^3 + 1)^{-5/3}\dfrac{d}{dx}(x^3 + 1) = -\tfrac{2}{3}(x^3 + 1)^{-5/3}3x$

$= -2x^2(x^3 + 1)^{-5/3}$

2.3.6. $\quad y = \sqrt{3x^4 - x^3 + 2x^2 + x - 5}$

$y' = \dfrac{\dfrac{d}{dx}(3x^4 - x^3 + 2x^2 + x - 5)}{2\sqrt{3x^4 - x^3 + 2x^2 + x - 5}},$

$y' = \dfrac{12x^3 - 3x^2 + 4x + 1}{2\sqrt{3x^4 - x^3 + 2x^2 + x - 5}}$

2.3.7. $f(x) = \dfrac{3x+1}{x^2-x+4}$

$) = \dfrac{(x^2-x+4)\dfrac{d}{dx}(3x+1)-(3x+1)\dfrac{d}{dx}(x^2-x+4)}{(x^2-x+4)^2}$

$) = \dfrac{(x^2-x+4)3-(3x+1)(2x-1)}{(x^2-x+4)^2} = \dfrac{-3x^2-2x+13}{(x^2-x+4)^2}$

2.3.8. $y = u^3 + 3u - 1$ and $u = x^2 + x - 4$. Find $\dfrac{dy}{dx}$.

Solution. $\dfrac{dy}{dx} = \dfrac{dy}{du}\dfrac{du}{dx} = \dfrac{d}{du}(u^3+3u-1)\dfrac{d}{dx}(x^2+x-4)$

$= (3u^2+3)(2x+1) = [3(x^2+x-4)^2+3](2x+1)$

2.3.9. $y = t^3(t^2+1)^{3/2}$

$' = t^3 \tfrac{3}{2}(t^2+1)^{1/2}2t + (t^2+1)^{3/2}3t^2 = 3t^2(t^2+1)^{1/2}(2t^2+1)$

2.3.10. $\quad y = \sqrt{3x}$

$y' = \dfrac{\dfrac{d}{dx}(3x)}{2\sqrt{3x}} = \dfrac{3}{2\sqrt{3x}}$

2.3.11. $\quad y = \sqrt{25-x^2}$

$\dfrac{dy}{dx} = \dfrac{\dfrac{d}{dx}(25-x^2)}{2\sqrt{25-x^2}} = \dfrac{-x}{\sqrt{25-x^2}}$

2.3.12. $y = \sqrt{1+\sqrt{1+\sqrt{1+x}}} = \{1+[1+(1+x)^{1/2}]^{1/2}\}^{1/2}$

$= \tfrac{1}{2}\{1+[1+(1+x)^{1/2}]^{1/2}\}^{-1/2}\dfrac{d}{dx}\{1+[1+(1+x)^{1/2}]^{1/2}\}$

$= \tfrac{1}{2}\{1+[1+(1+x)^{1/2}]^{1/2}\}^{-1/2}\tfrac{1}{2}[1+(1+x)^{1/2}]^{-1/2}$

$\qquad\qquad\qquad\qquad\qquad \dfrac{d}{dx}[1+(1+x)^{1/2}]$

$= \tfrac{1}{8}\{1+[1+(1+x)^{1/2}]^{1/2}\}^{-1/2}[1+(1+x)^{1/2}]^{-1/2}(1+x)^{-1/2}$

2.3.13. $y = (x^3 + 1)^2(x^2 - 4)^3$

$y' = (x^3 + 1)^2 3(x^2 - 4)^2 2x + (x^2 - 4)^3 2(x^3 + 1)3$

$ = 6x(x^3 + 1)(x^2 - 4)^2(2x^3 - 4x + 1)$

2.3.14. $y = \dfrac{2x + 1}{\sqrt{x + 2}}$

$y' = 2\dfrac{\sqrt{x + 2} - (2x + 1)\frac{1}{2}(x + 2)^{-1/2}}{x + 2}$

$ = \dfrac{2(x + 2) - \frac{1}{2}(2x + 1)}{(x + 2)^{3/2}} = \dfrac{x + \frac{7}{2}}{(x + 2)^{3/2}}$

This problem may also be differentiated as a product thu

$y = (2x + 1)(x + 2)^{-1/2}$

$y' = (2x + 1)(-\frac{1}{2})(x + 2)^{-3/2} + (x + 2)^{-1/2}2 = \dfrac{x + \frac{7}{2}}{(x + 2)^{3/2}}$

2.3.15. $y = (x^2 + 1)(x^2 + 2)(x^2 + 3)$

$ = (x^2 + 1)[(x^2 + 2)(x^2 + 3)]$

$y' = (x^2 + 1)\dfrac{d}{dx}[(x^2 + 2)(x^2 + 3)]$

$ + [(x^2 + 2)(x^2 + 3)]\dfrac{d}{dx}(x^2 + 1)$

$ = (x^2 + 1)[(x^2 + 2)2x + (x^2 + 3)2x]$

$ + [(x^2 + 2)(x^2 + 3)]2x$

$ = 2x[(x^2 + 1)(x^2 + 2) + (x^2 + 1)(x^2 + 3)$

$ + (x^2 + 2)(x^2 + 3)]$

2.3.16. $y = \dfrac{z}{\sqrt{2z + 7}} - \dfrac{\sqrt{2z + 7}}{z}$

$y' = \dfrac{\sqrt{2z + 7} - z\dfrac{2}{2\sqrt{2z + 7}}}{(2z + 7)} - \dfrac{z\dfrac{2}{2\sqrt{2z + 7}} - \sqrt{2z + 7}}{z^2}$

$ = \dfrac{(z + 7)(z^2 + 2z + 7)}{z^2(2z + 7)^{3/2}}$

Derivatives of higher order. Since the derivative itself [is a] function, it may also be differentiated giving us a derivative [of] a derivative, which is called a second derivative and is indicated by $\frac{d^2y}{dx^2}$ or $f''(x)$ or y''. The derivative of the second derivative is a third derivative, indicated by $\frac{d^3y}{dx^3} = f'''(x) = y'''$,

A freely falling body obeys the law $s = 16.1t^2$, where s [is] the distance fallen in feet and t is the time of the fall in seconds. [Th]e velocity of the fall is $\frac{ds}{dt}$. The time rate of change of velocity [is] $\frac{d^2s}{dt^2}$, which is called the acceleration.

2.4.1. Find the acceleration of a freely falling body.

Solution

$s = 16.1t^2$; $\quad s' = 32.3t$ ft per sec; $\quad s'' = 32.2$ ft per sec^2

Notice that the acceleration is constant and does not change with time.

2.4.2. A particle moves in a straight line according to the [la]w $s(t) = t^3 - 6t^2 + 20t - 5$, where $s(t)$ is distance in feet from [the] initial point, and t is time in seconds. Find the velocity of the [pa]rticle when its acceleration is zero.

Solution

$$s'(t) = 3t^2 - 12t + 20; \quad s''(t) = 6t - 12 = 0.$$

[H]ence $t = 2$, and

$$s'(2) = 3(2^2) - 12(2) + 20 = 8 \text{ ft per sec}$$

2.4.3. A bomb is dropped from a height of 6440 ft. When [a]nd at what velocity will it strike the ground?

Solution

$s(t) = 16.1t^2$; $\quad 16.1t^2 = 6440$; $\quad t = 20$ sec to reach the ground.

$s'(t) = 32.2t$; $\quad s'(20) = 32.2(20) = 644$ ft per sec

2.4.4. Find y', y'', y''', y^{IV}, if $y = \dfrac{t}{1+t}$.

Solution

$$y' = \frac{(1+t)(1) - t(1)}{(1+t)^2} = (1+t)^{-2}; \qquad y'' = -2(1+t)^{-3};$$

$$y''' = 6(1+t)^{-4}; \qquad y^{IV} = -24(1+t)^{-5}$$

2.4.5. If $f(x) = (x^2 + 1)^{3/2}$, find $f'(x)$, $f''(x)$, $f'''(x)$, $f^{IV}($

Solution

$$f'(x) = \tfrac{3}{2}(x^2+1)^{1/2} 2x = 3x(x^2+1)^{1/2}$$

$$f''(x) = 3x(\tfrac{1}{2})(x^2+1)^{-1/2} 2x + (x^2+1)^{1/2} 3$$
$$= 3x^2(x^2+1)^{-1/2} + 3(x^2+1)^{1/2}$$

$$f'''(x) = 3x^2(-\tfrac{1}{2})(x^2+1)^{-3/2} 2x + (x^2+1)^{-1/2} 6x$$
$$+ 3(\tfrac{1}{2})(x^2+1)^{-1/2} 2x = -3x^3(x^2+1)^{-3/2}$$
$$+ 9x(x^2+1)^{-1/2}$$

$$f^{IV}(x) = -3x^3(-\tfrac{3}{2})(x^2+1)^{-5/2} 2x + (x^2+1)^{-3/2}(-9x^2)$$
$$+ 9x(-\tfrac{1}{2})(x^2+1)^{-3/2} 2x + (x^2+1)^{-1/2} 9$$
$$= 9x^4(x^2+1)^{-5/2} - 18x^2(x^2+1)^{-3/2}$$
$$+ 9(x^2+1)^{-1/2}$$

2.4.6. Find higher derivatives of $y = uv$, where u and and therefore y, are functions of x.

Solution

$$y' = uv' + u'v$$

$$y'' = uv'' + u'v' + u'v' + u''v = uv'' + 2u'v' + u''v$$

$$y''' = uv''' + u'v'' + 2u'v'' + 2u''v' + u''v' + u'''v$$
$$= uv''' + 3u'v'' + 3u''v' + u'''v$$

Similarly,

$$y^{IV} = uv^{IV} + 4u'v''' + 6u''v'' + 4u'''v' + u^{IV}v$$

1, in general,

$$\ldots = uv^{(n)} + nu'v^{(n-1)} + \frac{n(n-1)}{2!} u''v^{(n-2)} + \cdots$$
$$+ \frac{n(n-1)\cdots(n-r+1)}{r!} u^{(r)}v^{(n-r)} + \cdots + u^{(n)}v$$

is last result is known as Leibnitz's formulas after the German thematician Gottfried Wilhelm von Leibnitz (1646–1716).

. Implicit differentiation. For an equation such as

$$x^3 y - 4xy^2 = y + x^2$$

find $\frac{dy}{dx}$ by differentiating both members of the equation by formulas of 2.3, treating x as the independent variable. then solve the resulting equation for $\frac{dy}{dx}$.

$$x^3 \frac{dy}{dx} + y \cdot 3x^2 - 4x \cdot 2y \frac{dy}{dx} - 4y^2 = \frac{dy}{dx} + 2x$$

$$(x^3 - 8xy - 1)\frac{dy}{dx} = -3x^2 y + 4y^2 + 2x$$

$$\frac{dy}{dx} = \frac{-3x^2 y + 4y^2 + 2x}{x^3 - 8xy - 1}$$

2.5.1. Find $\frac{dy}{dx}$ if $x^2 + y^2 = a^2$.

Solution

$$2x + 2y\frac{dy}{dx} = 0; \qquad \frac{dy}{dx} = -\frac{x}{y}$$

2.5.2. Find y' if $xy = (x+y)^3$.

Solution

$$xy' + y = 3(x+y)^2(1+y'); \qquad y' = \frac{3(x+y)^2 - y}{x - 3(x+y)^2}$$

2.5.3. Find y' and y'' if $x^{2/3} + y^{2/3} = a^{2/3}$.

Solution

$$\tfrac{1}{3}x^{-1/3} + \tfrac{1}{3}y^{-1/3}y' = 0; \qquad y' = -\frac{y^{1/3}}{x^{1/3}}.$$

Differentiating both members of this last equation with respect to x, remembering that y is a function of x, then

$$y'' = -\frac{x^{1/3}(\tfrac{1}{3})y^{-2/3}y' - y^{1/3}(\tfrac{1}{3})x^{-2/3}}{x^{2/3}}.$$

Replacing y' by its value above and simplifying,

$$y'' = -\frac{x^{1/3}(\tfrac{1}{3})y^{-2/3}\left(-\dfrac{y^{1/3}}{x^{1/3}}\right) - y^{1/3}(\tfrac{1}{3})x^{-2/3}}{x^{2/3}}$$

$$= \tfrac{1}{3}(x^{-2/3}y^{-1/3} + x^{-4/3}y^{1/3})$$

2.5.4. Given $xy = 1$. Find y''' by (a) differentiating implicitly and (b) solving first for y.

Solution

(a) $xy' + y = 0;\qquad y' = -\dfrac{y}{x};\qquad y'' = -\dfrac{xy' - y}{x^2}$

$$= \frac{y - x\left(-\dfrac{y}{x}\right)}{x^2} = \frac{2y}{x^2}, \quad y''' = \frac{x^2(2y') - 2y \cdot 2x}{x^4}$$

$$= \frac{x^2\left(-\dfrac{2y}{x}\right) - 4xy}{x^4} = -\frac{6xy}{x^4} = -\frac{6y}{x^3}$$

(b) $y = \dfrac{1}{x} = x^{-1};\quad y' = -x^{-2};\quad y'' = 2x^{-3};\quad y''' = -6x^{-4}$

To reduce the result in (a) to that in (b), we may replace y by $\dfrac{1}{x}$.

$$y''' = -\frac{6y}{x^3} = -\frac{6\left(\dfrac{1}{x}\right)}{x^3} = -6x^{-4}$$

2.5.5. Given $x^3 - xy = y^3$, show that $\dfrac{dy}{dx} = \dfrac{1}{\dfrac{dx}{dy}}$.

Solution. Differentiating implicitly with respect to x

$$3x^2 - x\frac{dy}{dx} - y = 3y^2 \frac{dy}{dx}; \qquad \frac{dy}{dx} = \frac{3x^2 - y}{3y^2 + x}$$

Differentiating implicitly with respect to y,

$$3x^2 \frac{dx}{dy} - x - y\frac{dx}{dy} = 3y^2; \qquad \frac{dx}{dy} = \frac{3y^2 + x}{3x^2 - y}$$

Hence

$$\frac{dy}{dx} = \frac{3x^2 - y}{3y^2 + x} = \frac{1}{\dfrac{3y^2 + x}{3x^2 - y}} = \frac{1}{\dfrac{dx}{dy}}$$

2.6. Differentiation of parametric equations. If x and y are given as functions of a parameter t, we have the following formula:

$$\frac{dy}{dx} = \frac{\dfrac{dy}{dt}}{\dfrac{dx}{dt}} \qquad\qquad (2.6.1)$$

2.6.1. $y = t^3 + 2t^2 + 3t + 1;\qquad x = t^2 + t + 3$

$$\frac{dy}{dx} = \frac{\dfrac{d}{dt}(t^3 + 2t^2 + 3t + 1)}{\dfrac{d}{dt}(t^2 + t + 3)} = \frac{3t^2 + 4t + 3}{2t + 1}$$

2.6.2. $y = \dfrac{\theta + 2}{\theta - 2};\qquad x = \theta^3 + 1$

$$\frac{dy}{dx} = \frac{\dfrac{(\theta - 2)1 - (\theta + 2)1}{(\theta - 2)^2}}{3\theta^2} = \frac{-4}{3\theta^2(\theta - 2)^2}$$

2.6.3. $y = \sqrt{\dfrac{\theta^2 + 1}{\theta - 1}}; \qquad x = \theta^2 - 2\theta + 5$

$$\dfrac{dy}{d\theta} = \dfrac{1}{2}\left(\dfrac{\theta^2 + 1}{\theta - 1}\right)^{-1/2} \dfrac{(\theta - 1)2\theta - (\theta^2 + 1)1}{(\theta - 1)^2}$$

$$= \dfrac{\theta^2 - 2\theta - 1}{2(\theta^2 + 1)^{1/2}(\theta - 1)^{3/2}}$$

$\dfrac{dx}{d\theta} = 2\theta - 2 = 2(\theta - 1); \qquad \dfrac{dy}{dx} = \dfrac{\dfrac{dy}{d\theta}}{\dfrac{dx}{d\theta}} = \dfrac{\theta^2 - 2\theta - 1}{4(\theta^2 + 1)^{1/2}(\theta - 1)^{5/2}}$

2.6.4. $y = t^3 + 1; \qquad x = t^2 + 3$

$\dfrac{dy}{dx} = \dfrac{3t^2}{2t} = \dfrac{3}{2}t; \qquad \dfrac{d^2y}{dx^2} = \dfrac{d}{dt}\left(\dfrac{3}{2}t\right)\dfrac{dt}{dx}$

$$= \dfrac{3}{2}\dfrac{1}{\dfrac{dx}{dt}} = \dfrac{3}{2}\dfrac{1}{2t} = \dfrac{3}{4t} \text{ by using 2.3.8}$$

$\dfrac{d^3y}{dx^3} = \dfrac{d}{dt}\left(\dfrac{3}{4t}\right)\dfrac{dt}{dx} = \dfrac{d}{dt}\left(\dfrac{3}{4}t^{-1}\right)\dfrac{1}{\dfrac{dx}{dt}} = -\dfrac{3}{4}t^{-2}\dfrac{1}{2t} = -\dfrac{3}{8t^3}$

2.6.5. $y = \dfrac{1}{u - 1}; \qquad x = \dfrac{u}{u^2 - 1}$

$\dfrac{dy}{du} = -(u - 1)^{-2}$

$\dfrac{dx}{du} = \dfrac{(u^2 - 1)\cdot 1 - u\cdot 2u}{(u^2 - 1)^2} = -\dfrac{u^2 + 1}{(u^2 - 1)^2}$

$\dfrac{dy}{dx} = \dfrac{-(u - 1)^{-2}}{-\dfrac{u^2 + 1}{(u^2 - 1)^2}} = \dfrac{(u + 1)^2}{u^2 + 1}$

$\dfrac{d^2y}{dx^2} = \dfrac{d}{du}\dfrac{(u + 1)^2}{u^2 + 1}\dfrac{du}{dx}$

$= \dfrac{(u^2 + 1)2(u + 1) - (u + 1)^2 2u}{(u^2 + 1)^2} \dfrac{1}{\dfrac{dx}{du}}$ (by using 2.3.8)

$= -2\dfrac{u^2 - 1}{(u^2 + 1)^2}\left(-\dfrac{(u^2 - 1)^2}{u^2 + 1}\right) = 2\dfrac{(u^2 - 1)^3}{(u^2 + 1)^3}$

Chapter 3

APPLICATIONS OF DIFFERENTIATION

3.1. Tangent and normal to a curve. From analytic geometry $y - y_0 = m(x - x_0)$ and $(y - y_0) = -\dfrac{1}{m}(x - x_0)$ are perpendicular (or normal) lines through (x_0, y_0) with the slopes m and $-\dfrac{1}{m}$, respectively. The slope of a curve $y = f(x)$ at any point (x, y) on the curve is $f'(x)$, and at a specific point (x_0, y_0), the slope would be $f'(x_0)$. Combining these results, we have for the equations of the tangent and normal, respectively, to the curve $y = f(x)$ at (x_0, y_0):

$$y - y_0 = f'(x_0)(x - x_0) \qquad (3.1.1)$$

$$y - y_0 = \dfrac{-1}{f'(x_0)}(x - x_0) \qquad (3.1.2)$$

Find the equations of the tangent and normal to each of the following curves at the points indicated.

3.1.1. $y = 2x^3 + 3x^2 + 1$ at $P(-2, -3)$
Solution. $[y']_P = [6x^2 + 6x]_P = 12$, where the notation means that the derivative is evaluated at the point $P(-2, -3)$.

Tangent
$$y + 3 = 12(x + 2); \qquad 12x - y = -21$$

Normal
$$y + 3 = \dfrac{-1}{12}(x + 2); \qquad x + 12y = -38$$

3.1.2. $y^3 = x^2 + 7$ at $P(1, 2)$
$$y = (x^2 + 7)^{1/3}; \qquad [y']_P = [\tfrac{1}{3}(x^2 + 7)^{-2/3} 2x]_P = \tfrac{1}{6}$$
or, using the method of 2.5,
$$3y^2 y' = 2x; \qquad [y']_P = \left[\dfrac{2x}{3y^2}\right]_P = \dfrac{1}{6}$$

Tangent
$$y - 2 = \tfrac{1}{6}(x - 1); \qquad x - 6y = -11$$

Normal
$$y - 2 = -6(x - 1); \quad 6x + y = 8$$

3.1.3. $y = x^{-2/3} - 3x^2 + x^{1/4} + 1$ at $P(1, 0)$
Solution
$$[y']_P = [-\tfrac{2}{3}x^{-5/3} - 6x + \tfrac{1}{4}x^{-3/4}]_P = -\tfrac{77}{12}$$

Tangent
$$y = -\tfrac{77}{12}(x - 1); \quad 77x + 12y = 77$$

Normal
$$y = \tfrac{12}{77}(x - 1); \quad 12x - 77y = 12$$

3.1.4. $y = \sqrt[5]{x^3 + 5}$ at $P(3, 2)$
Solution
$$[y']_P = [\tfrac{1}{5}(x^3 + 5)^{-4/5} 3x^2]_P = \tfrac{27}{80}$$

Tangent
$$y - 2 = \tfrac{27}{80}(x - 3); \quad 27x - 80y = -79$$

Normal
$$y - 2 = -\tfrac{80}{27}(x - 3); \quad 80x + 27y = 294$$

3.1.5. Obtain the equations of the tangent and normal to $y = 2x^2 - 6x + 7$ at the point P where the slope of the tangent is 2.
Solution
$$y' = 4x - 6 = 2; \quad x = 2; \quad [y']_P = 2;$$
$$[y]_P = [2 \cdot 2^2 - 6 \cdot 2 + 7] = 3$$

Tangent
$$y - 3 = 2(x - 2); \quad 2x - y = 1$$

Normal
$$y - 3 = \frac{-1}{2}(x - 2); \quad x + 2y = 8$$

3.1.6. Find the point in which the normal to $y = x^2 - x$ at $P(-2, 6)$ intersects the line $2x + 5y = 5$.

Solution. Let x be the width of the top, and A be the area of the section. A must be maximized. The depth of the gutter is

$$\sqrt{a^2 - \left(\frac{x-a}{2}\right)^2} = \tfrac{1}{2}\sqrt{3a^2 + 2ax - x^2};$$

$$A = \frac{x+a}{2} \cdot \frac{\sqrt{3a^2 + 2ax - x^2}}{2}$$

$$\frac{dA}{dx} = \frac{1}{4}\left[\frac{(x+a)(2a-2x)}{2\sqrt{3a^2 + 2ax - x^2}} + \sqrt{3a^2 + 2ax - x^2}\right]$$

$$= \frac{(2a-x)(a+x)}{2\sqrt{3a^2 + 2ax - x^2}} = 0.$$

Since x must be positive, $x = 2a$ is the only solution. If $x < 2a$, then $\frac{dA}{dx} > 0$. If $x > 2a$, then $\frac{dA}{dx} < 0$. Hence $x = 2a$ for a maximum.

3.5. Rate problems.

3.5.1. Two ships leave port simultaneously, one sailing due south at 20 miles per hour, the other sailing due east at 30 miles per hour. How fast are the ships separating at the end of 3 hours?

Solution. Let z be the distance between the ships and t the elapsed time in hours since they left port. Then $z^2 = (20t)^2 + (30t)^2$; $z^2 = 1300 t^2$; $z = \sqrt{1300}\, t$; $\frac{dz}{dt} = \sqrt{1300} = 36.06$ miles per hour approximately.

3.5.2. Water is running at the rate of 2 cu in. per minute into an inverted, right circular cone whose altitude is twice its diameter. How fast is the surface of the water rising when the water is 10 in. deep in the cone?

Solution. Let V be the volume of the water in the cone, h its depth, r its top radius, t the variable time in minutes. Then $V = \tfrac{1}{3}\pi r^2 h$, where $h = 4r$, so that

$$V = \frac{1}{3}\pi \left(\frac{h}{4}\right)^2 h = \frac{\pi}{48} h^3; \qquad \frac{dV}{dt} = \frac{d}{dh}\left(\frac{\pi h^3}{48}\right)\frac{dh}{dt};$$

$$\frac{dV}{dt} = \frac{\pi}{16} h^2 \frac{dh}{dt}.$$

Replacing known quantities by their values, we have $2 = \frac{\pi}{16} 10^2 \frac{dh}{dt}$. Hence $\frac{dh}{dt} = \frac{0.32}{\pi} = 0.102$ in. per min approximately.

3.5.3. A stone is dropped into still water. The radius of the outer ripple increases at the rate of 4 ft per second when the radius is 10 ft. How fast is the area of the circle of disturbed water increasing?

Solution

$$A = \pi r^2; \quad \frac{dA}{dt} = \frac{d}{dr}(\pi r^2)\frac{dr}{dt} = 2\pi r \frac{dr}{dt} = 2\pi \cdot 10 \cdot 4 = 80\pi$$

$$= 251.33 \text{ sq ft per sec approximately.}$$

3.5.4. A kite, 80 ft above level ground, is moving horizontally away from the boy who is flying it at the rate of 4 ft per sec. How fast is the boy releasing the string at the instant when he has released 100 ft?

Solution. Let x be the horizontal distance to the kite, z the length of string released, and t the time in seconds. Then $z^2 = x^2 + 80^2$, whence $2z\frac{dz}{dt} = 2x\frac{dx}{dt}$. By the Pythagorean theorem $x = 60$ when $z = 100$. Substituting the known values, after dividing out 2, gives $100\frac{dz}{dt} = 60(4)$. Therefore $\frac{dz}{dt} = 2.4$ ft per sec.

3.5.5. A point moves on the upper branch of the semicubical parabola $y^2 = x^3$ in such a way as to cause its abscissa to increase 5 units per second when $x = 4$. How fast is its ordinate changing?

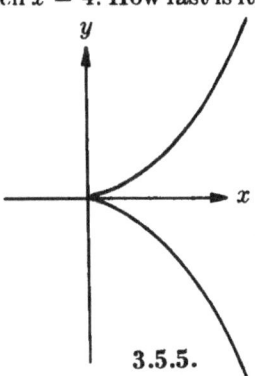

3.5.5.

Solution. Differentiating $y^2 = x^3$ with respect to the time t gives $2y\dfrac{dy}{dt} = 3x^2\dfrac{dx}{dt}$. When $x = 4$, $y = 8$. Substituting these values gives $2(8)\dfrac{dy}{dt} = 3(4^2)\cdot 5$; $\dfrac{dy}{dt} = 15$ units per second.

3.5.6. A 10-ft ladder is leaning against a wall on level ground. If the bottom of the ladder is dragged away from the wall at the rate of 4 ft per sec, how fast will the top descend at the instant when it is 8 ft from the ground?

Solution. Let y be the height of the top of the ladder above the ground and x the distance of the bottom of the ladder from the wall. When $y = 8$, $x = 6$. Also $x^2 + y^2 = 100$; $2x\dfrac{dx}{dt} + 2y\dfrac{dy}{dt} = 0$. Substituting known values, after dividing out 2, gives $6(4) + 8\dfrac{dy}{dt} = 0$; $\dfrac{dy}{dt} = -3$ ft per sec.

3.5.7. A man 6 ft tall is walking 3 miles an hour away from a light 15 ft above level ground. When his horizontal distance from the light is 12 ft, how fast is his shadow lengthening?

3.5.7.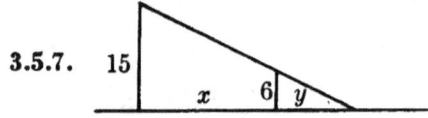

Solution. Let x be the man's horizontal distance from the light and y the length of his shadow. By similar triangles, $\dfrac{x+y}{15} = \dfrac{y}{6}$; $2x = 3y$; $2\dfrac{dx}{dt} = 3\dfrac{dy}{dt}$. Substituting, $2(3) = 3\dfrac{dy}{dt}$ then $\dfrac{dy}{dt} = 2$ miles per hour $= 2.933$ ft per sec.

3.5.8. In **3.5.7** how fast is the end of the man's shadow farthest from the light moving along the ground?

Solution. We need to find how fast $x + y$ is changing. Using the results of **3.5.7**, $\dfrac{d}{dt}(x + y) = \dfrac{dx}{dt} + \dfrac{dy}{dt} = 3 + 2 = 5$ miles per hour.

3.5.9. A man on a pier is pulling in a boat by means of a rope. His hands are 10 ft above the ring on the boat to which the rope is tied. The boat is 24 ft away measured over the surface of the water. If the man pulls in the rope at the rate of 3 ft per sec, how fast does the boat approach the pier?

3.5.9.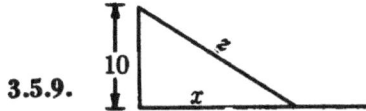

Solution. Let x be the distance to the boat over the surface of the water and z the length of the rope. When $x = 24$, $z = 26$. Then $\frac{dz}{dt} = -3$, and $x^2 + 100 = z^2$, and $2x\frac{dx}{dt} = 2z\frac{dz}{dt}$. Dividing out the 2 and substituting known values gives $24\frac{dx}{dt} = 26(-3)$, so that $\frac{dx}{dt} = -3.25$ ft per sec.

3.6. Plane curvilinear motion. If a point $P(x, y)$ moves in a plane along a curve C, then the coordinates x and y are functions of the time t. The derivatives $\frac{dx}{dt}$ and $\frac{dy}{dt}$ are the x and y components of the velocity of P, also denoted by v_x and v_y. The magnitude of the tangential velocity vector is given by

$$v = \sqrt{v_x^2 + v_y^2} \qquad (3.6.1)$$

The angle τ this vector makes with the x axis is given by

$$\tan \tau = \frac{v_y}{v_x} \qquad (3.6.2)$$

Also $\qquad v_x = v \cos \tau; \qquad v_y = v \sin \tau \qquad (3.6.3)$

The x and y components of acceleration are $a_x = \frac{dv_x}{dt} = \frac{d^2x}{dt^2}$; $a_y = \frac{dv_y}{dt} = \frac{d^2y}{dt^2}$. The resultant acceleration has magnitude

$$a = \sqrt{a_x^2 + a_y^2} \qquad (3.6.4)$$

and inclination ϕ.

$$\tan \phi = \frac{a_y}{a_x}; \qquad a_x = a \cos \phi; \qquad a_y = a \sin \phi \qquad (3.6.5)$$

The tangential component of acceleration has magnitude

$$a_t = \frac{dv}{dt} = \frac{v_x a_x + v_y a_y}{v} \qquad (3.6.6)$$

The magnitude of the normal component is

$$a_n = \frac{v_x a_y - v_y a_x}{v} \qquad (3.6.7)$$

3.6.1. As a particle moves on a plane curve, its coordinates are given by $x = t^2$ and $y = 2t - 1$, where t denotes time. Find the equation of the path and the direction and magnitude of both the velocity and acceleration when $t = 1$.

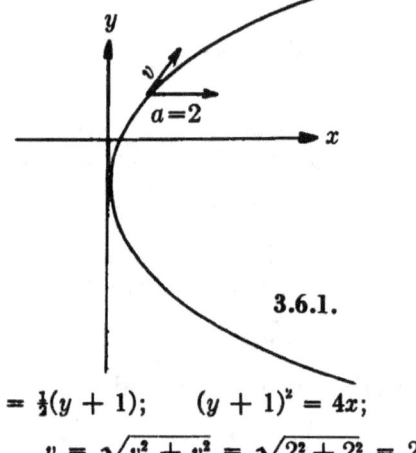

3.6.1.

Solution. $t = \frac{1}{2}(y + 1);$ $(y + 1)^2 = 4x;$ $v_x = 2t = 2;$

$v_y = 2;$ $v = \sqrt{v_x^2 + v_y^2} = \sqrt{2^2 + 2^2} = 2\sqrt{2};$

$\tan \tau = \frac{v_y}{v_x} = \frac{2}{2} = 1;$ $\tau = \frac{\pi}{4};$ $a_x = 2;$ $a_y = 0;$

$a = \sqrt{a_x^2 + a_y^2} = 2;$ $\tan \phi = \frac{a_y}{a_x} = \frac{0}{2} = 0;$ $\phi = 0$

3.6.2. In **3.6.1** find the tangential component a_t and the normal component a_n of the acceleration.
Solution. Using 3.6.6,

$$a_t = \frac{dv}{dt} = \frac{d}{dt}\sqrt{(2t)^2 + 2^2} = \frac{2(2t)2}{2\sqrt{(2t)^2 + 2^2}} = \sqrt{2},$$

Using the other part of 3.6.6, we may check thus:
$$a_t = \frac{2(2) - 2(0)}{2\sqrt{2}} = \sqrt{2}.$$

Using 3.6.7, $a_n = \dfrac{2(0) - 2(2)}{2\sqrt{2}} = -\sqrt{2}.$

3.6.3. The same as **3.6.1** and **3.6.2** but with $x = t^2$, $y = t^3$, $t = 2$.

Solution. $t = x^{1/2}$, $y = x^{3/2}$. Then $y^2 = x^3$ is the path, which is a semicubical parabola. See the figure of **3.5.5**. $v_x = 2t$;
$$v_y = 3t^2; \qquad v = \sqrt{v_x^2 + v_y^2} = \sqrt{4^2 + 12^2} = 4\sqrt{10};$$
$$\tan \tau = \frac{v_y}{v_x} = \frac{12}{4} = 3; \qquad \tau = 71°34'; \qquad a_x = 2; \qquad a_y = 6t;$$
$$a = \sqrt{a_x^2 + a_y^2} = \sqrt{2^2 + 12^2} = 2\sqrt{37};$$
$$\tan \phi = \frac{a_y}{a_x} = \frac{12}{2} = 6; \qquad \phi = 80°32'; \qquad a_t = \frac{dv}{dt}$$
$$= \frac{d}{dt}\sqrt{4t^2 + 9t^4} = \frac{8t + 36t^3}{2\sqrt{4t^2 + 9t^4}} = 3.8\sqrt{10} \text{ when } t = 2.$$

Then $a_n = \dfrac{v_x a_y - v_y a_x}{v} = \dfrac{4(12) - 12(2)}{4\sqrt{10}} = 0.6\sqrt{10}.$

3.6.4. A particle moves clockwise around the circle $x^2 + y^2 = 25$ at a constant speed of 5 ft per sec. At the instant that the particle passes through $(3, 4)$, find v_x, v_y, τ, a_x, a_y, a, ϕ, a_t, a_n.

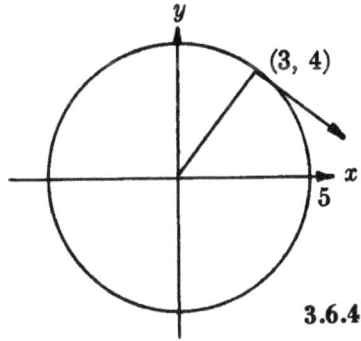

3.6.4.

Solution

$$2xv_x + 2yv_y = 0; \quad v_x = -\frac{y}{x}v_y; \quad v_x^2 + v_y^2 = v^2;$$

$$\left(-\frac{y}{x}v_y\right)^2 + v_y^2 = 25; \quad v_y^2 = \frac{25}{\left(-\frac{y}{x}\right)^2 + 1} = \frac{25}{(-\frac{4}{3})^2 + 1} = 9;$$

$$v_y = -3.$$

(The reason for the minus sign is clear from the figure.)

$$v_x = -\frac{y}{x}v_y = \frac{-4}{3}(-3) = 4; \quad \tan \tau = \frac{v_y}{v_x} = \frac{-3}{4};$$

$\tau = -36°52'$. Differentiating $xv_x + yv_y = 0$ with respect to t, we obtain $xa_x + v_x^2 + ya_y + v_y^2 = 0$. At (3, 4) this becomes $3a_x + 4a_y = -25$. By 3.6.6 $a_t = 0$, or $a_t = \frac{v_x a_x + v_y a_y}{v} = 0$. At (3, 4) this becomes $4a_x - 3a_y = 0$. Solving this simultaneously with $3a_x + 4a_y = -25$ gives $a_x = -3$ and $a_y = -4$. Hence $a = \sqrt{a_x^2 + a_y^2} = 5; \quad \tan \phi = \frac{a_y}{a_x} = \frac{-4}{-3}, \phi = -126°52'$. The acceleration is toward the center through (3, 4). Since $a_t = 0$, then $a_n = a = 5$.

3.6.5. A particle moves to the right along $y = e^x$ with $v_x = 2$ ft per sec and $a_x = 3$ ft per sec^2 at the instant it passes through (0, 1). At this instant find v_y, v, τ, a_y, a, ϕ, a_t, a_n.

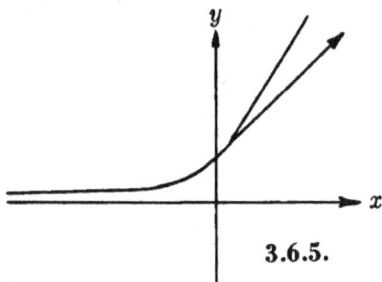

3.6.5.

Solution. Differentiating $y = e^x$ twice with respect to t gives $v_y = e^x v_x$ and $a_y = e^x a_x + v_x^2 e^x$. Substituting known values gives $v_y = v_x = 2$, and $a_y = 3 + 4 = 7$. Hence $v = \sqrt{v_x^2 + v_y^2} = \sqrt{2^2 + 2^2} = 2\sqrt{2}$.

$$\tan \tau = \frac{v_y}{v_x} = \frac{2}{2}; \quad \tau = 45°; \quad a = \sqrt{a_x^2 + a_y^2} = \sqrt{3^2 + 7^2}$$
$$= \sqrt{58} = 7.62^- \text{ ft per sec}^2$$
$$\tan \phi = \frac{a_y}{a_x} = \frac{7}{3}; \quad \phi = 66°48'; \quad a_t = \frac{v_x a_x + v_y a_y}{v}$$
$$= \frac{2(3) + 2(7)}{2\sqrt{2}} = 5\sqrt{2} = 7.07 \text{ ft per sec}^2$$

approximately. Then

$$a_n = \frac{v_x a_y - v_y a_x}{v} = \frac{2(7) - 2(3)}{2\sqrt{2}} = 2\sqrt{2} = 2.83^- \text{ ft per sec}^2.$$

Notice that differentiating twice the equation of the path together with $v_x^2 + v_y^2 = v^2$ and $a_x^2 + a_y^2 = a^2$ gives four equations in v_x, v_y, v, a_x, a_y, a. Hence, if two of these quantities are known, it may be possible to solve for the remaining four. Then a_t and a_n are expressible in terms of the first five of these quantities.

3.6.6. A particle moves to the left along $y = \sin x$ with $v_y = -2$ ft per sec and $a_x = 0$ at the instant that it passes through $\left(\frac{\pi}{4}, \frac{\sqrt{2}}{2}\right)$. At this instant find v_x, v, τ, a_y, a, ϕ, a_t, a_n.

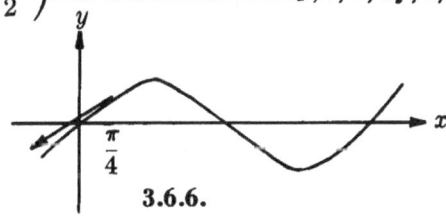

3.6.6.

Solution. Differentiating $y = \sin x$ twice, $v_y = (\cos x) v_x$, and $a_y = (\cos x) a_x - v_x^2 \sin x$. Substituting known values, $-2 = \frac{\sqrt{2}}{2} v_x$, and $a_y = \frac{\sqrt{2}}{2} 0 - 4 \frac{\sqrt{2}}{2}$.

Hence $v_x = -2\sqrt{2}$, and $v = \sqrt{(-2\sqrt{2})^2 + (-2)^2} = 2\sqrt{3}$;

$$\tan \tau = \frac{v_y}{v_x} = \frac{-2}{-2\sqrt{2}}; \quad \tau = -144°44'; \quad a_y = -2\sqrt{2};$$
$$a = 2\sqrt{2}; \quad \phi = -90°;$$

$$a_t = \frac{v_x a_x + v_y a_y}{v} = \tfrac{2}{3}\sqrt{6} = 1.63 \text{ ft per sec}^2 \text{ approximately;}$$

$$a_n = \frac{v_x a_y - v_y a_x}{v} = \frac{-2\sqrt{2}(-2\sqrt{2}) - (-2)0}{2\sqrt{3}} = \tfrac{2}{3}\sqrt{3}$$

$= 2.31^-$ ft per sec^2.

3.7. Differentials. We shall define dx as being identical to the increment Δx, where x is an independent variable. Then the following equation defines dy:

$$dy = f'(x)\, dx \qquad (3.7.1)$$

Here dx and dy are called, respectively, differential of x and differential of y. We see that dy is an approximation to Δy, the increment of y, and the smaller $\Delta x = dx$, the closer this approximation.

A formula for a derivative becomes a formula for a differential by application of 3.7.1. Thus, from the formulas for derivatives in 2.3, we obtain the following:

$$d(k) = 0 \qquad (3.7.2)$$

$$d(ku) = k\, du \qquad (3.7.3)$$

$$d(u \pm v) = du \pm dv \qquad (3.7.4)$$

$$d(u \cdot v) = u\, dv + v\, du \qquad (3.7.5)$$

$$d(u^n) = nu^{n-1}\, du \qquad (3.7.6)$$

$$d\left(\frac{u}{v}\right) = \frac{v\, du - u\, dv}{v^2} \qquad (3.7.7)$$

Find the differentials of the following functions y:

3.7.1. $\qquad y = x^4 + 3x^2 + 17x + 1$

Solution

$$dy = (4x^3 + 6x + 17)\, dx$$

3.7.2. $\qquad y = x^{3/2} + 4x^{1/3} + 6x^7$

Solution

$$dy = \left(\tfrac{3}{2} x^{1/2} + \tfrac{4}{3} x^{-2/3} + 42x^6\right) dx$$

3.7.3. $\qquad y = x^2 y^3 + \dfrac{3x}{y} - 4$

Solution

$$dy = x^2(3y^2\,dy) + y^3(2x\,dx) + 3\,\frac{y\,dx - x\,dy}{y^2};$$

$$y^2\,dy = 3x^2y^4\,dy + 2xy^5\,dx + 3y\,dx - 3x\,dy;$$

$$(y^2 - 3x^2y^4 + 3x)\,dy = (2xy^5 + 3y)\,dx;$$

$$dy = \frac{2xy^5 + 3y}{y^2 - 3x^2y^4 + 3x}\,dx$$

3.7.4. $\quad y = (x^3 - 2x^2 + 2x + 1)^{5/3}$
Solution

$$dy = \tfrac{5}{3}(x^3 - 2x^2 + 2x + 1)^{2/3}(3x^2 - 4x + 2)\,dx$$

3.7.5. $\quad y = \dfrac{x^3 + x + 2}{x^2 + 2x - 1}$

Solution

$$dy = \frac{(x^2+2x-1)\,d(x^3+x+2) - (x^3+x+2)\,d(x^2+2x-1)}{(x^2+2x-1)^2}$$

$$= \frac{(x^2+2x-1)(3x^2+1)\,dx - (x^3+x+2)(2x+2)\,dx}{(x^2+2x-1)^2}$$

$$= \frac{x^4 + 4x^3 - 4x^2 - 4x - 5}{(x^2+2x-1)^2}\,dx$$

3.7.6. A positive error of 0.01 in. is made in measuring the radius of a sphere of radius 12 in. What error will this produce in the computed volume of the sphere? Obtain an exact answer and also an approximate answer using differentials.

Exact solution. Computed volume $= v + \Delta v = \tfrac{4}{3}\pi(r + \Delta r)^3 = \tfrac{4}{3}\pi(12.01)^3 = 7256.340$ cu in., where v and r are the true volume and the true radius, while Δv and Δr are the errors in the volume and in the radius. True volume $= v = \tfrac{4}{3}\pi \cdot 12^3 = 7238.229$ cu in. Error in the volume $= \Delta v = 18.111$ cubic inches.

Approximate solution: dv is an approximation to Δv, and from $v = \tfrac{4}{3}\pi r^3$, we have $dv = 4\pi r^2\,dr = 4\pi r^2 \Delta r = 4\pi(12^2)0.01 = 18.096$ cubic inches. The difference between the exact error and the approximate error, computed by differentials, is only 0.015 cu in., not an appreciable amount.

3.7.7. Obtain the approximate square root of 9.04, using differentials.

Solution. Let $y = x^{1/2}$, so that $dy = \frac{1}{2}x^{-1/2}\,dx$, and $y + dy = x^{1/2} + \frac{1}{2}x^{-1/2}\,dx$. Take $x = 9$, $y = 3$, $dx = 0.04$, and then $y + dy = \sqrt{9.04}$ approximately, so that $y + dy = 9^{1/2} + 9^{-1/2} \cdot 0.04 = 3.006667^-$. A more exact value of the square root of 9.04 is 3.00665928.

3.7.8. If $y = x^4 - 2x^2 + 7x + 1$, find an approximate value of y when $x = 0.997$.

Solution

$$y + dy = (x^4 - 2x^2 + 7x + 1) + (4x^3 - 4x + 7)\,dx.$$

Let $x + dx = 0.997 = 1 - 0.003$ so that $x = 1$ and $dx = -0.003$. Then $y + dy$ is an approximate value of y for $x + dx = 0.997$. Substituting values in $y + dy$, we have $7 + 7(0.003) = 7.021$.

3.7.9. Considering the area of a circular ring as an increment of the area of a circle, find approximately the area of a ring whose inner and outer radii are, respectively, 5 and 5.05 in.

Solution. The differential is an approximation to the increment; hence $A = \pi r^2$; $dA = 2\pi r\,dr$. If $r = 5$ and $dr = 0.05$, then $dA = 2\pi(5)0.05 = 0.5\pi = 1.571$ sq in.

3.7.10. By 2.1 a freely falling body will fall $s = 16.1 t^2$ feet in t seconds. Find the approximate change in its speed for each increase of 0.1 sec in time.

Solution. Differentiating with respect to t: $\dfrac{ds}{dt} = v = 32.2 t$; $dv = 32.2\,dt$. If $dt = 0.1$, $dv = 3.22$ ft per sec.

Chapter 4

INTEGRATION

4.1. Area under a curve. Let us consider the area generated (or swept out) by a moving ordinate drawn to the curve $y = x^2$. Imagine this ordinate starting at $x = 1$ and moving to the right, always remaining parallel to the y axis. It will then leave behind it a growing area bounded by itself, the curve $y = x^2$, the ordinate to this curve at $x = 1$, and the x axis. We wish to find the instantaneous rate of growth of this changing area with respect to x. As we shall see, this instantaneous rate of growth, or derivative, of the area will enable us to find the area

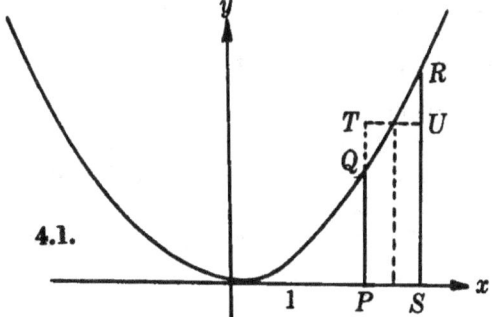

4.1.

itself. The process is called integration, the subject of this chapter. Let $A(x)$ represent the changing area since it is a function of x, the abscissa of the moving ordinate. If x takes on an increment Δx, then $A(x)$ takes on an increment ΔA represented by the area $PQRS$ in the figure. But $PQRS$ is equal to the area of the rectangle $PTUS$ whose altitude is the ordinate to $y = x^2$ at some point $x + \theta \Delta x$, where $0 < \theta < 1$. This altitude is then $(x + \theta \Delta x)^2$. Hence $\Delta A = (x + \theta \Delta x)^2 \Delta x$. Then $\dfrac{\Delta A}{\Delta x} = (x + \theta \Delta x)^2$. Allowing Δx to approach zero in both members, we have $\dfrac{dA}{dx} = x^2$. Since x^2 is the derivative of $A(x)$, then $A(x)$ itself must be $A(x) = \dfrac{x^3}{3} + c$, where c is any constant, this being the most general expression whose deriative is x^2. But when $x = 1$ our changing area was zero, since we imagined

ur moving ordinate as starting at $x = 1$. Hence $A(1) = \frac{1}{3} +$ $= 0$, and c must equal $-\frac{1}{3}$ in our case. Therefore $A(x) = \frac{x^3}{3} - \frac{1}{3}$.
his represents the area under the curve $y = x^2$ from the ordinate at one to the ordinate at x. If we want the area under the curve from $x = 1$ to $x = 3$, we need only substitute 3 for x thus: $A(3) = \frac{3^3}{3} - \frac{1}{3} = \frac{26}{3}$.

.2. Indefinite integrals: When, as in the last article, we nd an expression which has a given differential, we call this itegration or finding the indefinite integral. This function is ymbolized by a long s placed before the differential, thus: $\int 2x\, dx = x^2 + C$ where C is any constant, called, therefore, an rbitrary constant. An indefinite integral will always result in a unction with an arbitrary constant, since two functions which iffer only by a constant have the same differential. Below are hree formulas for integration, the student will notice that they re the same as certain differentiation formulas, only the notaion has been changed.

$$\int K\, du = Ku + C \quad \text{where } K \text{ is a constant} \quad (4.2.1)$$

$$\int (du \pm dv) = u \pm v + C \quad (4.2.2)$$

$$\int u^n\, dn = \frac{u^{n+1}}{n+1} + C \quad n \neq -1 \quad (4.2.3)$$

Evaluate the following indefinite integrals:

4.2.1. $I = \int (x^2 + 3x + 1)\, dx$

$$= \int x^2\, dx + 3\int x\, dx + \int dx$$

Solution

$$I = \frac{x^3}{3} + \frac{3x^2}{2} + x + C$$

4.2.2. $I = \int (x^2 + 1)^2 x \, dx = \frac{1}{2} \int (x^2 + 1)^2 \, 2x \, dx$

Solution. If $u = x^2 + 1$, then $du = 2x \, dx$. Hence $I = \frac{1}{2} \int u^2 \, du = \frac{1}{6} u^3 + C = \frac{1}{6}(x^2 + 1)^3 + C$

4.2.3. $I = \int (x^{3/2} - x^{-1/2} + 1) \, dx$

$$= \int x^{3/2} \, dx - \int x^{-1/2} \, dx + \int dx$$

Solution

$$I = \tfrac{2}{5} x^{5/2} - 2x^{1/2} + x + C$$

4.2.4. $I = \int (x^{3/2} - 1)^{-2/7} x^{1/2} \, dx = \frac{2}{3} \int (x^{3/2} - 1)^{-2/7} \tfrac{3}{2} x^{1/2} \, dx$

Solution

$$I = \frac{2}{3} \int (x^{3/2} - 1)^{-2/7} \, d(x^{3/2} - 1) = \tfrac{2}{3} \cdot \tfrac{7}{5}(x^{3/2} - 1)^{5/7} + C$$

$$= \tfrac{14}{15}(x^{3/2} - 1)^{5/7} + c$$

4.2.5. $I = \int \sqrt{(x + 5)} \, dx = \int (x + 5)^{1/2} \, d(x + 5)$

Solution

$$I = \tfrac{2}{3}(x + 5)^{3/2} + C$$

4.3. Definite integrals: It is readily seen that the arguments applied in 4.1 to $y = x^2$ may be applied to $y = f(x)$, where $f(x)$ is any continuous function which is greater than zero between $x = a$ and $x = b$. If $F(x)$ is any function whose differential is $f(x) \, dx$, then the area under the curve between a and b is given by $F(b) - F(a)$. This expression is called the definite integral of $f(x)$ from a to b, and is symbolized thus:

$$\int_a^b f(x) \, dx = F(b) - F(a)$$

4.3.1. $\int_2^5 (x^2 + 3x + 1) \, dx = \left(\frac{x^3}{3} + \frac{3x^2}{2} + x \right)_2^5$

$$= \left(\frac{5^3}{3} + \frac{3 \cdot 5^2}{2} + 5 \right) - \left(\frac{2^3}{3} + \frac{3 \cdot 2^2}{2} + 2 \right) = 73.5$$

4.3.2. $\int_1^4 (x^{-2} + 2x^{1/2} + 2)\, dx = \left(\dfrac{x^{-1}}{-1} + 2\cdot\tfrac{2}{3}x^{3/2} + 2x \right)\Big|_1^4$

$= \left(\dfrac{4^{-1}}{-1} + \tfrac{4}{3}4^{3/2} + 2\cdot 4 \right) - \left(\dfrac{1^{-1}}{-1} + \tfrac{4}{3}1^{3/2} + 2\cdot 1 \right) = \dfrac{193}{12}$

4.3.3. $\int_{-1}^2 \sqrt{x+1}\, dx = \int_{-1}^2 (x+1)^{1/2}\, d(x+1) = [\tfrac{2}{3}(x+1)^{3/2}]_{-1}^2$

$= [\tfrac{2}{3}(2+1)^{3/2}] - [\tfrac{2}{3}(-1+1)^{3/2}] = 2\sqrt{3}$

4.3.4. $\int_0^3 \sqrt{3x+1}\, dx = \dfrac{1}{3}\int_0^3 (3x+1)^{1/2}\, 3\, dx$

$= \dfrac{1}{3}\int_0^3 (3x+1)^{1/2}\, d(3x+1) = \tfrac{1}{3}\cdot\tfrac{2}{3}(3x+1)^{3/2}\Big|_0^3$

$= \tfrac{2}{9}(10^{3/2} - 1) = 6.805^+$

The symbol $\tfrac{1}{3}\cdot\tfrac{2}{3}(3x+1)^{3/2}\Big|_0^3$ is often used in place of $\tfrac{1}{3}\cdot\tfrac{2}{3}(3x+1)^{3/2}\Big|_0^3$.

4.3.5. $\int_{-3}^{-1} (x^3+1)^2 x^2\, dx = \dfrac{1}{3}\int_{-3}^{-1} (x^3+1)^2\, 3x^2\, dx$

$= \dfrac{1}{3}\int_{-3}^{-1} (x^3+1)^2\, d(x^3+1) = \dfrac{1}{9}(x^3+1)^3\Big|_{-3}^{-1}$

$= \tfrac{1}{9}[0 - (-26)^3] = 1952.89^-$

4.3.6. $\int_{-3}^{-1} (x^3+1)^2 x\, dx = \int_{-3}^{-1} (x^7 + 2x^4 + x)\, dx$

$= \left(\dfrac{x^8}{8} + \dfrac{2x^5}{5} + \dfrac{x^2}{2} \right)_{-3}^{-1} = \left(\dfrac{1}{8} - \dfrac{2}{5} + \dfrac{1}{2} \right)$

$- \left[\dfrac{(-3)^8}{8} + \dfrac{2(-3)^5}{5} + \dfrac{(-3)^2}{2} \right] = -727.2$

4.3.7. Find $\dfrac{dy}{dx}$ if (a) $y = \int \sqrt{x^3+1}\, dx$ and

(b) $y = \int_1^2 \sqrt{x^3+1}\, dx$.

Solution. (a) $\dfrac{dy}{dx} = \sqrt{x^3 + 1}$ since integration and differentiation are inverse processes, and one cancels the other.

(b) $\dfrac{dy}{dx} = 0$ since a definite integral is a constant, and the derivative of a constant is zero.

4.4. The fundamental theorem of integral calculus. We have obtained the area under the curve $y = f(x)$ from $x = a$ to $x = b$, but we have not defined that area. We shall now do

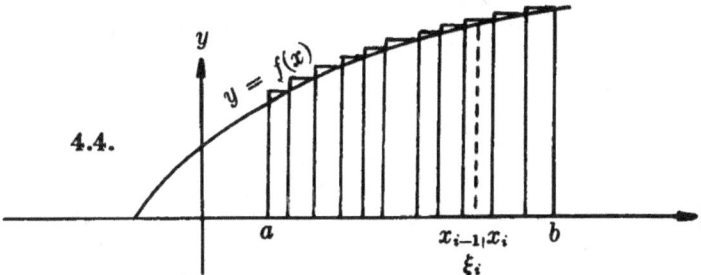

4.4.

that. We suppose that $f(x)$ is continuous and greater than zero. Let the interval from a to b be divided into n parts by the points $x_0 < x_1 < x_2 \ldots < x_n$, where $x_0 = a$ and $x_n = b$. Let ordinates be erected to the curve at these points. Next let x_{i-1} and x_i be any consecutive pair of such points and ξ_i any point between them, or even coinciding with either of them. Draw an ordinate to the curve at ξ_i. Its length will be $f(\xi_i)$. Then the area under the curve from x_{i-1} to x_i will be approximately equal to $f(\xi_i)\,\Delta x_i$ where $\Delta x_i = x_i - x_{i-1}$. Do the same for every strip into which we have divided the area under $y = f(x)$, then add together all the results. This sum is symbolized by

$$\sum_{i=1}^{i=n} f(\xi_i)\,\Delta x_i \qquad (4.4.1)$$

The Greek capital letter sigma, written $\sum$, is universally used in mathematics to indicate the sum of all terms that follow it. The $i = 1$ and the $i = n$ are called limits of the summation. Thus the whole symbol means the sum

$$f(\xi_1)\,\Delta x_1 + f(\xi_2)\,\Delta x_2 + f(\xi_3)\,\Delta x_3 + \cdots + f(\xi_n)\,\Delta x_n$$

and this sum approximates the area under $y = f(x)$ from a to
Let Δ represent the largest of the Δx_i, where $i = 1, 2, 3, \cdots$,

Now suppose that $n \to \infty$ while simultaneously $\Delta \to 0$. Let us suppose that, when this happens, 4.4.1 approaches a limit. This limit we *define* as the area under $y = f(x)$ from a to b. But we have seen that this area is given by $\int_a^b f(x)\, dx$. Hence we have

$$\lim_{\Delta \to 0} \sum_{i=1}^{i=n} f(\xi_i) \Delta x_i = \int_a^b f(x)\, dx \qquad (4.4.2)$$

This result is called the *fundamental theorem of integral calculus*. We may now divorce 4.4.2 from any geometric interpretation, or picture, whatsoever. Then 4.4.2 enables us to evaluate the limit in its left member no matter what was the source of this limit. If the solution of any problem may be expressed as the limit on the left of the equal sign, then that limit may be evaluated by using the definite integral on the right of the equal sign. Most surprising is the number of problems that can be solved in the form of the left member of 4.4.2, i.e., as the limit of a sum as the number of summands become infinite and each summand approaches zero. Most such limits can be evaluated by the fundamental theorem. Such problems make up a large part of the *integral calculus*. The student who has learned to use the fundamental theorem, has a useful tool. We next proceed to do this. But first let us illustrate briefly with a simple numerical example how the sum tends to approach the definite integral even when the number of summands is not large.

4.4.1. Find approximately the area bounded by $y = x^2$, $x = 1$, $x = 3$, and $y = 0$ by taking vertical rectangles and find the so-called "lower sum" and "upper sum." Compare the results with the exact area or integral, $\frac{26}{3} = 8.666\ldots$, as found in **4.1**.

Solution. Dividing the interval from $x = 1$ to $x = 3$ into 10 equal parts:

x	1	1.2	1.4	1.6	1.8	2.0	2.2	2.4	2.6	2.8	3.0	Sum
y	1	1.44	1.96	2.56	3.24	4.00	4.84	5.76	6.76	7.84	9.00	
Area lower rectangles	0.200	0.288	0.392	0.512	0.648	0.800	0.968	1.152	1.352	1.568		7.880
Area taller rectangles		0.288	0.392	0.512	0.648	0.800	0.968	1.152	1.352	1.568	1.800	9.480

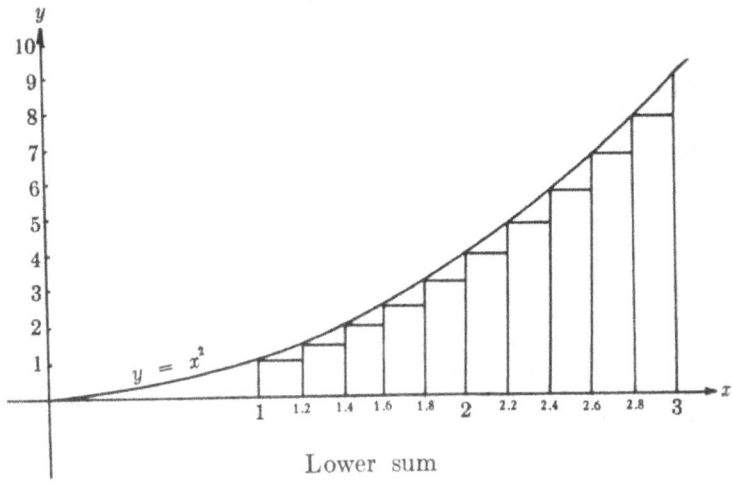

Lower sum

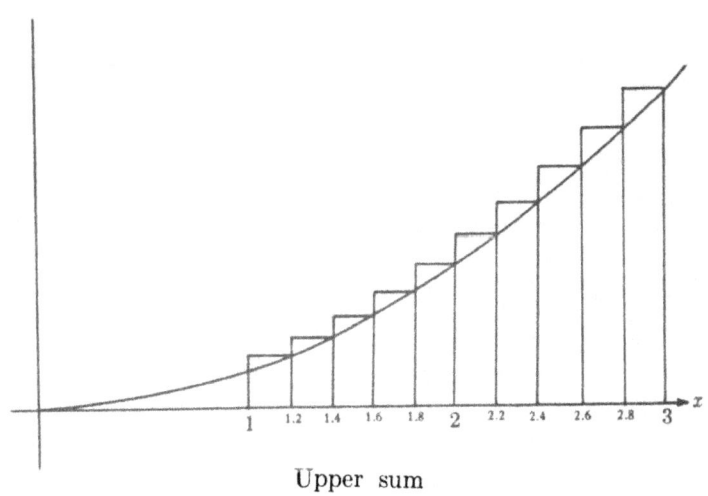

Upper sum

4.4.1.

The mean of the lower and upper sums is 8.680.
Dividing the interval $x = 1$ to $x = 3$ into 20 equal parts,

t	1.0	1.1	1.2	1.3	1.4	1.5	1.6	1.7	1.8	1.9	2.0
y	1.00	1.21	1.44	1.69	1.96	2.25	2.56	2.89	3.24	3.61	4.00
Area lower rectangles	0.100	0.121	0.144	0.169	0.196	0.225	0.256	0.289	0.324	0.361	0.400
Area taller rectangles		0.121	0.144	0.169	0.196	0.225	0.256	0.289	0.324	0.361	0.400
t	2.1	2.2	2.3	2.4	2.5	2.6	2.7	2.8	2.9	3.0	Sum
y	4.41	4.84	5.29	5.76	6.25	6.76	7.29	7.84	8.41	9.00	
Area lower rectangles	0.441	0.484	0.529	0.576	0.625	0.676	0.729	0.784	0.841		8.270
Area taller rectangles	0.441	0.484	0.529	0.576	0.625	0.676	0.729	0.784	0.841	0.900	9.070

The mean of the lower and upper sums is 8.670. Notice that both the lower and upper sums are closer to 8.666 ... , the exact area, when the number of rectangles is doubled.

4.4.2. Find the area under $y = x^3 + 3x^2 + 1$ from $x = -1$ to $x = 4$.

Solution. This area is $\lim_{\Delta \to 0} \sum_{i=1}^{i=n} f(\xi_i) \Delta x_i$, where $f(x) = x^3 + 3x^2 + 1$. By the fundamental theorem, the value of this limit is

$$\int_{-1}^{4} (x^3 + 3x^2 + 1) \, dx = \left[\frac{x^4}{4} + x^3 + x \right]_{-1}^{4}$$

$$= \left(\frac{4^4}{4} + 4^3 + 4 \right) - \left[\frac{(-1)^4}{4} + (-1)^3 - 1 \right] = 133.75$$

4.4.3. The line $y = 2x$ from $x = 0$ to $x = 4$ is revolved around the x axis thus generating a right circular cone. Find the volume of this cone.

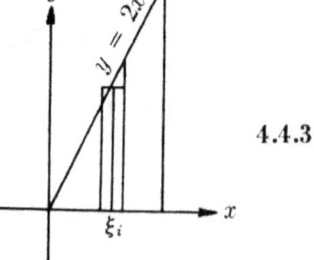

4.4.3.

Solution. Suppose this cone is cut into thin slices by planes perpendicular to the x axis. Let three consecutive such planes pass through x_{i-1}, ξ_i, x_i, where $x_{i-1} \leq \xi_i \leq x_i$. Then $\pi(2\xi_i)^2 \Delta x_i$, where $\Delta x_i = x_i - x_{i-1}$, is approximately the volume of the cone between the first and the last of these three planes, and the total volume of the cone is the limit of the sum of all such products as their number becomes infinite, and the thickness of each slice approaches zero. Then

$$\lim_{\Delta \to 0} \sum_{i=1}^{i=n} \pi(2\xi_i)^2 \Delta x_i = \int_0^4 \pi(2x)^2 \, dx = \frac{4\pi x^3}{3}\bigg|_0^4 = \frac{256\pi}{3}$$

The correctness of this result may be verified by the formula for the volume of a right circular cone from solid geometry, viz., $V = \frac{1}{3}\pi r^2 h = \frac{1}{3}\pi 8^2 4 = \frac{256\pi}{3}$. The argument for obtaining $\int_0^4 \pi(2x)^2 \, dx$ may be abbreviated as follows. The radius of a typical section of the cone perpendicular to the x axis is the value of y on the line $y = 2x$ or $2x$. The area of this typical section is then $\pi(2x)^2$. The thickness of the corresponding slice is dx, hence its volume is approximately $\pi(2x)^2 \, dx$. We need to take the limit of the sum of all such slices from $x = 0$ to $x = 4$. Hence these values of x are the limits of our definite integral, and we obtain $\int_0^4 \pi(2x)^2 \, dx$.

4.4.4. Find the volume generated by revolving the part of $y = x^2 + 1$ from $x = 1$ to $x = 3$ about the x axis.

Solution. The radius of a typical section is $x^2 + 1$, and its area is $\pi(x^2 + 1)^2$. The volume of a typical slice is $\pi(x^2 + 1)^2 \, dx$. The limit of the sum of these slices from $x = 1$ to $x = 3$ is, by the fundamental theorem,

$$\int_1^3 \pi(x^2 + 1)^2 \, dx = \pi \int_1^3 (x^4 + 2x^2 + 1) \, dx$$

$$= \pi \left(\frac{x^5}{5} + \frac{2x^3}{3} + x\right)_1^3 = 67.733$$

4.4.5. Find the volume generated by revolving $y = x^2 + 1$ from $y = 1$ to $y = 5$ about the y axis.

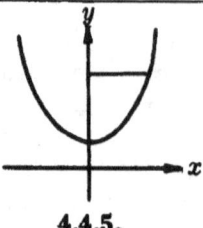

4.4.5.

Solution. The radius of a typical section is the value of x on the curve. Solve $y = x^2 + 1$ for x, and we obtain $x = \sqrt{y-1} = (y-1)^{1/2}$ as this radius so that the area of a typical section is $\pi[(y-1)^{1/2}]^2 = \pi(y-1)$. The volume of the corresponding slice is $\pi(y-1)\,dy$, and summing these, we have in the limit

$$\int_1^5 \pi(y-1)\,dy = \pi\left(\frac{y^2}{2} - y\right)_1^5 = 8\pi.$$

4.4.6. Solve **4.4.3** by using thin cylindrical shells.

Solution. A typical thin cylindrical shell will be generated by revolving the shaded area about the x axis. Its radius is the value of y on the curve; its circumference is then $2\pi y$. Its altitude is 4 minus the value of x on the curve, which is $\frac{y}{2}$. Its lateral area is then $2\pi y\left(4 - \frac{y}{2}\right)$. Since it is a *thin* cylindrical shell, its volume is approximately its lateral area times its thickness, or $2\pi y\left(4 - \frac{y}{2}\right)dy$. Summing all such shells from $y = 0$ to $y = 8$, and taking the limit of that sum as the number of shells becomes infinite, and the thickness of each shell approaches zero, we have

$$\int_0^8 2\pi y\left(4 - \frac{y}{2}\right)dy = \pi\left(4y^2 - \frac{y^3}{3}\right)_0^8 = \frac{256\pi}{3}.$$

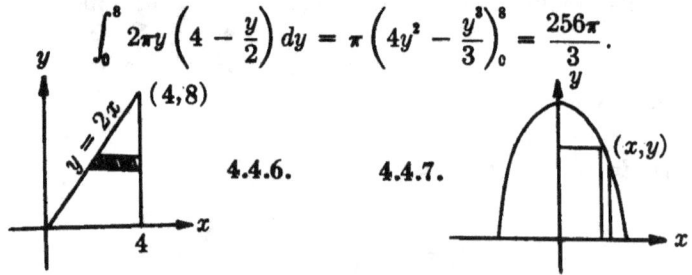

4.4.6.   4.4.7.

4.4.7. Find the volume generated by revolving about the y axis the part of $y + x^2 = 4$ which is above the x axis. Use (a) thin slices and (b) cylindrical shells.

Solution

(a) $V = \int_0^4 \pi x^2 \, dy = \int_0^4 \pi(4-y) \, dy = \pi\left(4y - \dfrac{y^2}{2}\right)_0^4 = 8\pi$

(b) $V = \int_0^2 2\pi xy \, dx = 2\pi \int_0^2 x(4-x^2) \, dx = \pi\left(4x^2 - \dfrac{x^4}{2}\right)$
$= 8\pi$

4.4.8. Find the volume of the prolate spheroid generated by revolving the ellipse $\dfrac{x^2}{a^2} + \dfrac{y^2}{b^2} = 1$, $a > b$, about the x axis.

Solution. Using thin slices, we obtain the volume from $x = 0$ to $x = a$ and double it on account of symmetry.

$$V = 2\int_0^a \pi y^2 \, dx = 2\int_0^a \pi b^2\left(1 - \dfrac{x^2}{a^2}\right) dx = 2\pi b^2\left[x - \dfrac{x^3}{3a^2}\right]_0^a$$

$$= 2\pi b^2\left(a - \dfrac{a}{3}\right) = \dfrac{4}{3}\pi ab^2$$

Using cylindrical shells such as the one generated by the shaded area,

$$V = 2\int_0^b 2\pi y \cdot x \cdot dy = 4\pi \int_0^b a\sqrt{1 - \dfrac{y^2}{b^2}}\, y \, dy$$

$$= 4\pi a\left(-\dfrac{b^2}{2}\right) \int_0^b \left(1 - \dfrac{y^2}{b^2}\right)^{1/2}\left(-\dfrac{2y}{b^2}\right) dy$$

$$= -2\pi ab^2 \int_0^b \left(1 - \dfrac{y^2}{b^2}\right)^{1/2} d\left(1 - \dfrac{y^2}{b^2}\right)$$

$$= -2\pi ab^2 \cdot \dfrac{2}{3}\left(1 - \dfrac{y^2}{b^2}\right)^{3/2}\bigg|_0^b = -2\pi ab^2 \cdot \tfrac{2}{3}(-1) = \tfrac{4}{3}\pi ab^2$$

4.4.9. Find the area bounded by $y = x^2$ and $x = y^2$.

Solution. These two parabolas intersect at $(0, 0)$ and $(1, 1)$. If the required area is divided into vertical strips, the length of a side of a typical strip is the value of y on $x = y^2$ minus the value of y on $y = x^2$. Hence the area of a typical strip is $(x^{1/2} - x^2)\,dx$. By the fundamental theorem, the required area is

$$\int_0^1 (x^{1/2} - x^2)\,dx = \left(\frac{2}{3} x^{3/2} - \frac{x^3}{3}\right)_0^1 = \frac{1}{3}.$$

4.4.10. Find the area bounded by the semicubical parabola $y^2 = 2x^3$ and the line $y = 2x$.

Solution. The curves intersect at $(0, 0)$ and $(2, 4)$. Using horizontal strips, the length of one side of a typical strip is the value of x on $y^2 = 2x^3$ minus the value of x on $y = 2x$. Hence the approximate area of a typical strip is

$$\left[\left(\frac{y^2}{2}\right)^{1/3} - \frac{y}{2}\right] dy.$$

By the fundamental theorem, the required area is

$$\int_0^4 \left[\left(\frac{y^2}{2}\right)^{1/3} - \frac{y}{2}\right] dy = \left(\frac{3}{5} \frac{y^{5/3}}{2^{1/3}} - \frac{y^2}{4}\right)_0^4 = \frac{4}{5}.$$

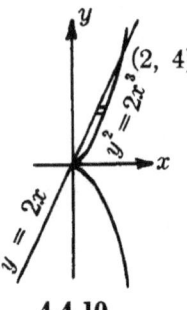

4.4.10.

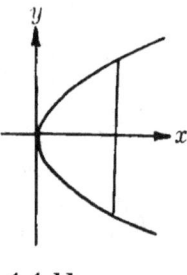

4.4.11.

4.4.11. Find the area A bounded by the parabola $y^2 = 4x$ and its latus rectum.

Solution. The focus of this parabola is at $(1, 0)$, and $x = 1$ is the equation of the line coinciding with its latus rectum. The upper branch of the parabola is $y = 2x^{1/2}$, while the lower branch is $y = -2x^{1/2}$. The area of a typical vertical strip is $[2x^{1/2} - (-2x^{1/2})]\,dx$. Hence $A = \int_0^1 4x^{1/2}\,dx = 4 \cdot \frac{2}{3} x^{3/2} \big|_0^1 = \frac{8}{3}.$

Or one could find the upper half of the required area and double it, thus: $A = 2\int_0^1 2x^{1/2}\,dx = \frac{8}{3}$.

4.4.12. Find the volume V of a segment of one base of a sphere of radius a, the altitude of the segment being h.

Solution. Let the sphere be generated by revolving the circle $x^2 + y^2 = a^2$ about the x axis. The base of the segment is then generated by the revolution of the line $x = a - h$. Using thin slices perpendicular to the x axis,

$$V = \int_{a-h}^{a} \pi y^2\,dx = \int_{a-h}^{a} \pi(a^2 - x^2)\,dx$$

$$= \pi\left(a^2 x - \frac{x^3}{3}\right)\Big|_{a-h}^{a} = \pi h^2\left(a - \frac{h}{3}\right)$$

4.4.12.

4.4.13.

4.4.13. A sphere of radius a has bored through its center a circular hole of radius h. Find the volume V of the remaining portion of the sphere.

Solution. This solid may be generated by revolving a segment of a circle about a diameter of the circle which is parallel to the chord of the segment. Our element of volume is a cylindrical shell generated by the revolution of the shaded area. The circle is

$$x^2 + y^2 = a^2;$$

$$V = \int_h^a 2\pi y(2x)\,dy = -2\pi \int_h^a (a^2 - y^2)^{1/2}(-2y\,dy)$$

$$= -2\pi \cdot \tfrac{2}{3}(a^2 - y^2)^{3/2}\Big|_h^a = \tfrac{4}{3}\pi(a^2 - h^2)^{3/2}$$

The result of this problem and that of 4.4.12 may be checked by adding together the volumes of the portion of the sphere removed and the portion of the sphere remaining.

4.4.14. A woodchopper notches a tree of radius a. The bottom of the notch is a horizontal semicircle, and the top makes an angle α with the bottom. Find the volume V of the wood removed.

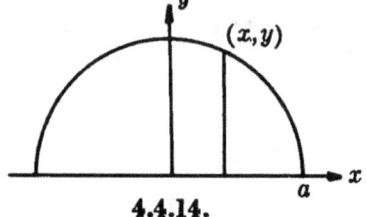

4.4.14.

Solution. The wood removed is a wedge. Divide it into thin slices by planes perpendicular to its edge. Each such plane makes a triangular section. The base of the triangle is $y = \sqrt{a^2 - x^2}$. The altitude of the triangle is $\sqrt{a^2 - x^2}\tan \alpha$. Hence the area of the triangle is $\tfrac{1}{2}\sqrt{a^2 - x^2}\sqrt{a^2 - x^2}\tan \alpha$.

$$V = 2\int_0^a \frac{1}{2}(a^2 - x^2)\tan \alpha\, dx$$

$$= \left(a^2 x - \frac{x^3}{3}\right)\tan \alpha \bigg|_0^a = \tfrac{2}{3}a^3 \tan \alpha$$

4.4.15. Find the area A between $x = 0$ and $x = 3$ which is bounded by the curves $y = 3x - x^2$ and $y = 3x^2 - x^3$. Designate these two curves as $y_1 = 3x - x^2$, and $y_2 = 3x^2 - x^3$.

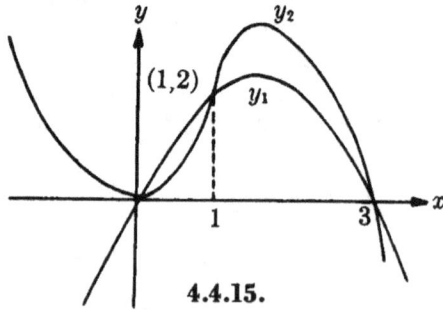

4.4.15.

Solution. Substitution shows that both curves pass through $(0, 0)$, $(1, 2)$, and $(3, 0)$. If $0 < x < 1$, then $y_1 > y_2$, and $(y_1 - y_2) \, dx$ represents a positive area as does $\int_0^1 (y_1 - y_2) \, dx$. If $1 < x < 3$, then $y_2 - y_1$ and $\int_1^3 (y_2 - y_1) \, dx$ are positive. Hence the required area is

$$A = \int_0^1 [(3x - x^2) - (3x^2 - x^3)] \, dx$$
$$+ \int_1^3 [(3x^2 - x^3) - (3x - x^2)] \, dx = \tfrac{5}{12} + \tfrac{32}{12} = \tfrac{37}{12}$$

Notice, for example, that $\int_0^3 [(3x - x^2) - (3x^2 - x^3)] \, dx = -\tfrac{27}{12}$, which is the area from 0 to 1 diminished by the area from 1 to 3.

4.4.16. A surface is generated by a line that always moves parallel to the xz plane and has points in common with the line $y + z = a$, $x = 0$, and the line $x = b$, $z = 0$. Find the volume in the first octant bounded by this surface.

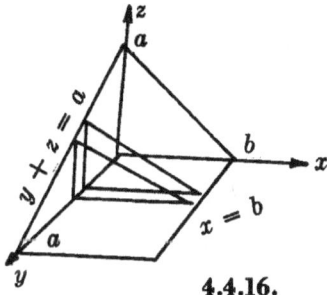

4.4.16.

Solution. Divide the required volume into thin slices by planes perpendicular to the y axis. A typical section is a right triangle whose base is b and whose altitude is $z = a - y$, and the approximate volume of the slice is $\tfrac{1}{2}b(a - y) \, dy$.

$$V = \int_0^a \tfrac{1}{2} b(a - y) \, dy = \tfrac{b}{2}\left(ay - \tfrac{y^2}{2}\right)\Big|_0^a = \tfrac{1}{4}a^2 b$$

It is instructive to notice that the element of volume is not exactly a triangular prism, but a wedge as is shown in the figure. However, the student can satisfy himself that the difference in volume between the wedge and the prism is $\frac{1}{2} b \Delta z \Delta y$ or $\frac{1}{2} b \overrightarrow{\Delta y}^2$. Since this difference is of the second order (Δy occurs to the second power) in the limiting case our calculations are correct.

4.5. Moments of areas and volumes. If the area bounded by $y = f(x)$, the x axis, $x = a$, and $x = b$ be divided into narrow strips by lines parallel to the y axis at $a = x_0 < x_1 < x_2 < \ldots < x_n = b$, let ξ_i be the abscissa of the center of gravity of the ith strip, and write $x_i - x_{i-1} = \Delta x_i$. The moment of an element of area about the y axis is $\xi_i f(\xi_i) \Delta x_i$, and by the fundamental theorem, the moment of the entire area, M_y, about the y axis is

$$M_y = \lim_{\Delta x \to 0} \sum_{i=1}^{n} \xi_i f(\xi_i) \Delta x_i = \int_a^b x f(x) \, dx = A \bar{x}$$

where A is $\int_a^b f(x) \, dx$, and $\bar{x}$ is the abcissa of the centroid of the area. The moment of inertia about the y axis is given by $I_y = \int_a^b x^2 y \, dx$.

4.5.1. Find the coordinates of the centroid of the area bounded by $y = x^2$, $x = 3$, $y = 0$.

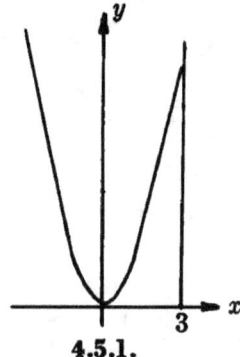

4.5.1.

Solution.

$$A = \int_0^3 x^2 \, dx = \frac{x^3}{3} \Big|_0^3 = 9; \quad \bar{x} = \frac{1}{9} \int_0^3 x \cdot x^2 \, dx;$$

$$\bar{x} = \frac{1}{9} \frac{x^4}{4} \Big|_0^3 = \frac{9}{4}$$

To find $\bar{y}$ we must take the first moment about the x axis. The area of a typical vertical strip may be thought of as $x^2\,dx$. The centroid of this strip is approximately midway of its length, or $\dfrac{x^2}{2}$ above the x axis. Hence its first moment about the x axis is approximately $\dfrac{x^4}{2}$. Summing all such elementary moments, in the limit we have

$$\bar{y} = \frac{1}{9}\int_0^3 \frac{x^4}{2}\,dx = \frac{1}{9}\frac{x^5}{10}\Big|_0^3 = 2.7$$

Another method for $\bar{y}$ is to use horizontal strips. Let (x, y) be a typical point of the curve. The approximate area of a typical horizontal strip is $(3 - x)\,dy = (3 - \sqrt{y})\,dy$, since $y = x^2$, or $x = \sqrt{y}$. The centroid of this strip is approximately at its distance y from the x axis. Hence its approximate moment about the x axis is $y(3 - \sqrt{y})\,dy$. Taking the limit of the sum of all such elementary moments from $y = 0$ to $y = 9$, in order to cover the whole area, we have

$$\bar{y} = \frac{1}{9}\int_0^9 y(3 - \sqrt{y})\,dy = \frac{1}{9}\int_0^9 (3y - y^{3/2})\,dy$$

$$= \tfrac{1}{9}(\tfrac{3}{2}y^2 - \tfrac{2}{5}y^{5/2})\Big|_0^9 = \tfrac{1}{9}(\tfrac{3}{2}9^2 - \tfrac{2}{5}9^{5/2}) - 0 = 2.7$$

4.5.2. Find the centroid of the area bounded by $y = -x^2 + 4$, $y = 0$.

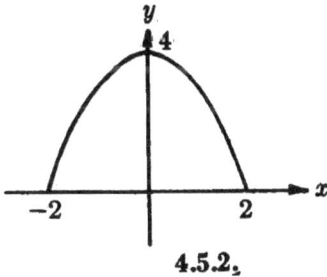

4.5.2.

Solution. By symmetry

$$\bar{x} = 0;$$

$$A = 2\int_0^2 y\,dx = 2\int_0^2 (-x^2 + 4)\,dx = 2\left(-\frac{x^3}{3} + 4x\right)\Big|_0^2 = \frac{32}{3}$$

Using vertical strips,

$$= 2 \cdot \frac{3}{32} \int_0^2 \frac{y}{2} y \, dx = \frac{3}{32} \int_0^2 y^2 \, dx = \frac{3}{32} \int_0^2 (-x^2 + 4)^2 \, dx$$

$$= \frac{3}{32} \int_0^2 (x^4 - 8x^2 + 16) \, dx = \frac{3}{32} \left(\frac{x^5}{5} - \frac{8x^3}{3} + 16x \right) \Big|_0^2 = 1.6$$

Using horizontal strips,

$$\bar{y} = \frac{3}{32} \int_0^4 2y \cdot x \, dy = \frac{3}{32} \int_0^4 2y(4-y)^{1/2} \, dy$$

Evaluation of the last integral requires more advanced technique than is available at the present stage so that this attempt to find $\bar{y}$ by using horizontal strips must be abandoned. However, it was good practice to set up the integral, and the point is illustrated that it is often much easier to solve a problem with one variable of integration than with another.

4.5.3. The area in 4.5.2 is rotated about the y axis to generate a paraboloid of revolution. Find the centroid of this volume.

Solution. Clearly, by symmetry, the centroid is on the y axis. Divide the volume into thin slices by planes perpendicular to the y axis.

$$V = \int_0^4 \pi x^2 \, dy = \int_0^4 \pi(4-y) \, dy = \pi \left(4y - \frac{y^2}{2} \right) \Big|_0^4 = 8\pi$$

$$\bar{y} = \frac{1}{V} \int_0^4 y \cdot \pi x^2 \, dy = \frac{1}{8\pi} \int_0^4 y\pi(4-y) \, dy = \frac{1}{8\pi} \pi \left(2y^2 - \frac{y^3}{3} \right) \Big|_0^4 = \frac{4}{3}$$

since the centroid of a slice is on the y axis at an approximate distance y from the origin.

4.5.4. Find the centroid of a right circular cone of radius r and altitude h.

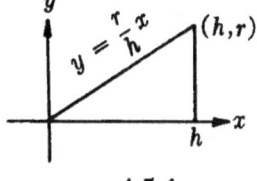

4.5.4.

Solution. Such a cone is generated by rotating the part of the line $y = \frac{r}{h}x$ from $x = 0$ to $x = h$ about the x axis.

$$V = \tfrac{1}{3}\pi r^2 h$$

$$\bar{x} = \frac{1}{V}\int_0^h x\pi y^2\,dx = \frac{3}{\pi r^2 h}\int_0^h x\pi \left(\frac{r}{h}x\right)^2 dx = \frac{3}{h^3}\cdot\frac{x^4}{4}\bigg|_0^h = \frac{3h}{4}$$

Clearly the centroid is on the x axis.

4.5.5. Find the centroid of a semicircular area of radius a.

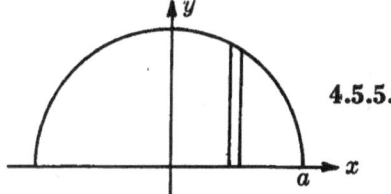

4.5.5.

Solution. Let $y = \sqrt{a^2 - x^2}$ be the equation of the semicircle. Using vertical strips, the area of a typical strip is $y\,dx = \sqrt{a^2 - x^2}\,dx$, and its moment about the x axis is

$$\frac{y}{2}\sqrt{a^2 - x^2}\,dx = \tfrac{1}{2}\sqrt{a^2 - x^2}\cdot\sqrt{a^2 - x^2}\,dx = \tfrac{1}{2}(a^2 - x^2)\,dx.$$

Then $A = \tfrac{1}{2}\pi a^2$;

$$A\bar{y} = 2\int_0^a \tfrac{1}{2}(a^2 - x^2)\,dx = \left(a^2 x - \frac{x^3}{3}\right)\bigg|_0^a = \tfrac{2}{3}a^3;$$

$$\bar{y} = \frac{\tfrac{2}{3}a^3}{\frac{\pi a^2}{2}} = \frac{4a}{3\pi}. \text{ By symmetry, } \bar{x} = 0.$$

4.5.6. Show that the centroid of a triangle is at the point of trisection of the medians.

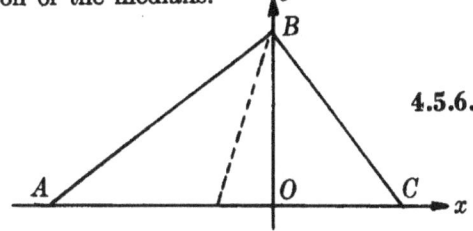

4.5.6.

Solution. Divide the triangle ABC into two right triangles by drawing an altitude to the base AC. Let this altitude and the base coincide, respectively, with the y axis and the x axis. Let the equation of BC be $\dfrac{x}{a} + \dfrac{y}{h} = 1$. Area of the triangle $OBC = \dfrac{ah}{2}$, and

$$\bar{y} = \frac{1}{A}\int_0^a \frac{y}{2} \cdot y\, dx = \frac{2}{ah}\int_0^a \tfrac{1}{2}h^2\left(1-\frac{x}{a}\right)^2 dx$$

$$= -h\int_0^a \left(1-\frac{x}{a}\right)^2 \left(-\frac{dx}{a}\right) = -\frac{h}{3}\left(1-\frac{x}{a}\right)^3\Big|_0^a = \frac{h}{3}$$

The same result would follow for the triangle AOB, and hence for triangle ABC, $\bar{y} = \dfrac{h}{3}$. The centroid of ABC is then on a line parallel to AC and passing through the point of trisection of the medians. (Recall from plane geometry that the medians of a triangle are concurrent at the point of trisection of each.) The same result would be true for either of the other sides of the triangle ABC. Hence the centroid of ABC must be at the point of trisection of the medians.

4.5.7. Find the centroid of the area bounded by $y = x^3$, $y = x^2$.

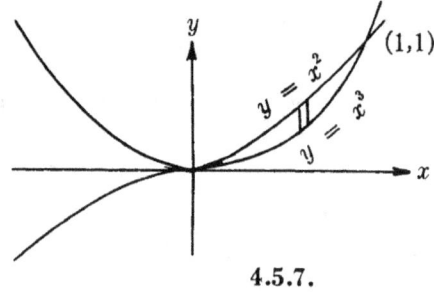

4.5.7.

Solution. The curves intersect at $(0, 0)$ and $(1, 1)$.

$$A = \int_0^1 (x^2 - x^3)\, dx = \left(\frac{x^3}{3} - \frac{x^4}{4}\right)_0^1 = \frac{1}{12};$$

$$A\bar{y} = \int_0^1 \frac{x^2 + x^3}{2}(x^2 - x^3)\,dx$$

$$= \frac{1}{2}\int_0^1 (x^4 - x^6)\,dx = \frac{1}{2}\left(\frac{x^5}{5} - \frac{x^7}{7}\right)\Big|_0^1 = \frac{1}{35};$$

$$\bar{y} = \tfrac{12}{35} = 0.343^-;$$

$$A\bar{x} = \int_0^1 x(x^2 - x^3)\,dx = \left(\frac{x^4}{4} - \frac{x^5}{5}\right)\Big|_0^1 = \frac{1}{20};$$

$$\bar{x} = \tfrac{12}{20} = 0.6$$

4.5.8. Find the moment of inertia and the radius of gyration of a rectangle a units by b units with respect to a side of a units.

Solution.

$$I_x = \int_0^b y^2(a\,dy) = \frac{ay^3}{3}\Big|_0^b = \frac{ab^3}{3} = M\frac{b^2}{3}$$

if we write M for the area.

$$R^2 M = I_x; \qquad R^2 = \frac{b^2}{3}; \qquad R = \frac{\sqrt{3}}{3}b$$

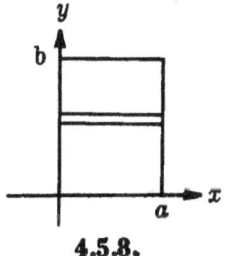

4.5.8.

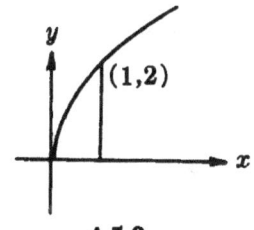

4.5.9.

4.5.9. Find (a) the moment of inertia and (b) the radius of gyration with respect to the x axis of the area bounded by $y^2 = 4x$, its latus rectum, and the x axis.

Solution.

(a) $$A = \int_0^2 (1 - x)\,dy = \int_0^2 \left(1 - \frac{y^2}{4}\right)dy = \left(y - \frac{y^3}{12}\right)_0^2 = \frac{4}{3};$$

$$I_x = \int_0^2 y^2(1 - x)\,dy = \int_0^2 y^2\left(1 - \frac{y^2}{4}\right)dy$$

$$= \left(\frac{y^3}{3} - \frac{y^5}{20}\right)_0^2 = \frac{16}{15}$$

To verify I_z, use vertical strips and the result of 4.5.8. The area of a strip is $y\,dx$, while its $R^2 = \frac{y^2}{3}$. Hence $I_z = \int_0^1 \frac{y^2}{3} y\,dx = \frac{1}{3}\int_0^1 8x^{3/2}\,dx = \frac{8}{3}\cdot\frac{2}{5}x^{5/2}\Big|_0^1 = \frac{16}{15}$ If $M = A = \frac{4}{3}$, we may write $I_z = (\frac{3}{4}\cdot\frac{16}{15})\frac{4}{3} = \frac{4}{5}M$.

(b) $R^2 A = I_z$; $R^2 \frac{4}{3} = \frac{16}{15}$; $R^2 = \frac{4}{5}$; $R = \frac{2}{5}\sqrt{5}$

4.5.10. Find the moment of inertia of the rectangle in 4.5.8 with respect to $y = \frac{b}{2}$.

Solution. Using the result of 4.5.8 to find I_z for each half of the rectangle above and below $y = \frac{b}{2}$ and adding the results,

$$I_z = \frac{a}{3}\left(\frac{b}{2}\right)^3 + \frac{a}{3}\left(\frac{b}{2}\right)^3 = \frac{ab^3}{12} = \frac{b^2}{12}M$$

4.5.11. Find the moment of inertia of a right circular cylinder with respect to its axis if its radius is a and its altitude h.

Solution. Let the cylinder be generated by revolving a rectangle about the x axis. A thin cylindrical shell with $I_z = y^2 2\pi y h\,dy$ is generated by the shaded area. Then for the cylinder, $I_z = \int_0^a 2\pi h y^3\,dy = \frac{\pi h a^4}{2} = \frac{Ma^2}{2}$

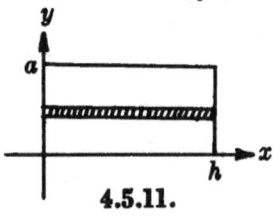
4.5.11.

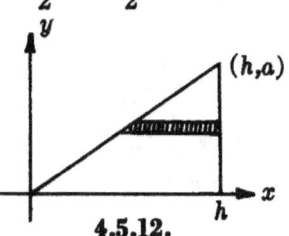
4.5.12.

4.5.12. (a) Find the moment of inertia of a right circular cone with respect to its axis if its radius is a and its altitude is h;

(b) find its radius of gyration.

Solution. (a) Let the cone be generated by revolving $y = \frac{a}{h}x$ from $x = 0$ to $x = h$ about the x axis. Use cylindrical shells generated by horizontal strips like the shaded one.

$$I_z = \int_0^a y^2 2\pi y(h-x)\,dy = 2\pi \int_0^a y^3\left(h - \frac{hy}{a}\right)dy$$
$$= 2\pi h\left(\frac{y^4}{4} - \frac{y^5}{5a}\right)_0^a = \frac{\pi}{10}ha^4;$$
$$M = V = \frac{\pi a^2 h}{3}; \quad I_z = \frac{\pi a^2 h}{3}\left(\frac{\pi h a^4}{10}\cdot \frac{3}{\pi a^2 h}\right) = \frac{3}{10}Ma^2;$$
$$R^2 M = I_z; \quad R^2 = \tfrac{3}{10}a^2; \quad R = \frac{\sqrt{30}a}{10}$$

4.5.13. Find the moment of inertia of the area bounded by $y^2 = x^3$, $x = 4$ with respect to: (a) the x axis, (b) $x = 4$, (c) the y axis.

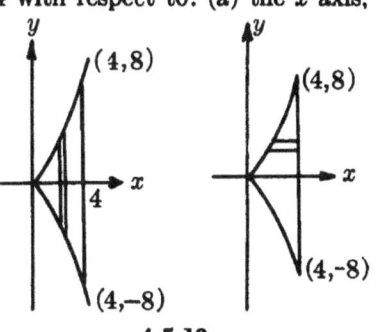

4.5.13.

Solution. (a) Using elements parallel to the x axis,
$$A = M = 2\int_0^8 (4-x)\,dy = 2\int_0^8 (4 - y^{2/3})\,dy$$
$$= 2(4y - \tfrac{3}{5}y^{5/3})\,|_0^8 = \frac{128}{5} = \frac{2^7}{5}$$

The moment of inertia for a typical element is
$$\Delta I_x = y^2(4-x)\,dy; \quad I_x = \int_{-8}^8 y^2(4-x)\,dy$$
$$= \int_{-8}^8 y^2(4 - y^{2/3})\,dy = (\tfrac{4}{3}y^3 - \tfrac{3}{11}y^{11/3})\Big|_{-8}^8$$
$$= \frac{2^{13}}{33} = \left(\frac{5}{2^7}\cdot\frac{2^{13}}{33}\right)\frac{2^7}{5} = \frac{320}{33}M$$

Using elements parallel to the y axis and the result of **4.5.8**,

$$\Delta I_z = \tfrac{1}{3}y^2 \cdot y\, dx; \qquad I_z = \frac{2}{3}\int_0^4 y^3\, dx$$

$$= \frac{2}{3}\int_0^4 x^{9/2}\, dx = \tfrac{2}{3}\cdot\tfrac{2}{11} x^{11/2}\Big|_0^4 = \frac{2^{13}}{33} = \frac{320}{33}M$$

(b) Using elements parallel to the y axis: The area of an element is $2y\, dx$ but now the arm is $4 - x$, so that $\Delta I_{x-4} = 2(4-x)^2 y\, dx$.

$$I_{x-4} = 2\int_0^4 (4-x)^2 y\, dx = 2\int_0^4 (4-x)^2 x^{3/2}\, dx$$

$$= 2\int_0^4 (16 x^{3/2} - 8x^{5/2} + x^{7/2})\, dx = 2\left(\tfrac{32}{5}x^{5/2} - \tfrac{16}{7}x^{7/2} + \tfrac{2}{9}x^{9/2}\right)\Big|_0^4$$

$$= \frac{2^{14}}{315} = \frac{128}{63}M$$

Using elements parallel to the x axis,

$$\Delta I_{x-4} = \tfrac{1}{3}(4-x)^2(4-x)\, dy; \qquad I_{x-4} = \frac{1}{3}\int_{-8}^{8}(4-x)^3\, dy$$

$$= \frac{1}{3}\int_{-8}^{8}(4 - y^{2/3})^3\, dy = \frac{1}{3}\int_{-8}^{8}(64 - 48x^{2/3} + 12x^{4/3} - x^2)\, dx$$

$$= \frac{1}{3}\left(64x - \frac{144}{5}x^{5/3} + \frac{36}{7}x^{7/3} - \tfrac{1}{3}x^3\right)\Big|_{-8}^{8} = \frac{2^{14}}{315} = \frac{128}{63}M$$

(c) Using elements parallel to the y axis, $\Delta I_y = x^2 \cdot 2y\, dx$.

$$I_y = 2\int_0^4 x^2 y\, dx = 2\int_0^4 x^2 \cdot x^{3/2}\, dx = 2\int_0^4 x^{7/2}\, dx$$

$$= 2\cdot \frac{2}{9}x^{9/2}\Big|_0^4 = \frac{2048}{9} = \frac{80}{9}M$$

Using elements parallel to the x axis. In this case we must take the difference of the inertias of the two rectangles shown in the figure.

$$\Delta I_y = \tfrac{1}{3}4^2\cdot 4\ dy - \tfrac{1}{3}x^2\cdot x\ dy; \qquad I_y = \int_{-8}^{8}\frac{64}{3}\ dy - \int_{-8}^{8}\frac{1}{3}x^3\ dy$$

$$= \frac{64}{3}\int_{-8}^{8} dy - \frac{1}{3}\int_{-8}^{8} y^2\ dy = \frac{64}{3}y\Big|_{-8}^{8} - \frac{1}{9}y^3\Big|_{-8}^{8}$$

$$= \frac{2^{10}}{3} - \frac{2^{10}}{9} = \frac{2^{11}}{9} = \frac{2048}{9} = \frac{80}{9}M$$

Notice that rectangles perpendicular to but removed from the axis of rotation give rise to the calculation of two separate moments of inertia.

Chapter 5

DIFFERENTIATION OF ELEMENTARY TRANSCENDENTAL FUNCTIONS

1. Differentiation of trigonometric functions. We use the four step rule to differentiate $\sin x$:

$$y = \sin x$$

$$y + \Delta y = \sin(x + \Delta x)$$

$$\Delta y = \sin(x + \Delta x) - \sin x$$

$$= 2 \cos\left(x + \frac{\Delta x}{2}\right) \sin \frac{\Delta x}{2}$$

by using a formula of trigonometry)

$$\frac{\Delta y}{\Delta x} = 2 \cos\left(x + \frac{\Delta x}{2}\right) \frac{\sin \frac{\Delta x}{2}}{\Delta x}$$

$$\lim_{\Delta x \to 0} \frac{\Delta y}{\Delta x} = \lim_{\Delta x \to 0} \cos\left(x + \frac{\Delta x}{2}\right) \lim_{\Delta x \to 0} \frac{\sin \frac{\Delta x}{2}}{\frac{\Delta x}{2}}$$

By example **1.2.4**, $\lim_{\Delta x \to 0} \frac{\sin \frac{\Delta x}{2}}{\frac{\Delta x}{2}} = 1$, and we have $\frac{dy}{dx} = \cos x$.

If $y = \sin u$, then $\frac{dy}{du} = \cos u$. If u is in turn a differentiable function of x, then y is a differentiable function of x and $\frac{dy}{dx} = \frac{dy}{du} \cdot \frac{du}{dx}$, which becomes

$$\frac{d}{dx} \sin u = \cos u \frac{du}{dx} \qquad (5.1.1)$$

The derivatives of the remaining trigonometric functions are given below.

$$\frac{d}{dx}\cos u = -\sin u \frac{du}{dx} \qquad (5.1.2)$$

$$\frac{d}{dx}\tan u = \sec^2 u \frac{du}{dx} \qquad (5.1.3)$$

$$\frac{d}{dx}\cot u = -\csc^2 u \frac{du}{dx} \qquad (5.1.4)$$

$$\frac{d}{dx}\sec u = \sec u \tan u \frac{du}{dx} \qquad (5.1.5)$$

$$\frac{d}{dx}\csc u = -\csc u \cot u \frac{du}{dx} \qquad (5.1.6)$$

It is useful to remember that the co-named functions (cosine cotangent, cosecant) all have minus signs in their derivatives. The same is true for the derivatives of the inverse trigonometric functions. The student should memorize the differentiation formulas of this and following sections.

Find $\frac{dy}{dx}$ for the following.

5.1.1. $y = \sin 2x$

Solution. Let $u = 2x$ so that $\frac{du}{dx} = 2$, then $\frac{dy}{dx} = 2\cos 2x$.

5.1.2. $y = \tan(x^2 + 2x + 3)$

Solution. Let $u = x^2 + 2x + 3$, so that $\frac{du}{dx} = 2x + 2$; then

$$\frac{dy}{dx} = [\sec^2(x^2 + 2x + 3)](2x + 2)$$

$$= 2(x + 1)\sec^2(x^2 + 2x + 3)$$

5.1.3. $y = \tan^2 x$

Solution. Let $u = \tan x$; then

$$\frac{du}{dx} = \sec^2 x; \qquad \frac{dy}{dx} = \frac{dy}{du} \cdot \frac{du}{dx}; \qquad \frac{dy}{dx} = 2\tan x \sec^2 x$$

5.1.4. $y = \sec(x + 1)$

Solution

$$\frac{dy}{dx} = \sec(x+1)\tan(x+1)$$

5.1.5. $y = \csc 2x$

Solution

$$\frac{dy}{dx} = -2\csc 2x \cdot \cot 2x$$

5.1.6. $y = \cot^3(2x+1)$
Solution. Let $\cot(2x+1) = v$ and $(2x+1) = u$, so that $= \cot u$. Hence $y = v^3$, and

$$\frac{dy}{dx} = 3v^2 \frac{dv}{du}\frac{du}{dx} = 3\cot^2(2x+1)[-\csc^2(2x+1)]2$$

$$= -6\cot^2(2x+1)\cdot\csc^2(2x+1)$$

5.1.7. $y = \sin x \cdot \tan x$

Solution

$$\frac{dy}{dx} = \sin x \frac{d}{dx}\tan x + \tan x \frac{d}{dx}\sin x$$

$$= \sin x \sec^2 x + \tan x \cos x$$

5.1.8. $y = \dfrac{\tan x + \sin x}{\cos x}$

Solution

$$y' = \frac{\cos x \dfrac{d}{dx}(\tan x + \sin x) - (\tan x + \sin x)\dfrac{d}{dx}\cos x}{\cos^2 x}$$

$$= \frac{\cos x(\sec^2 x + \cos x) - (\tan x + \sin x)(-\sin x)}{\cos^2 x}$$

$$= \sec^3 x + 1 + \sec x \tan^2 x + \tan^2 x$$

$$= \sec^3 x + \sec^2 x + \sec x \tan^2 x$$

Another solution of this example would be

$$y = \frac{\tan x + \sin x}{\cos x}$$

$$= \sec x \tan x + \tan x = \tan x (\sec x + 1)$$

$$y' = \tan x \frac{d}{dx}(\sec x + 1) + (\sec x + 1)\frac{d}{dx}\tan x$$

$$= \tan x \sec x \tan x + (\sec x + 1)\sec^2 x$$

$$= \sec x \tan^2 x + \sec^3 x + \sec^2 x$$

5.1.9. $y = \sqrt{\cot(4x+1)} = [\cot(4x+1)]^{1/2}$

Solution

$$y' = \frac{1}{2}[\cot(4x+1)]^{-1/2}\frac{d}{dx}\cot(4x+1)$$

$$= \tfrac{1}{2}[\cot(4x+1)]^{-1/2}[-\csc^2(4x+1)]4$$

$$= -2\sqrt{\tan(4x+1)}\csc^2(4x+1)$$

5.1.10. $y = x^n \cos x$

Solution

$$y' = x^n \frac{d}{dx}\cos x + \cos x \frac{d}{dx}x^n; \qquad = -x^n \sin x + nx^{n-1}\cos$$

5.1.11. $y = \sec^4 x^2$

Solution

$$y' = 4\sec^3 x^2 \frac{d}{dx}\sec x^2 = 4\sec^3 x^2 \sec x^2 \tan x^2 \frac{d}{dx}x^2$$

$$= 8x \sec^4 x^2 \tan x^2$$

5.1.12. $y = \dfrac{1 + \tan x}{1 - \tan x}$

Solution

$$f' = \frac{(1-\tan x)\dfrac{d}{dx}(1+\tan x) - (1+\tan x)\dfrac{d}{dx}(1-\tan x)}{(1-\tan x)^2}$$

$$= \frac{(1-\tan x)\sec^2 x - (1+\tan x)(-\sec^2 x)}{(1-\tan x)^2}$$

$$= \frac{2\sec^2 x}{(1-\tan x)^2}$$

5.1.13. $y = \sin 3x$

Solution

$$y' = 3\cos 3x; \qquad y'' = -3^2 \sin 3x;$$
$$y''' = -3^3 \cos 3x; \qquad y^{IV} = 3^4 \sin 3x$$

5.1.14. $y = \sin x$

Solution

$$y' = \cos x = \sin\left(\frac{\pi}{2} + x\right)$$

$$y'' = -\sin x = \sin\left(2\frac{\pi}{2} + x\right)$$

$$y''' = -\cos x = \sin\left(3\frac{\pi}{2} + x\right)$$

$$y^{IV} = \sin x = \sin\left(4\frac{\pi}{2} + x\right)$$

$$\cdots\cdots\cdots\cdots\cdots\cdots$$

$$y^{(n)} = \sin\left(n\frac{\pi}{2} + x\right)$$

5.2. Definition and differentiation of the inverse trigonometric functions. The symbols arc sin x and $\sin^{-1} x$ both mean an angle whose sine is x. Considered as a function of x, arc sin x is infinitely many-valued, i.e., for any value of x such that $-1 \leqslant x \leqslant 1$ there are an endless number of angles arc sin To make arc sin x a single-valued function of x and so give sense to its derivative, we agree to let it represent an ang between $-\frac{\pi}{2}$ and $\frac{\pi}{2}$. When this is done, we write $-\frac{\pi}{2} \leqslant A$ sin $x \leqslant \frac{\pi}{2}$ and say that Arc sin x is the principle branch of a sin x.

The other principal branches are:

$$0 \leqslant \text{Arc cos } x \leqslant \pi$$

$$-\frac{\pi}{2} < \text{Arc tan } x < \frac{\pi}{2}$$

$$0 < \text{Arc cot } x < \pi$$

$$-\pi \leqslant \text{Arc sec } x < -\frac{\pi}{2} \quad \text{or} \quad 0 \leqslant \text{Arc sec } x < \frac{\pi}{2}$$

$$-\pi < \text{Arc csc } x \leqslant -\frac{\pi}{2} \quad \text{or} \quad 0 < \text{Arc csc } x \leqslant \frac{\pi}{2}$$

5.2.

Below are listed the derivatives of these inverse trigonometric functions:

$$\frac{d}{dx} \text{Arc sin } u = \frac{1}{\sqrt{1 - u^2}} \frac{du}{dx} \qquad (5.2.1)$$

$$\frac{d}{dx} \text{Arc cos } u = \frac{-1}{\sqrt{1 - u^2}} \frac{du}{dx} \qquad (5.2.2)$$

$$\frac{d}{dx} \text{Arc tan } u = \frac{1}{1 + u^2} \frac{du}{dx} \qquad (5.2.3)$$

$$\frac{d}{dx} \operatorname{Arc\,cot} u = \frac{-1}{1+u^2} \frac{du}{dx} \qquad (5.2.4)$$

$$\frac{d}{dx} \operatorname{Arc\,sec} u = \frac{1}{u\sqrt{u^2-1}} \frac{du}{dx} \qquad (5.2.5)$$

$$\frac{d}{dx} \operatorname{Arc\,csc} u = \frac{-1}{u\sqrt{u^2-1}} \frac{du}{dx} \qquad (5.2.6)$$

Find $\dfrac{dy}{dx}$ for the following:

5.2.1. $y = \operatorname{Arc\,sin}(x^2+1)$

Solution. Let $u = x^2+1$ so that $\dfrac{du}{dx} = 2x$; then

$$\frac{dy}{dx} = \frac{1}{\sqrt{1-(x^2+1)^2}} \cdot 2x = \frac{2x}{\sqrt{-x^4-2x^2}}$$

5.2.2. $y = \operatorname{Arc\,sec} 3x$

Solution. Let $u = 3x$ so that $\dfrac{du}{dx} = 3$; then

$$\frac{dy}{dx} = \frac{1}{3x\sqrt{(3x)^2-1}} \cdot 3 = \frac{1}{x\sqrt{9x^2-1}}$$

5.2.3. $y = \operatorname{Arc\,cos}(4x+3)$

Solution. Let $u = 4x+3$ so that $\dfrac{du}{dx} = 4$; then

$$y' = -\frac{1}{\sqrt{1-(4x+3)^2}} \cdot 4 = -\frac{4}{\sqrt{-16x^2-24x-8}}$$

$$= -\frac{\sqrt{2}}{\sqrt{-2x^2-3x-1}}$$

5.2.4. $y = \operatorname{Arc\,tan}(3x-2)$

Solution. Let $u = 3x-2$, so that $\dfrac{du}{dx} = 3$; then

$$y' = \frac{1}{1+(3x-2)^2} \cdot 3 = \frac{3}{9x^2-12x+5}$$

5.2.5. $y = \operatorname{Arc\,cot} x^2$

Solution. Let $u = x^2$, so that $\dfrac{du}{dx} = 2x$; then $y' = \dfrac{-2x}{1+x^4}$

5.2.6. $y = \text{Arc csc}\,(x^3 + 1)$

Solution. Let $u = x^3 + 1$, so that $\dfrac{du}{dx} = 3x^2$, then

$$y' = \dfrac{-1}{(x^3+1)\sqrt{(x^3+1)^2 - 1}}\, 3x^2 = \dfrac{-3x^2}{(x^3+1)\sqrt{x^6 + 2x^3}}$$

5.2.7. $y = x^2\,\text{Arc tan}\,x^2$
Solution

$$y' = x^2\,\dfrac{2x}{1+x^4} + (\text{Arc tan}\,x^2)\,2x = \dfrac{2x^3}{1+x^4} + 2x\,\text{Arc tan}\,x^2$$

5.2.8. $y = \text{Arc tan}\,\dfrac{1+x}{1-x}$

Solution. Let $u = \dfrac{1+x}{1-x}$ so that

$$\dfrac{du}{dx} = \dfrac{(1-x)-(1+x)(-1)}{(1-x)^2} = \dfrac{2}{(1-x)^2};$$

$$y' = \dfrac{1}{1+\left(\dfrac{1+x}{1-x}\right)^2}\,\dfrac{2}{(1-x)^2} = \dfrac{1}{1+x^2}$$

5.2.9. $y = \dfrac{\text{Arc sec}\,x}{\text{Arc tan}\,x}$

Solution

$$y' = \dfrac{(\text{Arc tan}\,x)\,\dfrac{1}{x\sqrt{x^2-1}} - (\text{Arc sec}\,x)\,\dfrac{1}{1+x^2}}{(\text{Arc tan}\,x)^2}$$

$$y' = \dfrac{(1+x^2)\,\text{Arc tan}\,x - x\sqrt{x^2-1}\,\text{Arc sec}\,x}{x\sqrt{x^2-1}\,(1+x^2)\,(\text{Arc tan}\,x)^2}$$

5.2.10. $y = \text{Arc tan}\,(3\tan x)$

Solution. Let $u = 3 \tan x$ so that $\dfrac{du}{dx} = 3 \sec^2 x$.

$$y' = \frac{3 \sec^2 x}{1 + (3 \tan x)^2} = \frac{3}{\cos^2 x + 9 \sin^2 x}$$

5.2.11. $y^2 \tan x + y = \text{Arc} \tan x$
Solution

$$\sec^2 x + (\tan x) 2yy' + y' = \frac{1}{1 + x^2}; \qquad (2y \tan x + 1) y'$$

$$= \frac{1}{1 + x^2} - y^2 \sec^2 x; \qquad y' = \frac{1 - (1 + x^2) y^2 \sec^2 x}{(1 + x^2)(2y \tan x + 1)}$$

3. Differentiation of exponential and logarithmic functions. In the formulas listed below, e is the base of the natural logarithms, its approximate value being 2.71828 18284 .... The natural logarithm of a number N is written $\ln N$.

$$\frac{d}{dx} \ln x = \frac{1}{x} \qquad (5.3.1)$$

$$\frac{d}{dx} \log_a u = \frac{1}{u} \log_a e \frac{du}{dx} \qquad (5.3.2)$$

$$\frac{d}{dx} \ln u = \frac{1}{u} \frac{du}{dx} \qquad (5.3.3)$$

$$\frac{d}{dx} a^u = a^u \ln a \frac{du}{dx} \qquad (5.3.4)$$

$$\frac{d}{dx} e^u = e^u \frac{du}{dx} \qquad (5.3.5)$$

Find $\dfrac{dy}{dx}$ for the following:

5.3.1. $y = \ln (x^2 + 2)$
Solution. Let $u = x^2 + 2$ so that $\dfrac{du}{dx} = 2x$; then $\dfrac{dy}{dx} = \dfrac{2x}{x^2 + 2}$

5.3.2. $y = \ln (2x + 1)$
Solution. Let $u = 2x + 1$ so that $\dfrac{du}{dx} = 2$; then $\dfrac{dy}{dx} = \dfrac{2}{2x + 1}$

5.3.3. $y = \log_a (2x + 1)$

Solution
$$\frac{dy}{dx} = \frac{2}{2x+1} \log_a e$$

5.3.4. $y = \ln \sin x$

Solution. Let $u = \sin x$; then $\dfrac{du}{dx} = \cos x$;
$$y' = \frac{\cos x}{\sin x} = \cot x$$

5.3.5. $y = \ln \operatorname{Arc} \tan x$

Solution. Let $u = \operatorname{Arc} \tan x$; then $\dfrac{du}{dx} = \dfrac{1}{1+x^2}$;
$$y' = \frac{1}{(1+x^2) \operatorname{Arc} \tan}$$

5.3.6. $y = \log_{10}(2x^3 - 5)$
Solution
$$y' = \frac{6x^2}{2x^3 - 5} \log_{10} e$$

5.3.7. $y = \log_2 \sec x^3$
Solution

$u = \sec x^3$; $\dfrac{du}{dx} = 3x^2 \sec x^3 \tan x^3$;

$$y' = \frac{\log_2 e}{\sec x^3} 3x^2 \sec x^3 \tan x^3 = 3x^2 \log_2 e \tan x$$

5.3.8. $y = x(x+1)(x^2+1)$
Solution
$$\ln y = \ln x + \ln(x+1) + \ln(x^2+1)$$
$$\frac{1}{y}\frac{dy}{dx} = \frac{1}{x} + \frac{1}{x+1} + \frac{2x}{x^2+1}$$
$$\frac{dy}{dx} = y\left(\frac{1}{x} + \frac{1}{x+1} + \frac{2x}{x^2+1}\right)$$

5.3.9. $y = (x+1)(x^3 + 2x + 3)(x^2 + 1)^{1/2}$

Solution

$y = \ln(x+1) + \ln(x^3 + 2x + 3) + \tfrac{1}{2}\ln(x^2+1)$

$$\frac{dy}{dx} = \frac{1}{x+1} + \frac{3x^2+2}{x^3+2x+3} + \frac{x}{x^2+1}$$

$$\frac{dy}{dx} = y\left(\frac{1}{x+1} + \frac{3x^2+2}{x^3+2x+3} + \frac{x}{x^2+1}\right)$$

$$\frac{dy}{dx} = (x+1)(x^3+2x+3)(x^2+1)^{1/2}$$

$$\left(\frac{1}{x+1} + \frac{3x^2+2}{x^3+2x+3} + \frac{x}{x^2+1}\right)$$

5.3.10. $y = \log_{10}(\sin x + \cos x)^{3/2} = \tfrac{3}{2}\log_{10}(\sin x + \cos x)$

Solution

$$y' = \tfrac{3}{2}\log_{10} e \,\frac{\cos x - \sin x}{\sin x + \cos x}$$

5.3.11. $y = \ln \ln x$

Solution.

Let $u = \ln x$; then $\dfrac{du}{dx} = \dfrac{1}{x}$; $\dfrac{dy}{dx} = \dfrac{1}{\ln x}\cdot\dfrac{1}{x} = \dfrac{1}{x \ln x}$

5.3.12. $y = \ln \ln \ln x$

Solution. Let $u = \ln \ln x$; then $\dfrac{du}{dx} = \dfrac{1}{x \ln x}$ by 5.3.11

$$\frac{dy}{dx} = \frac{\frac{1}{x \ln x}}{\ln \ln x} = \frac{1}{x \ln x \cdot \ln \ln x}$$

5.3.13. $y = \ln^2 x$

Solution.

$$\frac{dy}{dx} = 2(\ln x)\frac{1}{x} = \frac{2\ln x}{x}$$

5.3.14. $y = \ln\sqrt{x^2-1} = \tfrac{1}{2}\ln(x^2-1)$
$= \tfrac{1}{2}\ln(x-1) + \tfrac{1}{2}\ln(x+1)$

Solution
$$y' = \frac{1}{2}\frac{1}{x-1} + \frac{1}{2}\frac{1}{x+1}$$

5.3.15. $y = \ln \text{Arc} \tan x$
Solution. Let $u = \text{Arc} \tan x$; then
$$\frac{du}{dx} = \frac{1}{1+x^2}; \quad \frac{dy}{dx} = \frac{1}{\text{Arc}\tan x} \cdot \frac{1}{1+x^2}$$

5.3.16. $y = e^{x^2}$
Solution. Let $u = x^2$ so that $\frac{du}{dx} = 2x$; then $\frac{dy}{dx} = e^{x^2} 2x$

5.3.17. $y = e^{\csc x}$
Solution. Let $u = \csc x$ so that $\frac{du}{dx} = -\csc x \cot x$; then
$$\frac{dy}{dx} = -e^{\csc x} \csc x \cot x$$

5.3.18. $y = 10^{x^2}$
Solution. Let $a = 10$, $u = x^2$, so that $\frac{du}{dx} = 2x$; then
$$\frac{dy}{dx} = (\ln 10)10^{x^2} \cdot 2x$$

Another solution depends on the identity $10 = e^{\ln 10}$, which follows from the definition of a natural logarithm.
$$y = 10^{x^2} = (e^{\ln 10})^{x^2} = e^{x^2 \ln 10}$$
Let $u = x^2 \ln 10$ so that $\frac{du}{dx} = 2x \ln 10$ and
$$\frac{dy}{dx} = e^{x^2 \ln 10} \cdot 2x \cdot \ln 10 = 2 \ln 10 \cdot 10^{x^2} x$$

5.3.19. $y = x^2 e^{3x}$
Solution
$$y' = x^2 \cdot e^{3x} \cdot 3 + e^{3x} \cdot 2x = e^{3x}(3x^2 + 2x)$$

5.3.20. $y = \dfrac{e^x - e^{-x}}{e^x + e^{-x}}$

Solution

$$y' = \frac{(e^x + e^{-x})(e^x + e^{-x}) - (e^x - e^{-x})(e^x - e^{-x})}{(e^x + e^{-x})^2} = \frac{4}{(e^x + e^{-x})^2}$$

5.3.21. $y = x^x$
Solution

$$\ln y = x \ln x; \qquad \frac{1}{y}\frac{dy}{dx} = x\frac{1}{x} + \ln x;$$

$$\frac{dy}{dx} = y(1 + \ln x) = x^x(1 + \ln x)$$

Since $x = e^{\ln x}$, a second method of solution is as follows: $y = x^x = (e^{\ln x})^x = e^{x \ln x}$. Let $u = x \ln x$ so that

$$y' = e^{x\ln x}\frac{d}{dx}(x \ln x) = e^{x \ln x}\left(x \cdot \frac{1}{x} + \ln x\right)$$
$$= x^x(1 + \ln x)$$

5.3.22. $y = e^x \sin x$

$y' = e^x \cos x + (\sin x)e^x = e^x(\cos x + \sin x)$

$y'' = e^x(-\sin x + \cos x) + (\cos x + \sin x)e^x = 2e^x \cos x$

$y''' = 2e^x(-\sin x) + (\cos x)2e^x = 2e^x(\cos x - \sin x)$

5.3.23. $y = e^{-x^2/2}$
Solution

$y' = e^{-x^2/2}(-x)$

$y'' = e^{-x^2/2}(-1) + (-x)e^{-x^2/2}(-x) = e^{-x^2/2}(x^2 - 1)$

$y''' = e^{-x^2/2}(2x) + (x^2 - 1)e^{-x^2/2}(-x) = e^{-x^2/2}(-x^3 + 3x)$

$y^{iv} = e^{-x^2/2}(-3x^2 + 3) + (-x^3 + 3x)e^{-x^2/2}(-x)$
$= e^{-x^2/2}(x^4 - 6x^2 + 3)$

5.3.24. $x^2 + \sqrt{\ln y + 1} = 2$
Solution

$$2x + \tfrac{1}{2}(\ln y + 1)^{-1/2}\frac{1}{y}y' = 0; \qquad 4xy + (\ln y + 1)^{-1/2}y' = 0;$$

$$y' = -4xy(\ln y + 1)^{1/2}$$

5.3.25. $y = e^{\ln x + 1} = ee^{\ln x} = ex$
Solution. $y' = e$

5.3.26. $y = \sqrt{e^x + 1} = (e^x + 1)^{1/2}$
Solution

$$y' = \tfrac{1}{2}(e^x + 1)^{-1/2} \cdot e^x = \frac{e^x}{2\sqrt{e^x + 1}}$$

5.3.27. $y = \text{Arc sin } e^x$

Solution. $y' = \dfrac{e^x}{\sqrt{1 - e^{2x}}}$

5.3.28. $y = \ln(e^{x^2} + 5)$

Solution. $y' = \dfrac{e^{x^2} \cdot 2x}{e^{x^2} + 5}$

5.3.29. $y = \tan^2 e^x$
Solution

$$y' = 2 \tan e^x \cdot \frac{d}{dx} \tan e^x$$

$$= 2 \tan e^x \cdot \sec^2 e^x \cdot \frac{d}{dx} e^x$$

$$= 2 e^x \cdot \tan e^x \cdot \sec^2 e^x$$

5.3.30. $y = \ln \text{Arc tan } e^x$
Solution

$$y' = \frac{1}{\text{Arc tan } e^x} \frac{d}{dx} \text{Arc tan } e^x,$$

$$= \frac{1}{\text{Arc tan } e^x} \frac{1}{1 + e^{2x}} \frac{d}{dx} e^x = \frac{e^x}{(1 + e^{2x}) \text{Arc tan } e^x}$$

5.3.31. $y = x^{\sin x}$
Solution

$$\ln y = (\sin x) \ln x;$$

$$\frac{1}{y} y' = (\sin x) \frac{1}{x} + (\ln x) \cos x;$$

$$y' = x^{\sin x} \left[\frac{\sin x}{x} + (\ln x) \cos x \right]$$

5.3.32. $y = e^{e^x}$

Solution. $y' = e^{e^x} \dfrac{d}{dx} e^x = e^{e^x} e^x$

5.3.33. $y = e^{1/\ln x}$

Solution

$$y' = e^{1/\ln x} \frac{d}{dx} \frac{1}{\ln x} = e^{1/\ln x} \frac{d}{dx} (\ln x)^{-1}$$

$$= e^{1/\ln x}(-1)(\ln x)^{-2} \frac{d}{dx} \ln x = e^{1/\ln x}(-1)(\ln x)^{-2} \cdot \frac{1}{x}$$

$$= \frac{e^{1/\ln x}}{-x(\ln x)^2}$$

5.3.34. $y = x^{x^x}$

Solution

$$\ln y = x^x \ln x;$$

$$\frac{1}{y} y' = x^x \frac{d}{dx} \ln x + (\ln x) \frac{d}{dx} x^x$$

By example **5.3.21**, $\dfrac{d}{dx} x^x = x^x(1 + \ln x)$. Using this, we have

$$y' = y \left[x^x \cdot \frac{1}{x} + (\ln x) x^x (1 + \ln x) \right]$$

$$= x^{x^x} \cdot x^x \left[(\ln x)^2 + \ln x + \frac{1}{x} \right]$$

Chapter 6

SOME PROPERTIES OF CURVES

6.1. Length of a curve. The differential of arc length can be found from the formulas

$$\frac{ds}{dx} = \sqrt{1 + \left(\frac{dy}{dx}\right)^2}; \qquad ds = \sqrt{1 + \left(\frac{dy}{dx}\right)^2}\, dx; \qquad (6.1.1)$$

$$ds = \sqrt{dx^2 + dy^2}$$

Using these formulas we can find the length of a curve,

$$s = \int_a^b \sqrt{1 + \left(\frac{dy}{dx}\right)^2}\, dx; \qquad s = \int_c^d \sqrt{1 + \left(\frac{dx}{dy}\right)^2}\, dy$$

6.1.1. Find the length of the semicubical parabola $y^2 = x^3$ from (0, 0) to (4, 8).

Solution

$$y = x^{3/2}; \qquad \frac{dy}{dx} = \frac{3}{2} x^{1/2};$$

$$s = \int_0^4 \sqrt{1 + \left(\frac{dy}{dx}\right)^2}\, dx = \int_0^4 \sqrt{1 + \frac{9x}{4}}\, dx$$

$$= \frac{4}{9} \int_0^4 \left(1 + \frac{9x}{4}\right)^{1/2} \left(\frac{9}{4}\, dx\right)$$

$$= \frac{4}{9} \cdot \frac{2}{3} \left(1 + \frac{9x}{4}\right)^{3/2} \Big|_0^4 = \frac{8}{27} (10^{3/2} - 1)$$

6.1.1.

6.1.2.

6.1.2. Find the length of the hypocycloid of four cusps whose equation is $x^{2/3} + y^{2/3} = a^{2/3}$.
Solution

$$\frac{2}{3}x^{-1/3} + \frac{2}{3}y^{-1/3}\frac{dy}{dx} = 0; \quad \frac{dy}{dx} = -\frac{y^{1/3}}{x^{1/3}}$$

Expressing the length as four times that of one arc,

$$s = 4\int_0^a \sqrt{1 + \frac{y^{2/3}}{x^{2/3}}}\,dx$$

$$= 4\int_0^a \sqrt{\frac{x^{2/3} + y^{2/3}}{x^{2/3}}}\,dx$$

$$= 4\int_0^a a^{1/3}x^{-1/3}\,dx = 4a^{1/3}\frac{3}{2}x^{2/3}\Big|_0^a = 6a$$

6.1.3. Find the length of the curve $y = \frac{x^2}{4} - \frac{1}{2}\ln x$ from $x = 1$ to $x = 4$.
Solution

$$y' = \frac{x}{2} - \frac{1}{2x} = \frac{1}{2}\left(x - \frac{1}{x}\right);$$

$$1 + (y')^2 = 1 + \frac{1}{4}\left(x^2 - 2 + \frac{1}{x^2}\right) = \frac{1}{4}\left(x + \frac{1}{x}\right)^2;$$

$$ds = \sqrt{1 + (y')^2}\,dx = \frac{1}{2}\left(x + \frac{1}{x}\right)dx;$$

$$s = \int_1^4 \frac{1}{2}\left(x + \frac{1}{x}\right)dx$$

$$= \left(\frac{x^2}{4} + \frac{1}{2}\ln x\right)_1^4 = \frac{15}{4} + \ln 2$$

6.1.4. Find the length of the curve $y^4 - 6xy + 3 = 0$ from $(\frac{2}{3}, 1)$ to $(\frac{19}{12}, 2)$.
Solution

$$x = \frac{y^4 + 3}{6y} = \frac{y^3}{6} + \frac{1}{2y};$$

$$\frac{dx}{dy} = \frac{y^2}{2} - \frac{1}{2y^2} = \frac{1}{2}\left(y^2 - \frac{1}{y^2}\right)$$

$$1 + \left(\frac{dx}{dy}\right)^2 = 1 + \frac{1}{4}\left(y^2 - \frac{1}{y^2}\right)^2 = \frac{1}{4}\left(y^2 + \frac{1}{y^2}\right)^2;$$

$$ds = \frac{1}{2}\left(y^2 + \frac{1}{y^2}\right)dy;$$

$$s = \int_1^2 \frac{1}{2}\left(y^2 + \frac{1}{y^2}\right)dy = \left(\frac{y^3}{6} - \frac{1}{2y}\right)_1^2 = \frac{17}{12}$$

6.1.5. Find the length of the curve $y = \frac{1}{\sqrt[3]{4}}(x-1)^{2/3}$ from $y = 0$ to $y = 1$.

Solution

$$(x-1)^{2/3} = 2^{2/3}y; \quad x - 1 = 2y^{3/2}; \quad \frac{dx}{dy} = 3y^{1/2};$$

$$s = \int_0^1 \sqrt{1 + \left(\frac{dx}{dy}\right)^2}\,dy = \int_0^1 \sqrt{1 + 9y}\,dy$$

$$= \frac{1}{9}\int_0^1 (1 + 9y)^{1/2}(9\,dy)$$

$$= \frac{1}{9} \cdot \frac{2}{3}(1 + 9y)^{3/2}\Big|_0^1 = \frac{2}{27}(10^{3/2} - 1)$$

6.2. Parametric representation of a curve

6.2.1. (a) Plot the curve having the parametric equations $x = 2t - 2$ and $y = t^2 - 1$. (b) Find the rectangular equation of the curve.

Solution. (a) We give the values in the following table to t and then compute the corresponding values of x and y.

t	-3	-2	-1	0	1	2	3
x	-8	-6	-4	-2	0	2	4
y	8	3	0	-1	0	3	8

Plotting the corresponding values of x and y as points and connecting these points by a smooth curve, we have the figure.

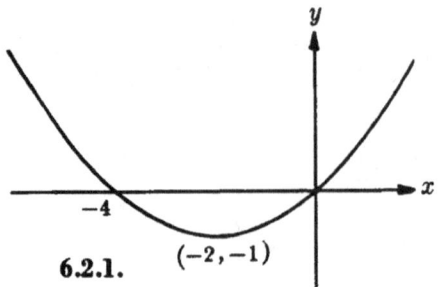

6.2.1. (−2,−1)

(b) Solving $x = 2t - 2$ for t gives $t = \frac{1}{2}(x + 2)$, which, substituted in $y = t^2 - 1$ yields the rectangular equation $y = \frac{1}{4}(x + 2)^2 - 1$. This can be written in the form $(x + 2)^2 = 4(y + 1)$, which shows that the curve is a parabola symmetrical with respect to the line $x + 2 = 0$ with its vertex at $(-2, -1)$ and opening upward.

6.2.2. Show that $x = a \cos \theta$ and $y = b \sin \theta$ are the parametric equations of an ellipse.

Solution. We obtain $\dfrac{x}{a} = \cos \theta$ and $\dfrac{y}{b} = \sin \theta$ so that

$$\frac{x^2}{a^2} + \frac{y^2}{b^2} = \sin^2 \theta + \cos^2 \theta; \qquad \frac{x^2}{a^2} + \frac{y^2}{b^2} = 1$$

6.2.3. Find the circumference of a circle of radius a.
Solution. This circle may be taken as $x^2 + y^2 = a^2$ so that

$$2x + 2y \frac{dy}{dx} = 0; \qquad \frac{dy}{dx} = -\frac{x}{y}.$$

Using 6.1.2,

$$ds = \sqrt{1 + \frac{x^2}{y^2}}\, dx = \sqrt{\frac{y^2 + x^2}{y^2}}\, dx$$

$$= \frac{a}{y}\, dx = \frac{a\, dx}{\sqrt{a^2 - x^2}}$$

and

$$s = 4 \int_0^a \frac{a\, dx}{\sqrt{a^2 - x^2}}$$

which is a definite integral that we cannot evaluate with the present state of our knowledge. However, if we place $b = a$ in the parametric equations of the ellipse given in **6.2.2**, we have parametric equations for our circle, viz., $x = a \cos \theta$ and

$y = a \sin \theta$. From these we obtain $dx = -a \sin \theta \, d\theta$ and $dy = a \cos \theta \, d\theta$, remembering that θ must be measured in radians. Substituting these values in 6.1.1,

$$ds = \sqrt{a^2 \sin^2 \theta \, d\theta^2 + a^2 \cos^2 \theta \, d\theta^2} = a \, d\theta.$$

As θ varies from 0 to 2π, the variable point (x, y) traces the complete circumference; hence $s = \int_0^{2\pi} a \, d\theta = a\theta \Big|_0^{2\pi} = 2\pi a$.

6.2.4. (a) Find the rectangular equation of the curve whose parametric equations are $x = t^2$ and $y = t^3$. (b) Use these parametric equations to find the length of this curve from $t = 0$ to $t = 2$.

Solution. (a) From $x = t^2$, $t = x^{1/2}$, and $y = t^3 = (x^{1/2})^3 = x^{3/2}$; $y^2 = x^3$.

(b) $dx = 2t \, dt$; $\quad dy = 3t^2 \, dt$. Substituting in 6 1.1,

$$ds = \sqrt{4t^2 \, dt^2 + 9t^4 \, dt^2} = \sqrt{4 + 9t^2} \, t \, dt = (4 + 9t^2)^{1/2} \, t \, dt.$$

Hence

$$s = \int_0^2 (4 + 9t^2)^{1/2} \, t \, dt = \frac{1}{18} \int_0^2 (4 + 9t^2)^{1/2} \cdot 18t \, dt$$

$$= \frac{1}{18} \cdot \frac{2}{3} [(4 + 9t^2)^{3/2}]_0^2 = \frac{8}{27} (10^{3/2} - 1).$$

Since the points $(0, 0)$ and $(4, 8)$ correspond to $t = 0$ and $t = 2$, respectively, this result verifies **6.1.1**

6.2.5. Find the rectangular equation of the curve $x = t + \frac{1}{t}$, $y = t - \frac{1}{t}$.

Solution. $x^2 = t^2 + 2 + \frac{1}{t^2}$; $\quad y^2 = t^2 - 2 + \frac{1}{t^2}$; $\quad x^2 - y^2 = 4$

6.2.6. Find parametric equations for the curve $x^3 + y^3 = xy$, which is called the folium of Descartes.

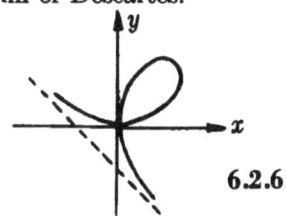

6.2.6.

Solution. Let $y = tx$ represent a family (or pencil) of lines through the origin with the variable slope t. Solve $y = tx$ simultaneously with

$$x^3 + y^3 = xy; \qquad x^3 + t^3 x^3 = xtx;$$

$$x^3(1 + t^3) = x^2 t; \qquad x = \frac{t}{1 + t^3}$$

Since $y = tx$, then $y = \dfrac{t^2}{1 + t^3}$.

3. Moments of a curve.
The first moment of arc of a curve given by

$$M_x = \int_a^b y\sqrt{1 + \left(\frac{dy}{dx}\right)^2}\, dx = \int_c^d y\sqrt{1 + \left(\frac{dx}{dy}\right)^2}\, dy;$$

$$M_y = \int_a^b x\sqrt{1 + \left(\frac{dy}{dx}\right)^2}\, dx = \int_c^d x\sqrt{1 + \left(\frac{dx}{dy}\right)^2}\, dy$$

Also the coordinates $(\bar{x}, \bar{y})$ of the centroid are defined by $\bar{x}s = M_y$ and $\bar{y}s = M_x$, respectively, where s is the length of the arc from (a, c) to (b, d). Similarly, the second moment of arc of a curve is given by

$$I_x = \int_a^b y^2\sqrt{1 + \left(\frac{dy}{dx}\right)^2}\, dx = \int_c^d y^2\sqrt{1 + \left(\frac{dx}{dy}\right)^2}\, dy;$$

$$I_y = \int_a^b x^2\sqrt{1 + \left(\frac{dy}{dx}\right)^2}\, dx = \int_c^d x^2\sqrt{1 + \left(\frac{dx}{dy}\right)^2}\, dy$$

Also the radii of gyration R about the x axis and the y axis are defined by $R^2 s = I_x$ and $R^2 s = I_y$, respectively.

6.3.1. For the arc from $(0, a)$ to $(a, 0)$ of the four-cusped hypocycloid $x^{2/3} + y^{2/3} = a^{2/3}$, find; (a) $(\bar{x}, \bar{y})$; (b) I_x and the corresponding R; (c) I_y and the corresponding R.

Solution. (a) By 6.1.2, $ds = a^{1/3} x^{-1/3}\, dx$, and the value of s from $(0, a)$ to $(a, 0)$ is $1.5a$.

$$M_y = \int_0^a x a^{1/3} x^{-1/3}\, dx = 0.6a^2; \qquad \bar{x}(1.5a) = 0.6a^2; \qquad \bar{x} = 0.4a$$

By symmetry, $\bar{y} = \bar{x} = 0.4a$.

(b) $I_y = \displaystyle\int_0^a x^2 a^{1/3} x^{-1/3}\, dx = \frac{3}{8} a^3; \quad R^2 \frac{3}{2} a = \frac{3}{8} a^3; \quad R = 0.5a$

(c) By symmetry, $I_y = I_x = \frac{2}{3}a^3$; $R = 0.5a$

6.3.2. Find the centroid of the upper half of the circle $x^2 + y^2 = a^2$.

Solution

$$2x + 2yy' = 0; \qquad y' = -\frac{x}{y};$$

$$M_x = \int_{-a}^{a} y\sqrt{1 + \left(-\frac{x}{y}\right)^2}\, dx = \int_{-a}^{a} a\, dx = 2a^2;$$

$$\bar{y}s = M_x; \qquad \bar{y}\pi a = 2a^2; \qquad \bar{y} = \frac{2a}{\pi} = 0.637\,a; \qquad \bar{x} = 0$$

by symmetry.

6.3.3. Find the radius of gyration about $x = 0$ for the arc from $x = 1$ to $x = 4$ of the curve $y = \frac{x^2}{4} - \frac{1}{2}\ln x$.

Solution. From 6.1.3, $ds = \frac{1}{2}\left(x + \frac{1}{x}\right) dx$ and $s = \frac{15}{4} + \ln 2$.

$$I_y = \int_1^4 x^2 \frac{1}{2}\left(x + \frac{1}{x}\right) dx = \frac{1}{2}\left(\frac{x^4}{4} + \frac{x^2}{2}\right)\Big|_1^4 = 35.625;$$

$$R^2 s = I_y; \qquad R^2\left(\frac{15}{4} + \ln 2\right) = 35.625;$$

$$R = 2\sqrt{\frac{35.625}{15 + \ln 16}} = 2.832^-$$

6.3.4. Find the moment of inertia with respect to $y = 0$ of the arc from $(\frac{2}{3}, 1)$ to $(\frac{13}{12}, 2)$ of the curve $y^4 - 6xy + 3 = 0$.

Solution. From 6.1.4,

$$ds = \frac{1}{2}\left(y^2 + \frac{1}{y^2}\right) dy; \qquad I_x = \int_1^2 y^2 \frac{1}{2}\left(y^2 + \frac{1}{y^2}\right) dy$$

$$= \frac{1}{2}\int_1^2 (y^4 + 1)\, dy = \frac{1}{2}\left(\frac{y^5}{5} + y\right)\Big|_1^2 = 3.6$$

6.4. Curvature, circle of curvature, and evolutes. The formula for curvature $K = \dfrac{d\alpha}{ds}$ is

$$K = \frac{\dfrac{d^2y}{dx^2}}{\left[1 + \left(\dfrac{dy}{dx}\right)^2\right]^{3/2}} = \frac{y''}{(1 + y'^2)^{3/2}} \qquad (6.4.1)$$

nce the sign of K is that of y'', K is positive if the curve is concave upward and negative if the curve is concave downward. he radius of curvature is given by

$$R = \frac{1}{K} = \frac{(1 + y'^2)^{3/2}}{y''} \qquad (6.4.2)$$

he coordinates of the center of curvature (x, y) are

$$x = x_1 - \frac{y_1'[1 + (y_1')^2]}{y_1''}; \qquad y = y_1 + \frac{1 + (y_1')^2}{y_1''} \qquad (6.4.3)$$

f we eliminate x_1 and y_1, we obtain an equation for the locus f the center of curvature called the evolute.

6.4.1. For $y = e^x$ at the point $(0, 1)$ find: (a) the curvature b) the radius of curvature, (c) coordinates of the center of urvature.

Solution. (a) $y' = e^x;$ $y_1' = e^0 = 1;$ $y'' = e^x;$ $y_1'' = e^0 = 1$

$$K = \frac{y_1''}{[1 + (y_1')^2]^{3/2}} = \frac{1}{(1 + 1^2)^{3/2}} = \frac{\sqrt{2}}{4}$$

(b) $\quad R = \dfrac{1}{K} = 2\sqrt{2} = 2.828^+$

(c) $\quad x = x_1 - \dfrac{y_1'[1 + (y_1')^2]}{y_1''}; \quad x = 0 - \dfrac{1(1 + 1^2)}{1} = -2;$

$\quad y = y_1 + \dfrac{1 + (y_1')^2}{y_1''}; \quad y = 1 + \dfrac{1 + 1^2}{1} = 3$

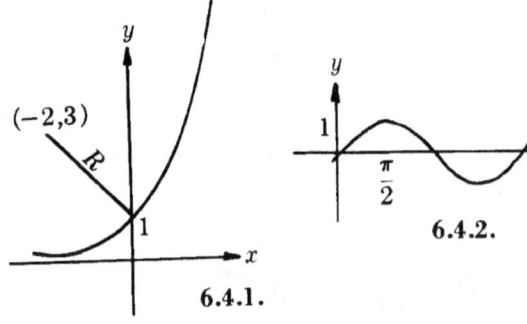

6.4.1.

6.4.2.

6.4.2. The same as 6.4.1 for $y = \sin x$ at the point $P\left(\dfrac{\pi}{2}, 1\right)$.

Solution. (a) $\quad y' = \cos x; \quad y'_1 = \cos \dfrac{\pi}{2} = 0;$

$y'' = -\sin x; \quad y''_1 = -\sin \dfrac{\pi}{2} = -1; \quad K = \dfrac{-1}{(1+0^2)^{3/2}} = -1$

(b) $\quad R = \dfrac{1}{K} = -1$

(c) $\quad x = \dfrac{\pi}{2} - \dfrac{0(1+0^2)}{-1} = \dfrac{\pi}{2}; \quad y = 1 + \dfrac{1+0^2}{-1} = 0$

6.4.3. The same as 6.4.1 for the folium of Descartes, $x^3 + y^3 = xy$. (See **6.2.6**) at the point $P(\tfrac{1}{2}, \tfrac{1}{2})$.

Solution

(a) $\quad 3x^2 + 3y^2 y' = xy' + y; \quad y' = \dfrac{y - 3x^2}{3y^2 - x}; \quad y'_1 = -1;$

$y'' = \dfrac{(3y^2 - x)(y' - 6x) - (y - 3x^2)(6yy' - 1)}{(3y^2 - x)^2};$

$y''_1 = \dfrac{(\tfrac{3}{4} - \tfrac{1}{2})(-1 - 3) - (\tfrac{1}{2} - \tfrac{3}{4})(-3 - 1)}{(\tfrac{3}{4} - \tfrac{1}{2})^2} = -32;$

$K = \dfrac{-32}{[1 + (-1)^2]^{3/2}} = -8\sqrt{2} = -11.314^-$

(b) $\quad R = \dfrac{1}{K} = -\dfrac{\sqrt{2}}{16} = -0.0884^-$

(c) $\quad x = \tfrac{1}{2} - \dfrac{(-1)[1 + (-1)^2]}{-32} = \dfrac{7}{16} = 0.4375;$

$y = \tfrac{1}{2} + \dfrac{1 + (-1)^2}{-32} = \dfrac{7}{16} = 0.4375$

Notice that y must equal x because of symmetry.

6.4.4. For the parabola $y^2 = x$ and the point $(1, 1)$, find: (a) the curvature, (b) the radius of curvature, (c) center of curvature, (d) the evolute of this parabola.

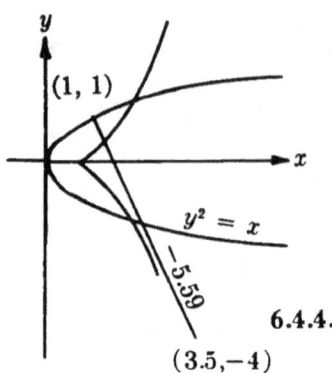

6.4.4.

Solution

(a) $\quad 2yy' = 1; \quad y' = \dfrac{1}{2y}; \quad y_1' = \dfrac{1}{2y_1} = \dfrac{1}{2};$

$yy'' + 2(y')^2 = 0; \quad y'' = -\dfrac{(y')^2}{y}; \quad y_1'' = -\dfrac{(\frac{1}{2})^2}{1} = -\dfrac{1}{4};$

$$K = \dfrac{-\frac{1}{4}}{[1 + (\frac{1}{2})^2]^{3/2}} = -\dfrac{2}{25}\sqrt{5} = -0.179^{-}$$

(b) $\quad R = \dfrac{1}{K} = -\dfrac{5}{2}\sqrt{5} = -5.590^{+}$

(c) $\quad x = 1 - \dfrac{\frac{1}{2}[1 + (\frac{1}{2})^2]}{-\frac{1}{4}} = 3.5;$

$\quad y = 1 + \dfrac{1 + (\frac{1}{2})^2}{-\frac{1}{4}} = -4$

(d) $\quad$ Eliminate x_1 and y_1 from the equations $y_1^2 = x_1$;

$x = x_1 - \dfrac{y_1'[1 + (y_1')^2]}{y_1''}; \quad y = y_1 + \dfrac{[1 + (y_1')^2]}{y_1''},$

or y_1 from the equivalent equations,

$$x = y_1^2 - \dfrac{\dfrac{1}{2y_1}\left[1 + \left(\dfrac{1}{2y_1}\right)^2\right]}{-\dfrac{1}{4y_1^3}} = 3y_1^2 + \dfrac{1}{2};$$

$$y = y_1 + \dfrac{1 + \left(\dfrac{1}{2y_1}\right)^2}{-\dfrac{1}{4y_1^3}} = -4y_1^3$$

which gives $27y^2 = 16(x - \frac{1}{3})^3$. This is a semicubical **parabo**(?) with its cusp at $(\frac{1}{3}, 0)$ and intersecting the parabola at $(2, \pm\sqrt{\ }$(?)

6.4.5. Solve 6.4.4 using the parametric equations $x = t^2$, $y = t$ of the parabola.

Solution

$$\frac{dy}{dx} = \frac{\frac{dy}{dt}}{\frac{dx}{dt}} = \frac{1}{2t}; \qquad \frac{dy^2}{dx^2} = \frac{d}{dt}\left(\frac{dy}{dx}\right)\frac{dt}{dx}$$

Differentiating $x = t^2$ with respect to x, we have $1 = 2t \frac{dt}{dx}$

so that $\dfrac{dt}{dx} = \dfrac{1}{2t}$. Then

$$\frac{d^2y}{dx^2} = \frac{d}{dt}\left(\frac{1}{2t}\right)\frac{1}{2t} = -\frac{1}{2t^2}\cdot\frac{1}{2t} = -\frac{1}{4t^3}.$$

The value $t = 1$ produces the point $(1, 1)$.

(a) $\qquad K = \dfrac{y_1''}{[1 + (y_1')^2]^{3/2}} = \dfrac{-\dfrac{1}{4t^3}}{\left[1 + \left(\dfrac{1}{2t}\right)^2\right]^{3/2}},\qquad$ which give(s)

$K = -\dfrac{2}{25}\sqrt{5}$ for the point $(1, 1)$ when t is given the value of (?)

(b) $\qquad R = \dfrac{1}{K} = -\dfrac{5}{2}\sqrt{5}$

(c) Let t_1 correspond to (x_1, y_1), then

$$x = x_1 - \frac{y_1'[1 + (y_1')^2]}{y_1''}$$

$$= t_1^2 - \frac{\dfrac{1}{2t_1}\left[1 + \left(\dfrac{1}{2t_1}\right)^2\right]}{-\dfrac{1}{4t_1^3}} = 3t_1^2 + \frac{1}{2};$$

$$y = y_1 + \frac{1 + (y_1')^2}{y_1''}$$

$$= t_1 + \frac{1 + \left(\dfrac{1}{2t_1}\right)^2}{-\dfrac{1}{4t_1^3}} = -4t_1^3$$

hese are parametric equations in the parameter t_1 of the evolute. The parameter t_1 can be eliminated like y_1 in 6.4.4 (d), with the identical result obtained there.

6.4.6. Find the evolute of the ellipse $\dfrac{x^2}{a^2} + \dfrac{y^2}{b^2} = 1$

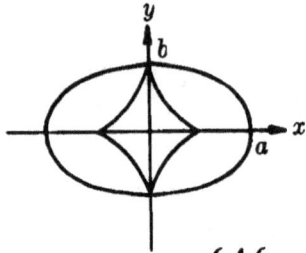

6.4.6.

Solution

$$\frac{2x}{a^2} + \frac{2yy'}{b^2} = 0; \qquad y' = -\frac{b^2 x}{a^2 y};$$

$$y'' = -\frac{b^2}{a^2}\frac{y - xy'}{y^2} = -\frac{b^2}{a^2}\frac{y - x\left(-\dfrac{b^2 x}{a^2 y}\right)}{y^2} = -\frac{b^4}{a^2 y^3};$$

$$x = x_1 - \frac{y_1'[1 + (y_1')^2]}{y_1''}$$

$$= x_1 - \frac{-\dfrac{b^2 x_1}{a^2 y_1}\left[1 + \left(-\dfrac{b^2 x_1}{a^2 y_1}\right)^2\right]}{-\dfrac{b^4}{a^2 y_1^3}} = \frac{(a^2 - b^2)x_1^3}{a^4};$$

$$y = y_1 + \frac{1 + (y_1')^2}{y_1''} = y_1 + \frac{1 + \left(-\dfrac{b^2 x_1}{a^2 y_1}\right)^2}{-\dfrac{b^4}{a^2 y_1^3}}$$

$$= -\frac{(a^2 - b^2)y_1^3}{b^4}$$

Solving for x_1 in terms of x, and y_1 in terms of y:

$$x_1 = \left(\frac{a^4 x}{a^2 - b^2}\right)^{1/3}; \qquad y_1 = \left(-\frac{b^4 y}{a^2 - b^2}\right)^{1/3}$$

Since (x_1, y_1) is a point on the ellipse, $\dfrac{x_1^2}{a^2} + \dfrac{y_1^2}{b^2} = 1$. Replacing x_1 and y_1 by their values,

$$\frac{1}{a^2}\left(\frac{a^4 x}{a^2 - b^2}\right)^{2/3} + \frac{1}{b^2}\left(-\frac{b^4 y}{a^2 - b^2}\right)^{2/3} = 1, \text{ or}$$

$$(ax)^{2/3} + (by)^{2/3} = (a^2 - b^2)^{2/3}$$

which is the distorted hypocycloid of four cusps represented in the figure.

6.4.7. Find parametric equations of the evolute of the orthocycloid $x = a(t - \sin t)$, $y = a(1 - \cos t)$.

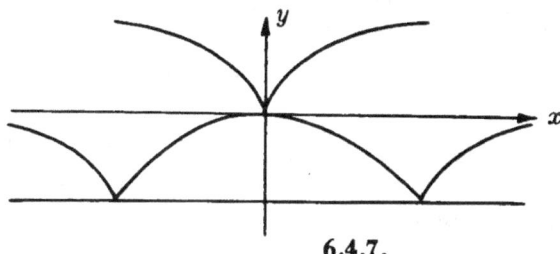

6.4.7.

Solution

$$\frac{dy}{dx} = \frac{\dfrac{dy}{dt}}{\dfrac{dx}{dt}} = \frac{a \sin t}{a(1 - \cos t)} = \frac{\sin t}{1 - \cos t}; \quad \frac{d^2 y}{dx^2} = \frac{d}{dt}\left(\frac{dy}{dx}\right)\frac{dt}{dx}$$

$$= \frac{(1 - \cos t)\cos t - \sin t \sin t}{(1 - \cos t)^2} \cdot \frac{1}{a(1 - \cos t)} = -\frac{1}{a(1 - \cos t)^2}$$

Let t_1 correspond to (x_1, y_1).

$$x = a(t_1 - \sin t_1) - \frac{\dfrac{\sin t_1}{1 - \cos t_1}\left[1 + \left(\dfrac{\sin t_1}{1 - \cos t_1}\right)^2\right]}{-\dfrac{1}{a(1 - \cos t_1)^2}}$$

$$= a(t_1 + \sin t_1)$$

$$y = a(1 - \cos t_1) + \frac{1 + \left(\dfrac{\sin t_1}{1 - \cos t_1}\right)^2}{-\dfrac{1}{a(1 - \cos t_1)^2}} = -a(1 - \cos t_1)$$

Here t_1 is the parameter. These are the equations of another orthocycloid shown below the first one in the figure.

6.4.8. A particle moves along $y = e^x$ at a constant speed of 5 units per second. At what point is its normal acceleration a maximum? See **6.4.1**.

Solution. Here $a_n = Kv^2$ so that a_n is a maximum when K is a maximum, v^2 being constant in our problem, and $y' = e^x$; $y'' = e^x$.

$$K = \frac{e^x}{(1 + e^{2x})^{3/2}} \qquad \frac{dK}{dx} = \frac{(1 + e^{2x})^{3/2} e^x - e^x \frac{3}{2}(1 + e^{2x})^{1/2} e^{2x} \cdot 2}{(1 + e^{2x})^3}$$

$$= \frac{(1 + e^{2x})e^x - 3e^{3x}}{(1 + e^{2x})^{5/2}} = \frac{e^x(1 - 2e^{2x})}{(1 + e^{2x})^{5/2}} = 0; \qquad 1 - 2e^{2x} = 0;$$

$$e^{2x} = \tfrac{1}{2}; \qquad x = -\tfrac{1}{2} \ln 2 = -0.347^-$$

If $x < -\tfrac{1}{2} \ln 2$, then $\dfrac{dK}{dx} > 0$. If $x > -\tfrac{1}{2} \ln$, then $\dfrac{dK}{dx} < 0$. Therefore $x = -0.347^-$ has maximum curvature and the particle has maximum normal acceleration.

CHAPTER 7

POLAR COORDINATES

7.1. Curves represented in polar coordinates. The locus (or place) of all points whose coordinates (r, θ) satisfy an equation in r and θ is, in general, a curve. Such a curve may be plotted pointwise by giving θ values and computing the corresponding values of r, or conversely, giving r values and computing the corresponding values of θ, whichever is the more convenient. While any pair of polar coordinates determines a unique point, the converse is not true, since any point may have an indefinite number of pairs of polar coordinates. For example, the point $\left(2, \dfrac{\pi}{4}\right)$ also has the coordinates

$$\left(2, -\frac{7\pi}{4}\right), \left(2, -\frac{15\pi}{4}\right), \left(-2, \frac{5\pi}{4}\right), \left(-2, -\frac{3\pi}{4}\right), \text{ etc.}$$

7.1.1. Plot $r = \sec \theta$.

	First quadrant					Second quadrant			
θ	0	$\dfrac{\pi}{6}$	$\dfrac{\pi}{4}$	$\dfrac{\pi}{3}$	$\dfrac{\pi}{2}$	$\dfrac{2\pi}{3}$	$\dfrac{3\pi}{4}$	$\dfrac{5\pi}{6}$	π
$\cos \theta$	1	$\dfrac{\sqrt{3}}{2}$	$\dfrac{\sqrt{2}}{2}$	$\dfrac{1}{2}$	0	$-\dfrac{1}{2}$	$-\dfrac{\sqrt{2}}{2}$	$-\dfrac{\sqrt{3}}{2}$	-1
$\sec \theta$	1	$\dfrac{2}{3}\sqrt{3}$	$\sqrt{2}$	2		-2	$-\sqrt{2}$	$-\dfrac{2}{3}\sqrt{3}$	-1
r	1.000	1.155	1.414	2		-2.000	-1.414	-1.155	-1.000

Following the same sequence of "common angles" through the third and fourth quadrants gives us no new points. All the points lie on a straight line perpendicular to the polar axis and one unit to the right of the pole. See **7.2.1**.

7.1.2. Plot $r = \sin \theta$.

	First quadrant				Second quadrant			
θ	0	$\dfrac{\pi}{6}$	$\dfrac{\pi}{4}$	$\dfrac{\pi}{3}$	$\dfrac{\pi}{2}$	$\dfrac{2\pi}{3}$	$\dfrac{3\pi}{4}$	$\dfrac{5\pi}{6}$	π
$\sin \theta$	0	$\dfrac{1}{2}$	$\dfrac{1}{2}\sqrt{2}$	$\dfrac{1}{2}\sqrt{3}$	1	$\dfrac{1}{2}\sqrt{3}$	$\dfrac{1}{2}\sqrt{2}$	$\dfrac{1}{2}$	0
r	0	0.500	0.707	0.866	1	0.866	1.414	0.5	0

Following the same sequence of "common angles" through the third and fourth quadrants gives us no new points. All the points lie on a circle of radius 0.5 tangent to the polar axis at the pole and lying above the polar axis. See 7.2.2.

7.1.3. Plot $r = 2\theta$.

First quadrant

θ	0	$\frac{\pi}{6}$	$\frac{\pi}{4}$	$\frac{\pi}{3}$	$\frac{\pi}{2}$
2θ	0	$\frac{\pi}{3}$	$\frac{\pi}{2}$	$\frac{2\pi}{3}$	π
r	0	1.047	1.571	2.094	3.142

Second quadrant

θ	$\frac{2\pi}{3}$	$\frac{3\pi}{4}$	$\frac{5\pi}{6}$	π
2θ	$\frac{4\pi}{3}$	$\frac{3\pi}{2}$	$\frac{5\pi}{3}$	2π
r	4.189	4.712	5.236	6.283

Third quadrant

θ	$\frac{7\pi}{6}$	$\frac{5\pi}{4}$	$\frac{4\pi}{3}$	$\frac{3\pi}{2}$	$\frac{5\pi}{3}$
2θ	$\frac{7\pi}{3}$	$\frac{5\pi}{2}$	$\frac{8\pi}{3}$	3π	$\frac{10\pi}{3}$
r	7.330	7.854	8.378	9.425	10.472

Fourth quadrant

θ	$\frac{7\pi}{4}$	$\frac{11\pi}{6}$	2π	$\frac{13\pi}{6}$
2θ	$\frac{7\pi}{2}$	$\frac{11\pi}{3}$	4π	$\frac{13\pi}{3}$
r	10.996	11.519	12.566	13.614

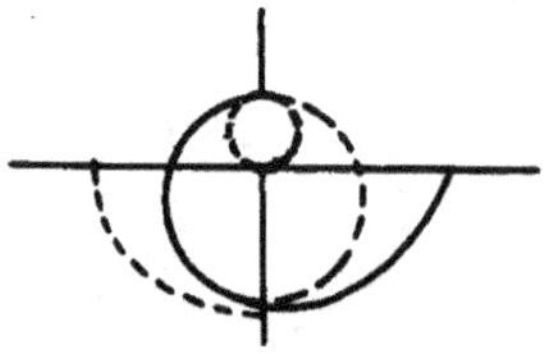

Changing the sign of θ changes the sign of r, giving the dotted branch in the figure.

7.2. Change of coordinate systems. Suppose that the polar axis is taken coincident with the positive x axis of a system of rectangular coordinates, the pole coinciding with the origin. Then very simple relations exist between the polar coordinates of any point P and the rectangular coordinates of the same point. From the figure these are seen to be

$$x = r \cos \theta; \qquad y = r \sin \theta \qquad (7.2.1)$$

$$r = \pm \sqrt{x^2 + y^2}; \qquad \theta = \arctan \frac{y}{x} \qquad (7.2.2)$$

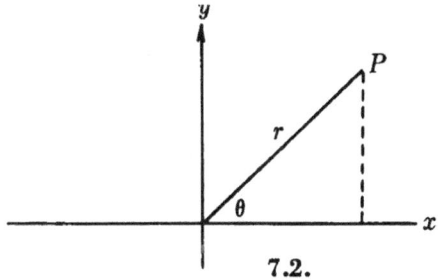

7.2.

By means of these relations an equation in rectangular coordinates may be transformed into one in polar coordinates, or vice versa. The curve may then be plotted from the simpler of its two equations.

7.2.1. Change to rectangular coordinates: $r = \sec\theta$.

Solution. $r = \sec\theta = \dfrac{1}{\cos\theta}$; $r\cos\theta = 1$; $x = 1$ by 7.2.1. See 7.1.1.

7.2.2. The same as 7.2.1 for $r = \sin\theta$.

Solution. $r^2 = r\sin\theta$; $x^2 + y^2 = y$; $x^2 + (y^2 - y + \tfrac{1}{4}) = \tfrac{1}{4}$; $x^2 + (y - \tfrac{1}{2})^2 = (\tfrac{1}{2})^2$, which is a circle of radius $\tfrac{1}{2}$ and center at $(0, \tfrac{1}{2})$. See 7.1.2.

7.2.3. The same as 7.2.1 for $r = 2\theta$.

Solution. $\pm\sqrt{x^2 + y^2} = 2\arctan\dfrac{y}{x}$ by 7.2.2.

This equation is much more difficult to plot than the polar equation. See 7.1.3.

7.2.4. The same as 7.2.1 for $r^2 \sin 2\theta = 20$.

Solution. $r^2 \cdot 2\sin\theta\cos\theta = 20$; $xy = 10$

7.2.5. The same as 7.2.1 for $r^2 - 6r\cos\left(\theta - \dfrac{5\pi}{4}\right) + 5 = 0$.

Solution

$$r^2 - 6r\left(\cos\theta\cos\dfrac{5\pi}{4} + \sin\theta\sin\dfrac{5\pi}{4}\right) + 5 = 0$$

$$r^2 - 6(r\cos\theta)\left(-\dfrac{\sqrt{2}}{2}\right) - 6(r\sin\theta)\left(-\dfrac{\sqrt{2}}{2}\right) + 5 = 0$$

$$x^2 + y^2 + 3\sqrt{2}(x + y) + 5 = 0$$

7.2.6. The same as 7.2.1 for $\cos\theta + \sin\theta = 1$.

Solution. $r\cos\theta + r\sin\theta = r$; $\quad x + y = \pm\sqrt{x^2 + y^2}$; $(x + y)^2 = x^2 + y^2$; $\quad xy = 0$, which represents both rectangular axes.

7.2.7. Change to polar coordinates: $ax + by + c = 0$.
Solution. $ar\cos\theta + br\sin\theta + c = 0$, by 7.2.1

7.2.8. The same as **7.2.7** for $y + 3 = 0$.
Solution. $r\sin\theta = -3$; $\quad r = -3\csc\theta$

7.2.9. The same as **7.2.7** for $x^2 + y^2 = a^2$.
Solution. $r^2 = a^2$, by 7.2.2; $\quad r = \pm a$

7.2.10. The same as **7.2.7** for $x^2 + y^2 - 4x + 4y = 0$.
Solution. $r^2 - 4r\cos\theta + 4r\sin\theta = 0$;
$r^2 - 4r(\cos\theta - \sin\theta) = 0$; $\quad r = 4(\cos\theta - \sin\theta)$

7.3. Angle from the radius vector to the tangent. In the figure let the points $P(r, \theta)$ and $Q(r + \Delta r, \theta + \Delta\theta)$ lie on the curve $r = f(\theta)$. Let TP be tangent to this curve at P, PS be perpendicular to OQ at S; angle $OQP = \psi'$; angle $OPT = \psi$.

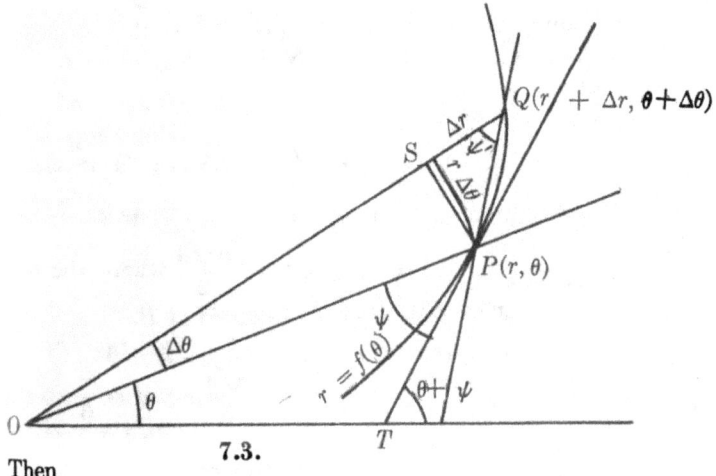

7.3.

Then

$$\tan\psi' = \frac{SP}{(r + \Delta r) - r\cos\Delta\theta} = \frac{SP}{r\,\Delta\theta} \cdot \frac{r\,\Delta\theta}{r(1 - \cos\Delta\theta) + \Delta r}$$

As in **6.1**, we shall assume that $\dfrac{SP}{r\,\Delta\theta} \to 1$ as $\Delta\theta \to 0$. This seems plausible. Also as $\Delta\theta \to 0$, OQ approaches OP; secant QP approaches tangent TP so that ψ' approaches ψ. Thus the above equation becomes in the limit

$$\tan \psi = r \frac{d\theta}{dr} \tag{7.3.1}$$

where ψ is the angle from the radius vector at P to the tangent at P. If α is the inclination of the tangent TP, we have readily from the figure

$$\alpha = \theta + \psi \tag{7.3.2}$$

7.3.1. For the cardioid $r = a(1 - \cos\theta)$ and the point $P\left(\dfrac{a}{2}, \dfrac{\pi}{3}\right)$ on it, find: (a) the angle from the radius vector to the tangent and (b) the inclination of the tangent.

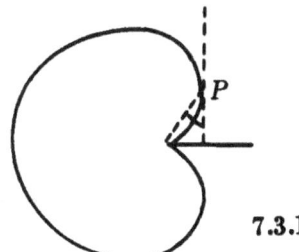

7.3.1.

Solution. (a) $\left[\dfrac{dr}{d\theta}\right]_P = [a \sin\theta]_P = \dfrac{a\sqrt{3}}{2}$ where the notation means that the derivative is evaluated at P.

$$[\tan\psi]_P = \left[r\frac{d\theta}{dr}\right]_P = \frac{a}{2} \cdot \frac{2}{a\sqrt{3}} = \frac{\sqrt{3}}{3}; \quad \psi = \frac{\pi}{6}$$

(b) $\qquad \alpha = \dfrac{\pi}{3} + \dfrac{\pi}{6} = \dfrac{\pi}{2}$

7.3.2. For the lituus $r^2\theta = 1$ and the point $P(\sqrt{2}, \frac{1}{2})$ on it, find: (a) the angle from the radius vector to the tangent and (b) the inclination of the tangent.

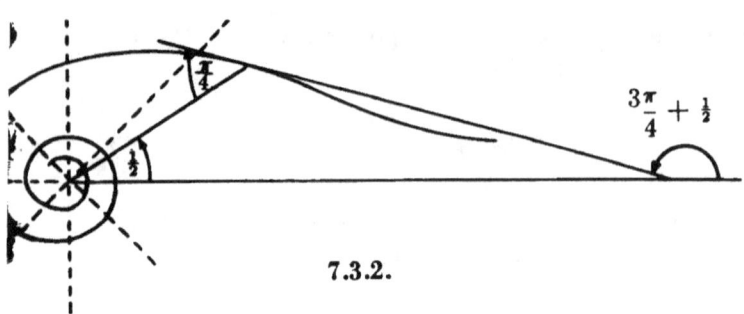

7.3.2.

Solution.

(a) $\quad r^2 \dfrac{d\theta}{dr} + \theta \cdot 2r = 0; \quad \left[\dfrac{d\theta}{dr}\right]_P = \left[-\dfrac{2\theta}{r}\right]_P = -\dfrac{1}{\sqrt{2}};$

$[\tan \psi]_P = \left[r \dfrac{d\theta}{dr}\right]_P = \sqrt{2}\left(-\dfrac{1}{\sqrt{2}}\right) = -1; \quad \psi = -\dfrac{\pi}{4}$

(b) $\quad \alpha = -\dfrac{\pi}{4} + \dfrac{1}{2}$

7.3.3. For the logarithmic spiral $r = e^{a\theta}$, show that the angle ψ is the same for all points on the curve thus justifying the name equiangular spiral which is sometimes given to this curve.

Solution

$$\dfrac{dr}{d\theta} = e^{a\theta} a; \quad \tan \psi = \dfrac{r}{\dfrac{dr}{d\theta}} = \dfrac{e^{a\theta}}{e^{a\theta} a} = \dfrac{1}{a}$$

7.4. Curvature. In **6.4** we defined the curvature K as $\dfrac{d\alpha}{ds}$. It is necessary to express this derivative in polar coordinates. We use 3.7.4 to obtain the following differentials from 7.2.1,

$dx = -r \sin \theta \, d\theta + \cos \theta \cdot dr; \quad dy = r \cos \theta \, d\theta + \sin \theta \cdot dr$

Substituting these results in 6.1.1, we obtain

$$ds = \sqrt{r^2 \, d\theta^2 + dr^2} = \left[r^2 + \left(\dfrac{dr}{d\theta}\right)^2\right]^{1/2} d\theta$$

Finally, we have the formula for curvature in terms of polar coordinates.

$$\frac{d\alpha}{ds} = K = \frac{r^2 + 2\left(\frac{dr}{d\theta}\right)^2 - r\frac{d^2r}{d\theta^2}}{\left[r^2 + \left(\frac{dr}{d\theta}\right)^2\right]^{3/2}} \qquad (7.4.$$

7.4.1. Find the curvature of $r = 2a \cos \theta$.
Solution

$$\frac{dr}{d\theta} = -2a \sin \theta; \qquad \frac{d^2r}{d\theta^2} = -2a \cos \theta$$

$$K = \frac{r^2 + 2(-2a \sin \theta)^2 - r(-2a \cos \theta)}{[r^2 + (-2a \sin \theta)^2]^{3/2}}$$

$$= \frac{(2a \cos \theta)^2 + 2(-2a \sin \theta)^2 - 2a \cos \theta(-2a \cos \theta)}{[(2a \cos \theta)^2 + (-2a \sin \theta)^2]^{3/2}} = \frac{1}{a}$$

which shows that the curvature is everywhere the same, so that we must be dealing with a circle. Multiplying both members of $r = 2a \cos \theta$ by r, we have $r^2 = 2ar \cos \theta$. Using 7.2.1 and 7.2.2 this becomes $x^2 + y^2 = 2ax$, or $(x - a)^2 + y^2 = a^2$. This is a circle of radius a whose curvature is $\frac{1}{a}$ by 6.4.2. We have thus verified 6.4.2 for the special case of this circle.

7.4.2. Find the radius of curvature of the spiral of Archimedes $r = a\theta$ at the point $P(\pi a, \pi)$. See **7.1.3**.
Solution

$$\frac{dr}{d\theta} = a; \qquad \frac{d^2r}{d\theta^2} = 0;$$

$$K = \frac{(\pi a)^2 + 2(a)^2 - \pi a \cdot 0}{[(\pi a)^2 + (a)^2]^{3/2}} = \frac{0.3312^+}{a}; \qquad R = \frac{1}{K} = 3.019^+ \, a$$

7.4.3. Find the curvature of $r = \sec \theta$. See **7.1.1** and **7.2.1**.
Solution

$$\frac{dr}{d\theta} = \sec \theta \tan \theta; \qquad \frac{d^2r}{d\theta^2} = \sec^3 \theta + \sec \theta \tan^2 \theta$$

$$K = \frac{(\sec \theta)^2 + 2(\sec \theta \tan \theta)^2 - \sec \theta(\sec^3 \theta + \sec \theta \tan^2 \theta)}{[(\sec \theta)^2 + (\sec \theta \tan \theta)^2]^{3/2}}$$

mplifying the numerator, we obtain

$$\sec^2 \theta (1 + 2 \tan^2 \theta - \sec^2 \theta - \tan^2 \theta) = 0$$

nce the denominator can not vanish, we have $K = 0$ for all ints. Hence we must be dealing with a straight line.

7.4.4. Find the curvature and the radius of curvature at the) of a leaf of the four-leaf rose $r = a \sin 2\theta$.

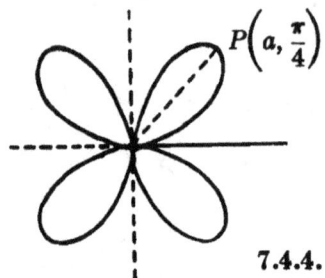

7.4.4.

Solution. The point $P\left(a, \dfrac{\pi}{4}\right)$ is at the tip of a leaf.

$$\left[\frac{dr}{d\theta}\right]_P = [2a \cos 2\theta]_P = 0; \quad \left[\frac{d^2r}{d\theta^2}\right]_P = [-4a \sin 2\theta]_P = -4a;$$

$$K = \frac{a^2 + 2 \cdot 0^2 - a(-4a)}{(a^2 + 0^2)^{3/2}} = \frac{5}{a}; \quad R = \frac{1}{K} = \frac{a}{5}$$

CHAPTER 8

INDETERMINATE FORMS

8.1. The mean value theorem. The mean value theorem ca be stated as follows: If a function $f(x)$ is a continuous functic of the variable x for the closed interval $a \leqq x \leqq b$, and has derivative at all points of the open interval $a < x < b$, then the is at least one value of $x = x_1$, $a < x_1 < b$, such that

$$\frac{f(b) - f(a)}{b - a} = f'(x_1)$$

8.1.

Geometrically this theorem means that there is a value o $x = x_1$ between a and b where the tangent to the curve is paralle to the chord joining $A = [a, f(a)]$ and $B = [b, f(b)]$. The slope of the tangent at x_1 is $f'(x_1)$, and of AB is $\frac{f(b) - f(a)}{b - a}$. The student should observe that this is an existence theorem; it tells us only that x_1 exists without locating it further than $a < x_1 < b$. It is called a mean value theorem, since it concerns an average value of the slope.

It is important that the derivative exist at *all* points of the interval. The function $y = (1 - x^{2/3})^{3/2}$ is continuous in the interval $-\frac{1}{8} < x < \frac{1}{8}$.

$$f\left(-\frac{1}{8}\right) = f\left(\frac{1}{8}\right) = \left(\frac{3}{4}\right)^{3/2}, \text{ so } \frac{f(b) - f(a)}{b - a} = 0.$$

However, at $x = 0$ the derivative does not exist, and there is no point between $-\frac{1}{8}$ and $\frac{1}{8}$ where the derivative vanishes.

In the special case where $f(b) = f(a) = 0$, we have Rolle's theorem that $f'(x_1) = 0$, $a < x_1 < b$.

The mean value theorem does not apply to $\dfrac{x^2 - 9}{x - 3}$ in any interval including $x = 3$, since the function is not continuous at that point (see **1.3.1**). It is true that $f(1) = 4$, and $f(5) = 8$, $\dfrac{f(5) - f(1)}{5 - 1} = 1$, and $f'(x_1)$ is 1 for $x = 2$, say, but this does not follow from the mean value theorem. Another example would be the step function (see **1.3.10**), where $f(\tfrac{3}{4}) = 2$, $f(\tfrac{1}{4}) = 1$, and $\dfrac{f(\tfrac{3}{4}) - f(\tfrac{1}{4})}{\tfrac{3}{4} - \tfrac{1}{4}} = 1$, but there is no point x_1, $\tfrac{1}{4} < x_1 < \tfrac{3}{4}$ where $f'(x_1) = 1$. However, the function is discontinuous at $x = 1$.

8.1.1. Find x_1 for $f(x) = x^2 + x$ if $a = 3$, $b = 6$.

Solution

$$f(3) = 12; \quad f(6) = 42; \quad \frac{f(6) - f(3)}{6 - 3} = \frac{30}{3} = 10$$

We must find x_1, so $f'(x_1) = 2x_1 + 1 = 10$; $\quad 3 < x_1 < 6$; $x_1 = 4\tfrac{1}{2}$

8.1.2. Find x_1 for $f(x) = \tfrac{1}{3}x^3 + x^2$ if $a = 1$, $b = 2$.

Solution

$$f(1) = \frac{4}{3}; \quad f(2) = \frac{20}{3}; \quad \frac{f(2) - f(1)}{2 - 1} = \frac{16}{3}$$

So $\quad x_1^2 + 2x_1 = \dfrac{16}{3} \quad$ or $\quad x_1 = \dfrac{-2 \pm \sqrt{4 + \tfrac{64}{3}}}{2}$

$$= -1 \pm \sqrt{\frac{19}{3}}.$$

We select $x_1 = -1 + \sqrt{\dfrac{19}{3}} = 1.52^-$, so $\quad 1 < x_1 < 2$.

8.1.3. Find x_1 for $f(x) = x^3 + 2x^2 - 1$, $a = -1$, $b = 4$.

Solution

$$f(-1) = 0; \quad f(4) = 95; \quad \frac{f(4) - f(-1)}{4 - (-1)} = 19$$

So $\quad 3x_1^2 + 4x_1 = 19; \quad x_1 = \frac{-4 \pm \sqrt{16 + 228}}{6}.$

We select $x_1 = \dfrac{-4 + \sqrt{244}}{6} = 1.94^-; \quad -1 < x_1 < 4.$

8.2. L'Hospital's rule. The second mean value theorem states that if $f(x)$ and $g(x)$ are continuous and each has a derivative $a < x < b$, and if in addition $g'(x) \neq 0$ for $a < x < b$, then there exists an x_1 such that

$$\frac{f(b) - f(a)}{g(b) - g(a)} = \frac{f'(x_1)}{g'(x_1)}.$$

From this mean value theorem can be derived L'Hospital's rule that states if $\lim\limits_{x \to a} \dfrac{f(x)}{g(x)}$ has the form $\tfrac{0}{0}$, then

$$\lim_{x \to a} \frac{f(x)}{g(x)} = \lim_{x \to a} \frac{f'(x)}{g'(x)}$$

providing the latter limit exists. Note that the form $\tfrac{0}{0}$ occurs when $\lim\limits_{x \to a} f(x) = 0$ and $\lim\limits_{x \to a} g(x) = 0$.

If the limit remains indeterminate, the rule may be applied a second time:

$$\lim_{x \to a} \frac{f(x)}{g(x)} = \lim_{x \to a} \frac{f'(x)}{g'(x)} = \lim_{x \to a} \frac{f''(x)}{g''(x)}$$

However, the student must be careful not to apply the rule after the numerator or denominator has a limit other than zero, since the reduction is no longer valid. The possibility of simplifying algebraically should be examined at each step. This rule can also be used if $\lim\limits_{x \to a} \dfrac{f(x)}{g(x)}$ has the form $\dfrac{\infty}{\infty}$, and the process stops when numerator or denominator has a limit other than ∞.

If $\lim_{x \to a} \frac{f'(x)}{g'(x)}$ is infinite, so is $\lim_{x \to a} \frac{f(x)}{g(x)}$.

8.2.1. $\lim_{x \to 2} \frac{x^4 - 2x^3 + 3x - 6}{x^2 - x - 2} = \lim_{x \to 2} \frac{4x^3 - 6x^2 + 3}{2x - 1} = \frac{11}{3}$

Notice $\lim_{x \to 2} \frac{x^4 - 2x^3 + 3x - 6}{x^2 - x - 2} = \frac{4x^3 - 6x^2 + 3}{2x - 1}$ is incorrect.
Numerator and denominator are differentiated separately.

8.2.2. $\lim_{x \to \infty} \frac{x^3 + 7x^2 - 4x}{2x^3 + 5x - 4} = \lim_{x \to \infty} \frac{3x^2 + 14x - 4}{6x^2 + 5}$

$$= \lim_{x \to \infty} \frac{6x + 14}{12x} = \lim_{x \to \infty} \frac{6}{12} = \frac{1}{2}$$

Here we had to apply the rule three times before the $\frac{\infty}{\infty}$ reduced to $\frac{1}{2}$.

8.2.3. $\lim_{x \to \pi/2} \frac{\cos x}{x - \pi/2} = \lim_{x \to \pi/2} \frac{-\sin x}{1} = -1$

8.2.4. $\lim_{\theta \to \pi/2} \frac{\cos \theta - \cot \theta}{(\theta - \pi/2)^2} = \lim_{\theta \to \pi/2} \frac{-\sin \theta + \csc^2 \theta}{2(\theta - \pi/2)}$

$$= \lim_{\theta \to \pi/2} \frac{-\cos \theta - 2 \csc^2 \theta \cot \theta}{2} = 0$$

8.2.5. $\lim_{x \to 0^+} \frac{\ln x}{e^{1/x}} = \lim_{x \to 0^+} \frac{\frac{1}{x}}{-\frac{1}{x^2} e^{1/x}} = \lim_{x \to 0^+} \frac{-x}{e^{1/x}} = 0$

Observe that $\lim_{x \to 0} \frac{\ln x}{e^{1/x}}$ has no meaning, since $\ln x$ does not exist for $x < 0$. If the algebraic simplification of multiplying numerator and denominator by x^2 had not been made, further application of the rule would have made no simplification. This multiplication is valid if $x \neq 0$, and since the limit is not concerned with the values at $x = 0$, the result is valid.

8.2.6. $\lim_{x \to \infty} \frac{x^2}{e^x} = \lim_{x \to \infty} \frac{2x}{e^x} = \lim_{x \to \infty} \frac{2}{e^x} = 0$

8.2.7. $\lim_{x \to \infty} \frac{x^n}{e^x} = \lim_{x \to \infty} \frac{nx^{n-1}}{e^x} = \lim_{x \to \infty} \frac{n!}{e^x} = 0$ (after n differentiations)

8.2.8. $\lim\limits_{x \to 0} \dfrac{\tan x - x}{x - \sin x} = \lim\limits_{x \to 0} \dfrac{\sec^2 x - 1}{1 - \cos x} = \lim\limits_{x \to 0} \dfrac{2 \sec^2 x \tan x}{\sin x}$

$= \lim\limits_{x \to 0} \dfrac{2 \sec^4 x + 4 \sec^2 x \tan^2 x}{\cos x} =$

We could also write by dividing

$$\lim\limits_{x \to 0} \dfrac{2 \sec^2 x \tan x}{\sin x} = \lim\limits_{x \to 0} 2 \sec^3 x = 2.$$

If $\lim\limits_{x \to a} f(x)g(x)$ is of the form $0 \cdot \infty$, we may write $f(x)g(x)$ in the form $\dfrac{f(x)}{\dfrac{1}{g(x)}}$ or $\dfrac{g(x)}{\dfrac{1}{f(x)}}$ and so obtain either $\dfrac{0}{0}$ or $\dfrac{\infty}{\infty}$, then apply the rules. If $\lim\limits_{x \to a} [f(x) - g(x)]$ has the form $\infty - \infty$, consider

$$f(x) - g(x) = \dfrac{\dfrac{1}{g(x)} - \dfrac{1}{f(x)}}{\dfrac{1}{f(x)g(x)}},$$

which will have the form, $\tfrac{0}{0}$ and apply the rule.

8.2.9. $\lim\limits_{x \to 0^+} x \ln x = \lim\limits_{x \to 0^+} \dfrac{\ln x}{\dfrac{1}{x}} = \lim\limits_{x \to 0^+} \dfrac{\dfrac{1}{x}}{-\dfrac{1}{x^2}} = \lim\limits_{x \to 0^+} - x = 0$

Here $x \ln x$ is in the form $0 \cdot \infty$, but $\dfrac{\ln x}{\dfrac{1}{x}}$ in the form $\dfrac{\infty}{\infty}$. Note the necessary algebraic manipulation.

8.2.10. $\lim\limits_{x \to -\infty} x^2 e^x = \lim\limits_{x \to -\infty} \dfrac{x^2}{e^{-x}} = \lim\limits_{x \to -\infty} \dfrac{2x}{-e^{-x}} = \lim\limits_{x \to -\infty} \dfrac{2}{e^{-x}} = 0$

Also $\lim\limits_{x \to -\infty} x^2 e^x = \lim\limits_{x \to -\infty} \dfrac{e^x}{\dfrac{1}{x^2}}$, but successive application of the rule is not helpful here.

8.2.11. $\lim\limits_{x \to \pi/2} \left(\dfrac{\pi}{2} - x\right) \tan x = \lim\limits_{x \to \pi/2} \dfrac{\dfrac{\pi}{2} - x}{\cot x}$

$= \lim\limits_{x \to \pi/2} \dfrac{-1}{-\csc^2 x} = 1$

8.2.12. $\lim\limits_{x \to \pi/2} (\sec x - \tan x) = \lim\limits_{x \to \pi/2} \dfrac{1}{\sec x + \tan x} = 0$

8.2.13. $\lim\limits_{x \to 1} \left(\dfrac{1}{\ln x} - \dfrac{1}{x-1} \right) = \lim\limits_{x \to 1} \dfrac{x - 1 - \ln x}{(x-1)\ln x}$

$= \lim\limits_{x \to 1} \dfrac{1 - \dfrac{1}{x}}{\ln x + \dfrac{x-1}{x}} = \lim\limits_{x \to 1} \dfrac{\dfrac{1}{x^2}}{\dfrac{1}{x} + \dfrac{1}{x^2}} = \dfrac{1}{2}$

8.2.14. $\lim\limits_{x \to 0} \left(\dfrac{1}{\sin x} - \dfrac{1}{x} \right) = \lim\limits_{x \to 0} \dfrac{x - \sin x}{x \sin x}$

$= \lim\limits_{x \to 0} \dfrac{1 - \cos x}{\sin x + x \cos x} = \lim\limits_{x \to 0} \dfrac{\sin x}{\cos x + \cos x - x \sin x} = 0$

8.2.15. $\lim\limits_{x \to \pi/4} (1 - \tan x) \sec 2x = \lim\limits_{x \to \pi/4} \dfrac{1 - \tan x}{\cos 2x}$

$= \lim\limits_{x \to \pi/4} \dfrac{-\sec^2 x}{-2\sin 2x} = 1$

If we wish to find $\lim\limits_{x \to a} [f(x)]^{g(x)}$, and $[f(x)]^{g(x)}$ takes any of the forms 1^∞, ∞^0, or 0^0, we take the logarithm of the limit, which is $\lim\limits_{x \to a} \ln[f(x)]^{g(x)}$ or $\lim\limits_{x \to a} g(x) \ln f(x)$. This now has the form $0 \cdot \infty$, and if this latter is l, the student must be careful to remember

$$e^l = e^{\ln \lim\limits_{x \to a} [f(x)]^{g(x)}} = \lim\limits_{x \to a} [f(x)]^{g(x)}$$

8.2.16. Find $\lim\limits_{x \to 0^+} x^x$

Solution. $\ln \lim\limits_{x \to 0^+} x^x = \lim\limits_{x \to 0^+} \ln x^x = \lim\limits_{x \to 0^+} x \ln x = \lim\limits_{x \to 0^+} \dfrac{\ln x}{\dfrac{1}{x}}$

$= \lim\limits_{x \to 0^+} \dfrac{\dfrac{1}{x}}{-\dfrac{1}{x^2}} = \lim\limits_{x \to 0^+} - x = 0; \quad \lim\limits_{x \to 0^+} x^x = e^0 = 1$

8.2.17. Find $\lim\limits_{x \to e} (\ln x)^{1/1 - \ln x}$

Solution

$$\ln [\lim_{x \to e} (\ln x)^{1/1-\ln x}] = \lim_{x \to e} \frac{1}{1 - \ln x} \ln \ln x = \lim_{x \to e} \frac{\frac{1}{x \ln x}}{-\frac{1}{x}}$$

$$= \lim_{x \to e} \frac{-1}{\ln x} = -1; \quad \lim_{x \to e} (\ln x)^{1/1-\ln x} = e^{-1}$$

8.2.18. Find $\lim_{x \to 0^+} (\csc x)^{\sin x}$

Solution

$$\ln \lim_{x \to 0^+} (\csc x)^{\sin x} = \lim_{x \to 0^+} \sin x \ln \csc x = \lim_{x \to 0^+} \frac{\ln \csc x}{\csc x}$$

$$= \lim_{x \to 0^+} \frac{\frac{-\csc x \cot x}{\csc x}}{-\csc x \cot x} = \lim_{x \to 0^+} \frac{1}{\csc x} = 0; \quad \lim_{x \to 0^+} (\csc x)^{\sin x} = e^0 = 1$$

8.2.19. Find $\lim_{x \to 0^+} \left(\frac{1}{x}\right)^x$

Solution

$$\ln \lim_{x \to 0^+} \left(\frac{1}{x}\right)^x = \lim_{x \to 0^+} - x \ln x = \lim_{x \to 0^+} \frac{-\ln x}{\frac{1}{x}} = \lim_{x \to 0^+} \frac{\frac{1}{x}}{\frac{1}{x^2}}$$

$$= \lim_{x \to 0^+} x = 0; \quad \lim_{x \to 0^+} \left(\frac{1}{x}\right)^x = e^0 = 1$$

8.2.20. Find $\lim_{x \to \pi/2} (\sin x)^{\tan x}$

Solution

$$\ln \lim_{x \to \pi/2} (\sin x)^{\tan x} = \lim_{x \to \pi/2} \tan x \ln \sin x = \lim_{x \to \pi/2} \frac{\ln \sin x}{\cot x}$$

$$= \lim_{x \to \pi/2} \frac{\frac{\cos x}{\sin x}}{-\csc^2 x} = 0; \quad \lim_{x \to \pi/2} (\sin x)^{\tan x} = e^0 = 1$$

8.2.21. $\lim_{x \to \infty} [\ln (2x + 1) - \ln (2x - 1)] = \lim_{x \to \infty} \ln \frac{2x + 1}{2x - 1}$

$$= \ln \lim_{x \to \infty} \frac{2x + 1}{2x - 1} = \ln \lim_{x \to \infty} \frac{2}{2} = \ln 1 = 0$$

CHAPTER 9

CURVE TRACING

9.1. Symmetry and asymptotes. A curve is symmetric with respect to the y axis if x can be replaced by $-x$ to give an equivalent equation. Below are given four other substitutions which are tests for their respective symmetries:

The x axis,	y replaced by $-y$
The origin,	x replaced by $-x$, y by $-y$
The line $y = x$,	x replaced by y, y by x
The line $y = -x$,	x replaced by $-y$, y by $-x$

An asymptote is a line which a curve approaches closely without actually reaching; it is a limiting position of the tangent under certain conditions. Vertical asymptotes occur when $\lim_{x \to a} f(x) = +\infty$ the asymptote being $x = a$. Horizontal asymptotes occur when $\lim_{x \to \pm\infty} f(x) = l$, the asymptote being $y = l$.

9.1.1. Sketch $y = \dfrac{x^2}{x^2 - 9}$

<u>Symmetry</u>: By applying our tests, we see the curve is symmetric with respect to the y axis.

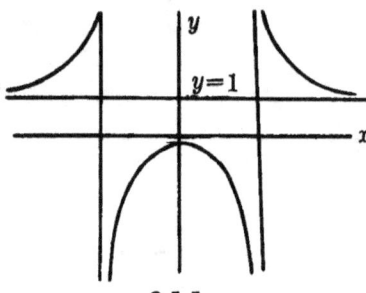

9.1.1.

<u>Asymptotes</u>: By L'Hospital's rule, $\lim_{x \to \pm\infty} \dfrac{x^2}{x^2 - 9} = 1$, and $y = 1$ is a horizontal asymptote. Since $y > 1$ for numerically large x, the curve should approach the asymptote from above. $\lim_{x \to \pm 3} \dfrac{x^2}{x^2 - 9} = \pm\infty$, and $x = 3$, $x = -3$ are asymptotes.

Notice that $\lim_{x \to 3^+} \frac{x^2}{x^2 - 9} = +\infty$, and $\lim_{x \to 3^-} \frac{x^2}{x^2 - 9} = -\infty$, so that y assumes large positive values to the right and numerically large negative values to the left of $x = 3$.

Critical points: $y' = \frac{-18x}{(x^2 - 9)^2}$

so that $x = 0$ is a critical point.

$$y'' = \frac{18(3x^2 + 9)}{(x^2 - 9)^3}$$

Then $x = 0$ is a relative maximum. There are no points of inflection.

9.1.2. Sketch $y = \frac{-x}{x^2 + 1}$

Symmetry: By substituting $-x$ for x and $-y$ for y, we find the curve symmetric with respect to the origin.

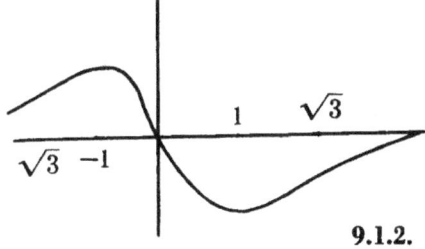

9.1.2.

Asymptotes: $\lim_{x \to \pm\infty} \frac{-x}{x^2 + 1} = 0$. The x axis is an asymptote. As $x \to +\infty$, y approaches 0 through negative values, and the curve lies below the x axis; as $x \to -\infty$ the curve lies above the axis. There are no vertical asymptotes.

Critical points: $y' = \frac{x^2 - 1}{(x^2 + 1)^2}$; $y'' = \frac{2x(3 - x^2)}{(x^2 + 1)^3}$

Then $x = -1$ is a maximum, $x = +1$ a minimum, $x = 0$, $\sqrt{3}, -\sqrt{3}$ are points of inflection.

9.1.3. Sketch $y = \frac{(x - 1)^2}{x + 1}$

Symmetry: The usual tests for symmetry fail.

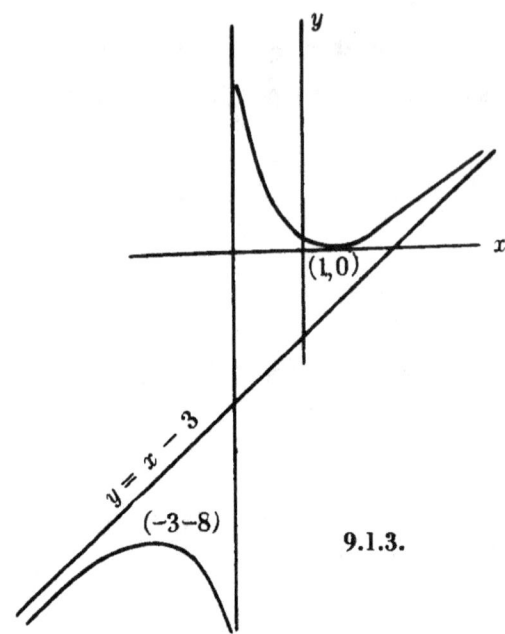

9.1.3.

Asymptotes: $\lim_{x \to -1} \dfrac{(x-1)^2}{x+1} = \pm \infty$. Notice that y is large and positive for $x > -1$ and large negative for $x < -1$. By division $\dfrac{(x-1)^2}{x+1} = x - 3 + \dfrac{4}{x+1}$ so that for large x, $\dfrac{4}{x+1}$ is insignificant and $y = x - 3$ is an asymptote. Since $\dfrac{4}{x+1}$ is positive for large positive x and negative for large negative x, the curve will lie above $y = x - 3$ as $x \to +\infty$ and below as $x \to -\infty$.

Critical values: $y' = \dfrac{(x-1)(x+3)}{(x+1)^2}$; $y'' = \dfrac{8}{(x+1)^3}$

Then $x = 1$ is a relative minimum, $x = -3$ a relative maximum. There are no points of inflection. When $x = 1$, $y = 0$. The student will recognize the curve as an hyperbola.

9.1.4. Sketch $y = \dfrac{x^2}{(x-1)^2}$

This curve passes through the origin but does not cross

either axis at any other point. None of the usual tests for symmetry work.

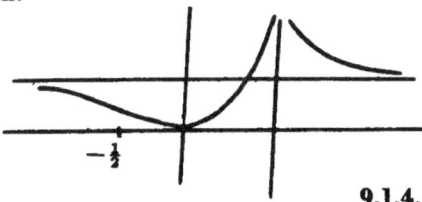

9.1.4.

Asymptotes: $\lim\limits_{x \to \pm\infty} \dfrac{x^2}{(x-1)^2} = 1$ by L'Hospital's rule, and $y = 1$ is an asymptote. y is always positive (quotient of two squares). $\lim\limits_{x \to 1} \dfrac{x^2}{(x-1)^2} = +\infty$. Hence $x = 1$ is an asymptote. Then y is a large positive quantity on both sides of the asymptote. Observe that as $x \to +\infty$, $\dfrac{x^2}{(x-1)^2} > 1$, and as $x \to -\infty$, $\dfrac{x^2}{(x-1)^2} < 1$.

Critical values: $y' = \dfrac{-2x}{(x-1)^3}$; $y'' = \dfrac{4x+2}{(x-1)^4}$
Now $x = 0$ is a minimum, $x = -\tfrac{1}{2}$ is a point of inflection.

9.1.5. Sketch $y = \ln \sin x$

The usual tests for symmetry fail. Notice that when $\sin x$ is negative, $(-\pi < x < 0, \pi < x < 2\pi$ etc.$)$, y does not exist.

Asymptotes: $\lim\limits_{x \to n\pi} \ln \sin x = -\infty$, and the lines $x = \pm n\pi$ are vertical asymptotes, where n has integral values.

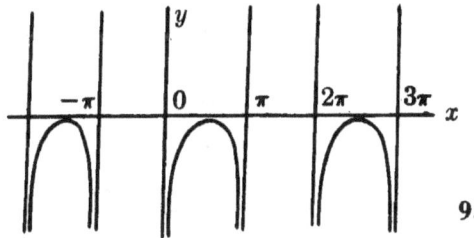

9.1.5.

Critical values: $y' = \cot x$; $y'' = -\csc^2 x$
Here $x = \dfrac{\pi}{2}, \dfrac{5\pi}{2} \cdots$ are maxima, $y'' \ne 0$, and there are no

points of inflection; $y' = 0$ at $\frac{-\pi}{2}, \frac{3\pi}{2}, \ldots$, but the curve does not exist at those points.

9.1.6. Sketch $y = \dfrac{1}{1+x^2}$

<u>Symmetry</u>: The curve is symmetric with respect to the y axis and y is always positive.

<u>Asymptotes</u>: $\lim\limits_{x \to \pm\infty} \dfrac{1}{1+x^2} = 0$, and the x axis is an asymptote.

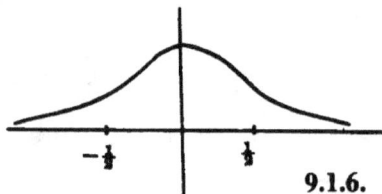

9.1.6.

<u>Critical values</u>: $y' = \dfrac{-2x}{(x^2+1)^2}$; $y'' = \dfrac{6x^2-2}{(x^2+1)^3}$.

Now $x = 0$ is a maximum, $x = \dfrac{\sqrt{3}}{3}$ and $x = -\dfrac{\sqrt{3}}{3}$ are points of inflection.

We wish now to introduce a more exact procedure for finding oblique asymptotes than we have used hitherto. In the first place, it is evident that if an oblique asymptote exists, $y \to \infty$ as $x \to \infty$ but y', the slope of the tangent, assumes a limiting value as $x \to \infty$. This limiting value is the slope of the asymptote; the tangent, and hence the curve, are closely approximated by this limiting position for large x. For the curve we have just examined in **9.1.3**

$$y = \dfrac{(x-1)^2}{x+1}; \qquad y' = \dfrac{(x-1)(x+3)}{(x+1)^2}$$

and by L'Hospital's rule, we see that $\lim\limits_{x \to \infty} \dfrac{(x-1)(x+3)}{(x+1)^2} = 1$, which is the slope of the asymptote. There remains the problem of finding some point through which the asymptote passes so that we may write its equation. Denote by (x_1, y_1) a point on the curve, and by y_1' the value of the derivative at the point.

The equation of the tangent at that point is

$$\frac{y - y_1}{x - x_1} = y_1' \quad \text{or} \quad y = y_1'x - x_1y_1' + y_1$$

If the y intercept $y_1 - x_1y_1'$ or $f(x_1) - x_1f'(x_1)$ has a limit as $x_1 \to \infty$, this will be the y intercept of the asymptote. For our example:

$$f(x) - xf'(x) = \frac{(x-1)^2}{x+1} - x\frac{(x-1)(x+3)}{(x+1)^2}$$

$$= \frac{-3x^2 + 2x + 1}{(x+1)^2}$$

By L'Hospital's rule

$$\lim_{x \to \infty} \frac{-3x^2 + 2x + 1}{(x+1)^2} = -3$$

and the equation of the asymptote is $y = x - 3$ as before. This method will be useful in our subsequent discussion.

9.2. Vertical tangents. The function $y = (x - 1)^{2/3}$ has the derivative $y' = \frac{2}{3}(x-1)^{-1/3}$. At $x = 1$, $y = 0$, but y' is infinite at $x = 1$ so that the tangent at that point is vertical. Now y' is negative to the left of $x = 1$ and positive to the right, the function is decreasing to the left of this point and increasing to the right, so the point must be a minimum. Since the tangent is vertical we have a cusp as shown. The best way to determine the nature of points such as this is to examine the derivative to the left and right of the critical point.

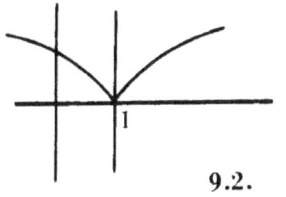

9.2.

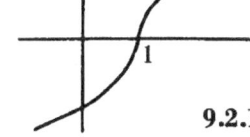

9.2.1.

9.2.1. Sketch $y = (x - 1)^{1/3}$

<u>Symmetry</u>: The curve is symmetric with respect to the point $(1, 0)$.

<u>Critical values</u>: $y' = \frac{1}{3}(x-1)^{-2/3}$; $\quad y' \to \infty$

as $x \to 1$. Now y' has the same sign on either side of $x = 1$, so that this is a vertical point of inflection.

9.2.2. Sketch $y = (x - 2)^{2/3}(x + 1)^{1/3}$

$$y' = \tfrac{2}{3}(x - 2)^{-1/3}(x + 1)^{1/3} + \tfrac{1}{3}(x - 2)^{2/3}(x + 1)^{-2/3}$$

$$= \frac{x}{(x - 2)^{1/3}(x + 1)^{2/3}}$$

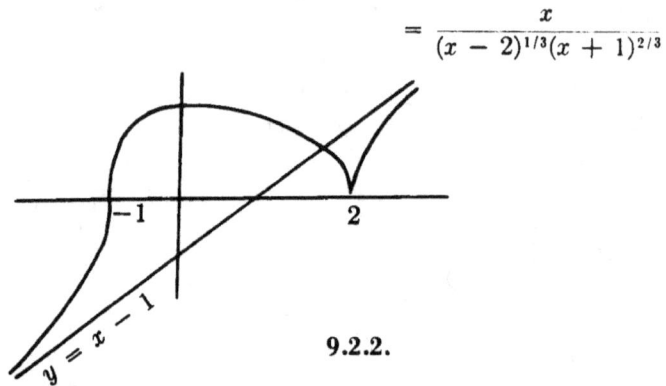

9.2.2.

Critical values: y' being 0 at $x = 0$, and $y' \to \infty$ as $x \to -1$ and $x \to 2$, then -1, 0, and 2 are the critical values. Slightly to the left and right of $x = -1$, y' has the same sign (positive) and this must be a point of inflection. From left to right of $x = 0$ our sign changes from plus to minus, so $x = 0$ is a maximum. Finally as we move from left to right through $x = 2$ our derivative changes from minus to plus and this is a relative minimum.

$$y'' = \frac{-2}{(x - 2)^{4/3}(x + 1)^{5/3}}$$

There are no other points of inflection.

Asymptotes:

$$\lim_{x \to \pm\infty} \frac{x}{(x - 2)^{1/3}(x + 1)^{2/3}} = \lim_{x \to \pm\infty} \frac{1}{\left(1 - \frac{2}{x}\right)^{1/3}\left(1 + \frac{1}{x}\right)^{2/3}} = 1.$$

An asymptote exists of slope 1.

$$f(x) - xf'(x) = (x-2)^{2/3}(x+1)^{1/3} - \frac{x^2}{(x-2)^{1/3}(x+1)^{2/3}}$$

$$= \frac{-x-2}{(x-2)^{1/3}(x+1)^{2/3}} \cdot \lim_{x \to \pm\infty} - \frac{x+2}{(x-2)^{1/3}(x+1)^{2/3}}$$

$$= \lim_{x \to \pm\infty} - \frac{1 + \frac{2}{x}}{\left(1 - \frac{2}{x}\right)^{1/3}\left(1 + \frac{1}{x}\right)^{2/3}} = -1$$

Now $y = x - 1$ is an asymptote. Notice the limit is true for $\pm\infty$, so it is an asymptote on both ends.

9.2.3. Sketch $y = x^{5/3} + x + 1$

<u>Symmetry</u>: There is no symmetry.

<u>Critical values</u>: $y' = \tfrac{5}{3}x^{2/3} + 1$

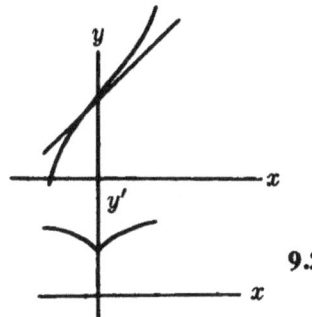

9.2.3.

Now $y' = 0$ implies $x^{2/3} = -\tfrac{3}{5}$, or $x^2 = \frac{-27}{125}$, so that there are no critical points. Then $y'' = \tfrac{10}{9}x^{-1/3}$. The second derivative is infinite for $x = 0$, and since y'' is negative to the left of $x = 0$ and positive to the right, y' must reach a minimum value at $x = 0$. However, if y' has a minimum value for a value of x, this in turn implies that the curve has a point of inflection at that point (see 3.4). Hence $(0, 1)$ is a point of inflection, and the slope of the tangent at that point is 1. Also $y'' > 0$ if $x > 0$, $y'' < 0$ if $x < 0$, and the curve is concave upward on the right and concave downward on the left of $x = 0$.

9.3. Double-valued functions. In this section we shall be concerned for the most part with functions of the form $y^2 = f(x)$.

the first place it is evident all such curves are symmetric
th respect to the x axis, and that the locus exists only when
$r) \geq 0$. Usually we shall consider the two branches $y = \sqrt{f(x)}$
id $y = -\sqrt{f(x)}$, which together make up the locus; each of
ese branches may be inspected by the analysis developed in
.e preceding sections of this chapter. Wherever the curve
osses the x axis, it does so perpendicularly save for some
:ceptional cases. To see this,

$$y^2 = f(x); \qquad 2yy' = f'(x); \qquad y' = \frac{f'(x)}{2y}$$

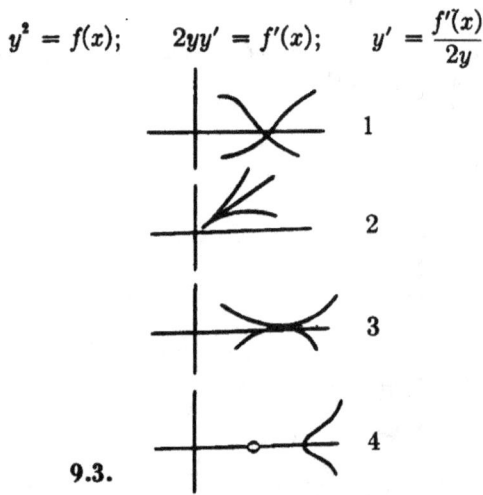

9.3.

f we set $y = 0$, we see that y' is infinite and that the tangent to
he curve is perpendicular to the x axis. The only exceptions
xccur when $f'(x) = 0$, when $y = 0$; then y' assumes the inde-
:erminate form $\frac{0}{0}$. These are called singular points. If this limit
s different for each branch, the two branches cross at that
point in a *node* as in figure 1. If the limit is the same for each
oranch, the two branches form a *cusp*, figure 2, or a *tacnode*,
figure 3. If the limit is imaginary, the point will be an isolated
point as in figure 4. This may also occur when the limits are the
same.

9.3.1. Sketch $x^4 + x^2y^2 = 4y^2$

We rewrite this in the form $y = \dfrac{x^2}{\pm\sqrt{4 - x^2}}$ and consider the
branch corresponding to the positive root.

Extent: The locus exists only if $4 - x^2 > 0$, or $-2 < x < 2$.
Critical values:

$$y' = \frac{2x\sqrt{4-x^2} - x^2 \dfrac{-x}{\sqrt{4-x^2}}}{4-x^2} = \frac{8x - 2x^3 + x^3}{(4-x^2)^{3/2}}$$

$$= \frac{x(8-x^2)}{(4-x^2)^{3/2}} \quad (9.3.1.)$$

$$y'' = \frac{4(x^2+8)}{(4-x^2)^{5/2}}$$

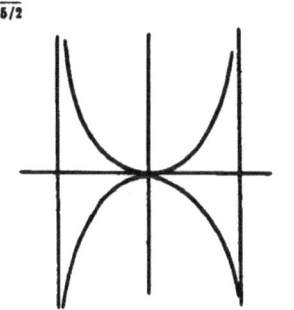

9.3.1.

Now $x = 0$ is a minimum, and there are no points of inflection. The student may well ask why, in view of the fact that $y = 0$ when $x = 0$, the derivative does not assume the form $\frac{0}{0}$ at $x = 0$. Actually it does.

$$y^2 = \frac{x^4}{4-x^2}; \quad 2yy' = \frac{16x^3 - 2x^5}{(4-x^2)^2}; \quad y' = \frac{\dfrac{16x^3 - 2x^5}{(4-x^2)^2}}{2\dfrac{\sqrt{4-x^2}}{x^2}}$$

It is precisely here that y' assumes the form $\frac{0}{0}$, and to reduce the derivative to the form 9.3.1, we must justify the step by the limiting process. This will be true of most of the examples here; though as above the step may be hidden. The procedure will simply be to find y' for those values of x which are roots of $f(x) = 0$ (if y' still has the form $\frac{0}{0}$, L'Hospital's rule must be used), and to use the analysis outlined in the preceding paragraph. In the present case, since the curve exists on both sides of $x = 0$, and since y' is the same for both branches, we have a tacnode.

__Asymptotes__: Since $\lim\limits_{x\to\pm 2} \dfrac{x^2}{\sqrt{4-x^2}} = \infty$, then $x = \pm 2$
are vertical asymptotes.

9.3.2. Sketch $x^3 + xy^2 + y^2 - 3x^2 = 0$

On the positive branch one has $y = \sqrt{\dfrac{3x^2 - x^3}{x+1}}$.

__Extent__: $3x^2 - x^3 \geqslant 0$ and $x + 1 \geqslant 0$ implies $x \geqslant -1$ and $x \leqslant 3$. Notice $3x^2 - x^3 < 0$ and $x + 1 < 0$ simultaneously is not possible; hence $-1 < x < 3$.

__Critical values__:

$$y' = \frac{6x - 2x^2}{(x+1)^2} \frac{1}{2}\left(\frac{3x^2 - x^3}{x+1}\right)^{-1/2} = \frac{3 - x^2}{(x+1)^{3/2}(3-x)^{1/2}}$$

Now $y' = 0$ when $x = \pm\sqrt{3}$. Since $-\sqrt{3}$ is excluded, we examine $\sqrt{3}$ only, and the curve has a relative maximum at this point.

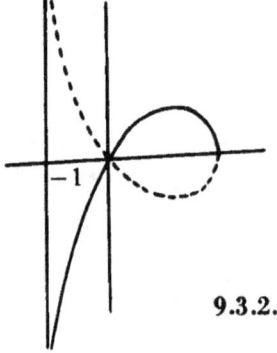

9.3.2.

__Singular points__: $y = 0$ for $x = 0$ and $x = 3$. Notice that y' is indeterminate at $x = 0$ in its first form, but by using theorems on limits, we can justify substitution in the second. We find $y' = \sqrt{3}$ for $x = 0$ on the positive branch. On the negative branch we use $-\dfrac{1}{2}\left(\dfrac{3x^2 - x^3}{x+1}\right)^{-1/2}$ in the above formula, and $y' = -\sqrt{3}$ at $x = 0$. Hence the point is a node. The curve crosses the axis perpendicularly at $x = 3$.

__Asymptotes__: Since $y \to \infty$ as $x \to -1$, the line $x = -1$ is a vertical asymptote. In the figure the positive branch is represented by a heavy line, the negative by a dotted line.

9.3.3. Sketch $xy^2 = (x+1)^3$

__Symmetry__: With respect to the x axis.

__Extent__: On the positive branch, $y = \sqrt{\dfrac{(x+1)^3}{x}}$, and w
must have $x + 1 > 0, x > 0$, which means $x > 0$; or $x + 1 <$ (
$x < 0$, which is true if $x < -1$.

__Critical values__:
$$y' = \frac{(2x-1)(x+1)^{1/2}}{2x^{3/2}}; \quad y'' = \frac{3}{4x^{5/2}(x+1)^{1/2}}$$

9.3.3.

Now $x = -1$ is a singular point; $x = \frac{1}{2}$ is a relative **minimum**.

Singular points: $y = 0$ at $x = -1$. Since $y' = 0$ for both branches at $x = -1$, this must be a cusp or tacnode. Since the curve does not exist immediately to the right of $x = -1$, it is a cusp.

__Asymptotes__: $\lim\limits_{x \to 0} \sqrt{\dfrac{(x+1)^3}{x}} = \infty$, and the y axis is a vertical asymptote. Since
$$\lim_{x \to \pm\infty} \frac{(2x-1)(x+1)^{1/2}}{2x^{3/2}} = \lim_{x \to \pm\infty} \frac{1}{2}\left(2 - \frac{1}{x}\right)\left(1 + \frac{1}{x}\right)^{1/2} = 1$$

there is an oblique asymptote of slope 1.

$$f(x) - xf'(x) = \sqrt{\frac{(x+1)^3}{x}} - \frac{(2x-1)(x+1)^{1/2}}{2x^{1/2}}$$

$$= \frac{3}{2}\sqrt{\frac{x+1}{x}} \cdot \lim_{x \to \pm\infty} \frac{3}{2}\sqrt{\frac{x+1}{x}} = \frac{3}{2}$$

Then $y = x + \frac{3}{2}$ is an asymptote. For the negative branch $y = -x - \frac{3}{2}$ is an asymptote. Again the negative branch is represented by a dotted line.

9.3.4. Sketch $y^2 = x^2(x-1)$

<u>Extent</u>: On the positive branch $y = x\sqrt{x-1}$ and $x \geq 1$.

<u>Critical values</u>: $y' = \dfrac{3x-2}{2\sqrt{x-1}}$; $y'' = \dfrac{3x-4}{4(x-1)^{3/2}}$

9.3.4.

At first glance, it would seem that $x = \frac{2}{3}$ is a critical point since $y' = 0$ there. However, y is imaginary for that point, and so $x = \frac{2}{3}$ does not enter into our considerations. Then $y'' = 0$ when $x = \frac{4}{3}$, and this is a point of inflection since y'' changes sign as x passes through this value.

<u>Singular points:</u> $y = 0$ when $x = 0$ or $x = 1$, and $x = 0$ is an isolated point since y' is imaginary for $x = 0$. Also $y' \to \infty$ as $x \to 1$, so the curve crosses the x axis perpendicularly at this point.

<u>Asymptotes:</u> There are no oblique asymptotes; y' has no limit as $x \to \infty$. The dotted portion represents the negative branch.

Chapter 10

INTEGRATION BY ELEMENTARY SUBSTITUTIONS

10.1. Elementary algebraic substitutions. In solving integrals, it is often useful to substitute u for one of the functions and so reduce the integral to the form $\int u^n \, du$. With this method, the du is as important as the u, and the alert student should have his eyes open for a suitable du as well as u.

<u>Example</u>: In the integral $\int \dfrac{\cos \theta \, d\theta}{\sin^5 \theta}$,

we choose $\sin \theta = u$, for then $\cos \theta \, d\theta$ fits in nicely as du.

Integrals can often be solved by multiplying the numerator and denominator of the integrand by the same quantity.

<u>Example</u>:
$$\int \frac{d\theta}{1 - \sin \theta} = \int \frac{(1 + \sin \theta) \, d\theta}{(1 + \sin \theta)(1 - \sin \theta)}$$
$$= \int \frac{(1 + \sin \theta) \, d\theta}{\cos^2 \theta} = \int (\sec^2 \theta + \sec \theta \tan \theta) \, d\theta$$
$$= \tan \theta + \sec \theta + c$$

Sometimes numerator and denominator are multiplied by the same quantity in order to supply a convenient du.

<u>Example</u>: $\int \dfrac{dx}{1 + e^{-x}} = \int \dfrac{e^x \, dx}{e^x + 1}$,

and now we can make the substitution $e^x + 1 = u$, $e^x \, dx = du$, so that

$$\int \frac{e^x \, dx}{e^x + 1} = \int \frac{du}{u} = \ln |u| + c = \ln |e^x + 1| + c$$

The student must beware of trying to make a single substitution do too much work. In **10.1.1**, $\int \sqrt{1 + \sin x} \cos x \, dx$, he might be tempted to let $u = \sqrt{1 + \sin x}$, or in the integral $\int \dfrac{e^x \, dx}{1 + e^{2x}}$, to let $u = 1 + e^{2x}$. (See **10.2.4.** below.)

10.1.1. $I = \int \sqrt{1 + \sin x} \cos x \, dx$

Let $1 + \sin x = u$, $\cos x \, dx = du$.

$$I = \int u^{1/2} \, du = \tfrac{2}{3} u^{3/2} + c = \tfrac{2}{3}(1 + \sin x)^{3/2} + c$$

10.1.2. $I = \int \dfrac{dx}{\sqrt{4 + 5x}}$

Let $4 + 5x = u$, $dx = \tfrac{1}{5} du$.

$I = \dfrac{1}{5} \int \dfrac{du}{u^{1/2}} = \dfrac{1}{5} \int u^{-1/2}\, du = \dfrac{2}{5} u^{1/2} + c = \dfrac{2}{5}(4 + 5x)^{1/2} + c$

10.1.3. $I = \int \dfrac{\sec^2 \theta\, d\theta}{2 + \tan \theta}$

Let $2 + \tan \theta = u$, $\sec^2 \theta\, d\theta = du$.

$I = \int \dfrac{du}{u} = \ln |u| + c = \ln |2 + \tan \theta| + c$

10.1.4. $I = \int (1 + \ln x) \dfrac{dx}{x}$

Let $1 + \ln x = u$, $\dfrac{dx}{x} = du$.

$I = \int u\, du = \dfrac{1}{2} u^2 + c = \dfrac{1}{2}(1 + \ln x)^2 + c$

Here we choose $du = \dfrac{dx}{x}$ because it fits in with the $\ln x$ in the integrand.

10.1.5 $\int (x^3 - 5)^2\, dx = \int (x^6 - 10x^3 + 25)\, dx$

$= \dfrac{1}{7} x^7 - \dfrac{10}{4} x^4 + 25x + c$

This is an example which is not amenable to substitution; there is no $x^2\, dx$, and multiplying numerator and denominator by x^2 would only unnecessarily complicate the problem.

10.1.6. $I = \int \sqrt{4 + 3e^x}\, e^x\, dx$

Let $4 + 3e^x = u$; $e^x\, dx = \tfrac{1}{3} du$.

$I = \dfrac{1}{3} \int \sqrt{u}\, du = \dfrac{2}{9} u^{3/2} + c = \dfrac{2}{9}(4 + 3e^x)^{3/2} + c$

10.1.7. $I = \int \dfrac{x^{1/3}\, dx}{(1 + x^{4/3})^5}$

Let $1 + x^{4/3} = u$, $x^{1/3}\, dx = \tfrac{3}{4} du$.

$I = \dfrac{3}{4} \int \dfrac{du}{u^5} = -\dfrac{3}{16} u^{-4} + c = -\dfrac{3}{16}(1 + x^{4/3})^{-4} + c$

10.1.8. $I = \int \dfrac{7x^3\,dx}{(1-x^4)^{1/2}}$

Let $1 - x^4 = u$, $x^3\,dx = -\tfrac{1}{4}\,du$.

$$I = -\frac{1}{4}\int \frac{7\,du}{u^{1/2}} = -\frac{7}{2}u^{1/2} + c = -\frac{7}{2}(1-x^4)^{1/2} + c$$

10.1.9. $I = \int (5x+8)^4\,dx$

Let $5x + 8 = u$, $dx = \tfrac{1}{5}\,du$.

$$I = \frac{1}{5}\int u^4\,du = \frac{1}{25}u^5 + c = \frac{1}{25}(5x+8)^5 + c$$

10.1.10 $I = \int \left(1 - \dfrac{1}{y}\right)^3 \dfrac{dy}{y^2}$

Let $1 - \dfrac{1}{y} = u$, $\dfrac{dy}{y^2} = du$.

$$I = \int u^3\,du = \frac{1}{4}u^4 + c = \frac{1}{4}\left(1 - \frac{1}{y}\right)^4 + c$$

10.1.11. $I = \int \sin^4 3x \cos 3x\,dx$

Let $\sin 3x = u$, $\cos 3x\,dx = \tfrac{1}{3}\,du$.

$$I = \frac{1}{3}\int u^4\,du = \frac{1}{15}u^5 + c = \frac{1}{15}\sin^5 3x + c$$

10.1.12. $I = \int \dfrac{\text{Arc}\tan x\,dx}{1+x^2}$

Let $\text{Arc}\tan x = u$, $\dfrac{dx}{1+x^2} = du$.

$$I = \int u\,du = \frac{1}{2}u^2 + c = \frac{1}{2}(\text{Arc}\tan x)^2 + c$$

10.1.13. $I = \int \dfrac{\sec^2 x\,dx}{\tan^2 x}$

Let $\tan x = u$, $\sec^2 x\,dx = du$.

$$I = \int \frac{du}{u^2} = -u^{-1} + c = -\frac{1}{\tan x} + c$$

10.1.14. $I = \int \tan \theta \cos \theta \, d\theta = \int \sin \theta \, d\theta = -\cos \theta + c.$

Here we changed to sines and cosines.

10.1.15. $I = \int \sec^3 \theta \tan \theta \, d\theta$

Let $\sec \theta = u$, $\sec \theta \tan \theta = du$.

$$I = \int \sec^2 \theta \cdot \sec \theta \tan \theta \, d\theta$$

$$= \int u^2 \, du = \frac{1}{3} u^3 + c = \frac{1}{3} \sec^3 \theta + c$$

10.1.16. $I = \int \frac{e^{2x} \, dx}{1 + e^{2x}}$

Let $1 + e^{2x} = u$, $e^{2x} \, dx = \frac{1}{2} du$.

$$I = \frac{1}{2} \int \frac{du}{u} = \frac{1}{2} \ln |u| + c = \frac{1}{2} \ln |1 + e^{2x}| + c$$

10.1.17. $I = \int \frac{dx}{x(1 - \ln x)}$

Let $1 - \ln x = u$, $\frac{dx}{x} = -du$.

$$I = -\int \frac{du}{u} = -\ln |u| + c = -\ln |1 - \ln x| + c$$

10.1.18. $I = \int \tan x \, dx = \int \frac{\sin x \, dx}{\cos x} = -\ln |\cos x| + c$

10.1.19. $I = \int \sec x \, dx = \int \frac{\sec x (\sec x + \tan x)}{\sec x + \tan x} \, dx$

Let $\sec x + \tan x = u$, $\sec^2 x + \sec x \tan x = du$.

$$I = \int \frac{du}{u} = \ln |u| + c = \ln |\sec x + \tan x| + c$$

A different type of substitution is illustrated by the following examples.

10.1.20. $I = \int \sin 3x \, dx$

Let $3x = u$, $dx = \frac{1}{3} du$.

$$I = \frac{1}{3} \int \sin u \, du = -\frac{1}{3} \cos u + c = -\frac{1}{3} \cos 3x + c$$

10.1.21. $I = \int e^{5x+1} \, dx$

Let $5x + 1 = u$, $dx = \tfrac{1}{5} du$.

$$I = \frac{1}{5}\int e^u \, du = \frac{1}{5} e^u + c = \frac{1}{5} e^{5x+1} + c$$

10.1.22. $I = \int \dfrac{\sec^2 x \, dx}{e^{\tan x}}$

Let $\tan x = u$, $\sec^2 x \, dx = du$.

$$I = \int e^{-u} \, du = -e^{-u} + c = -e^{-\tan x} + c$$

10.1.23. $I = \int \dfrac{\sin e^{-x}}{e^x} \, dx$

Let $e^{-x} = u$, $-e^{-x} dx = du$.

$$I = -\int \sin u \, du = \cos u + c = \cos e^{-x} + c$$

10.1.24. $I = \int -\dfrac{1}{x^2} \sin \dfrac{1}{x} \, dx$

Let $\dfrac{1}{x} = u$, $-\dfrac{dx}{x^2} = du$.

$$I = \int \sin u \, du = -\cos u + c = -\cos \frac{1}{x} + c$$

10.1.25. $I = \int e^{\cos x} \sin x \, dx$

Let $\cos x = u$, $-\sin x \, dx = du$.

$$I = -\int e^u \, du = -e^u + c = -e^{\cos x} + c$$

10.1.26. $I = \int \sin^2 (5x + 1) \cos (5x + 1) \, dx$

Let $5x + 1 = u$, $dx = \tfrac{1}{5} du$.

$$I = \frac{1}{5} \int \sin^2 u \cos u \, du$$

Now let $\sin u = v$, $\cos u \, du = dv$.

$$I = \frac{1}{5} \int v^2 \, dv = \frac{1}{15} v^3 + c = \frac{1}{15} \sin^3 u + c = \frac{1}{15} \sin^3(5x + 1) + c$$

10.1.27. $I = \int \ln^2 \cos x \tan x \, dx$
Let $\ln \cos x = u$, $-\tan x \, dx = du$.
$$I = -\int u^2 \, du = -\frac{1}{3} u^3 + c = -\frac{1}{3} \ln^3 \cos x + c$$

10.1.28. $I = \int x^3 \cos 5x^4 \, dx$
Let $5x^4 = u$.
$$I = \frac{1}{20} \int \cos u \, du = \frac{1}{20} \sin u + c = \frac{1}{20} \sin 5x^4 + c$$

10.1.29. $I = \int \frac{\ln \ln x \, dx}{x \ln x}$
Let $\ln x = u$, $\frac{dx}{x} = du$.
$$I = \int \frac{\ln u}{u} \, du, \quad \text{and now we let } \ln u = v, \frac{du}{u} = dv.$$
$$I = \int v \, dv = \frac{1}{2} v^2 + c = \frac{1}{2} (\ln u)^2 + c = \frac{1}{2} (\ln \ln x)^2 + c$$

10.2. Integrals leading to inverse trigonometric functions

10.2.1. $I = \int \frac{dx}{25 + x^2}$
Let $x = 5u$, $dx = 5 \, du$.
$$I = 5 \int \frac{du}{25 + 25u^2} = \frac{1}{5} \int \frac{du}{1 + u^2} = \frac{1}{5} \operatorname{Arc\,tan} u + c$$
$$= \frac{1}{5} \operatorname{Arc\,tan} \frac{x}{5} + c$$

10.2.2. $I = \int \frac{dx}{\sqrt{9 - 7x^2}}$
Let $\sqrt{7} x = 3u$, $dx = \frac{3}{\sqrt{7}} \, du$.
$$I = \frac{3}{\sqrt{7}} \int \frac{du}{\sqrt{9 - 9u^2}} = \frac{1}{\sqrt{7}} \int \frac{du}{\sqrt{1 - u^2}} = \frac{1}{\sqrt{7}} \operatorname{Arc\,sin} u + c$$
$$= \frac{1}{\sqrt{7}} \operatorname{Arc\,sin} \frac{\sqrt{7} x}{3} + c$$

10.2.3. $I = \int \dfrac{dx}{x\sqrt{x-1}}$

Let $x = u^2$, $dx = 2u\,du$.

$$I = 2\int \dfrac{u\,du}{u^2\sqrt{u^2-1}} = 2\int \dfrac{du}{u\sqrt{u^2-1}} = 2\,\text{Arc sec}\, u + c$$
$$= 2\,\text{Arc sec}\,\sqrt{x} + c$$

10.2.4. $I = \int \dfrac{e^x\,dx}{1+e^{2x}}$

Let $e^x = u$, $e^x\,dx = du$.

$$I = \int \dfrac{du}{1+u^2} = \text{Arc tan}\, u + c = \text{Arc tan}\, e^x + c$$

10.2.5. $I = \int \dfrac{\sec^2 x\,dx}{\sqrt{1-\tan^2 x}}$

Let $\tan x = u$, $\sec^2 x\,dx = du$.

$$I = \int \dfrac{du}{\sqrt{1-u^2}} = \text{Arc sin}\, u + c = \text{Arc sin}\,(\tan x) + c$$

In the following examples methods are developed for certain integrands containing $ax^2 + bx + c$ where $ax^2 + bx + c$ can be written in the form $(a_1 x + b_1)^2 + c_1^2$, or $\sqrt{ax^2 + bx + c}$ where $\sqrt{ax^2 + bx + c}$ can be written $\sqrt{c_1^2 - (a_1 x + b_1)^2}$.

10.2.6. $I = \int \dfrac{dx}{3x^2 + 6x + 5}$

Let $\sqrt{3}\,(x+1) = u\sqrt{2}$, $\sqrt{3}\,dx = \sqrt{2}\,du$.

$$I = \int \dfrac{dx}{3(x+1)^2 + 2} = \int \dfrac{\sqrt{\tfrac{2}{3}}\,du}{2u^2 + 2} = \dfrac{1}{\sqrt{6}}\int \dfrac{du}{u^2+1}$$
$$= \dfrac{1}{\sqrt{6}}\,\text{Arc tan}\, u + c = \dfrac{1}{\sqrt{6}}\,\text{Arc tan}\,\sqrt{\dfrac{3}{2}}\,(x+1) + c$$

Once the denominator has been made the sum of two squares, the substitution is found by making $3(x+1)^2 + 2 = 2(u^2+1)$.

10.2.7. $I = \int \dfrac{dx}{\sqrt{2+6x-3x^2}}$

Let $\sqrt{3}(x-1) = \sqrt{5}\,u$, $dx = \sqrt{\dfrac{5}{3}}\,du$.

$$I = \int \frac{dx}{\sqrt{5 - 3(1 - 2x + x^2)}} = \int \frac{dx}{\sqrt{5 - 3(x - 1)^2}}$$

$$= \frac{\sqrt{5}}{\sqrt{3}} \int \frac{du}{\sqrt{5 - 5u^2}} = \frac{1}{\sqrt{3}} \int \frac{du}{\sqrt{1 - u^2}}$$

$$= \frac{1}{\sqrt{3}} \text{Arc sin } u + c = \frac{1}{\sqrt{3}} \text{Arc sin } \sqrt{\frac{3}{5}}(x - 1) + c$$

Also, $1 - 2x + x^2 = (1 - x)^2$ and the problem could be solved by making the substitution $\sqrt{3}(1 - x) = \sqrt{5}u$, and this would lead to the answer $\frac{1}{\sqrt{3}} \text{Arc cos } \sqrt{\frac{3}{5}}(1 - x) + c$. Notice that the difference between the inverse trigonometric functions in the two forms of the answer is the constant $\frac{\pi}{2\sqrt{3}}$, which is absorbed in the arbitrary constant.

10.2.8. $I = \int \frac{e^x \, dx}{e^{2x} + 2e^x + 3}$

Let $e^x = u$, $e^x \, dx = du$.

$$I = \int \frac{du}{u^2 + 2u + 3} = \int \frac{du}{(u + 1)^2 + 2}.$$

Let $u + 1 = \sqrt{2}v$, $du = \sqrt{2} \, dv$.

$$I = \sqrt{2} \int \frac{dv}{2v^2 + 2} = \frac{1}{\sqrt{2}} \int \frac{dv}{v^2 + 1} = \frac{1}{\sqrt{2}} \text{Arc tan } v + c$$

$$= \frac{1}{\sqrt{2}} \text{Arc tan } \frac{u + 1}{\sqrt{2}} + c = \frac{1}{\sqrt{2}} \text{Arc tan } \frac{e^x + 1}{\sqrt{2}} + c$$

10.2.9. $I = \int \frac{dx}{(2x + 1)\sqrt{4x^2 + 4x - 1}}$

$$= \int \frac{dx}{(2x + 1)\sqrt{(2x + 1)^2 - 2}}$$

Let $2x + 1 = \sqrt{2}u$, $dx = \frac{1}{\sqrt{2}} du$.

$$I = \frac{1}{\sqrt{2}} \int \frac{du}{\sqrt{2}u \sqrt{2u^2 - 2}} = \frac{1}{2\sqrt{2}} \int \frac{du}{u\sqrt{u^2 - 1}}$$

$$= \frac{1}{2\sqrt{2}} \text{Arc sec } u + c = \frac{1}{2\sqrt{2}} \text{Arc sec } \frac{2x + 1}{\sqrt{2}} + c$$

10.2.10. $I = \int \dfrac{(x+1)\,dx}{x^2+1} = \int \dfrac{x\,dx}{x^2+1} + \int \dfrac{dx}{x^2+1}$

$= \tfrac{1}{2} \ln(x^2+1) + \text{Arc tan } x + c$

10.3. Integration by parts. The formula $\int u\,dv = uv - \int v\,du$ where u and v are functions of x, can often be used in solving integrals. Repeated application can often solve such integrals as shown below.

Example (a). $\int x^2 \cos x\,dx$

Choose $u = x^2$, $dv = \cos x\,dx$.

$$I = x^2 \sin x - 2 \int x \sin x\,dx$$

and choose $u = x$, $dv = \sin x\,dx$.

$$I = x^2 \sin x - 2\left[-x \cos x + \int \cos x\,dx \right]$$
$$= x^2 \sin x + 2x \cos x - 2 \sin x + c$$

Example (b). $I = \int x^3 \sqrt{x+1}\,dx$

Choose $u = x^3$, $\sqrt{x+1}\,dx = dv$.

$$I = \tfrac{2}{3}(x^3)(x+1)^{3/2} - 2 \int x^2 (x+1)^{3/2}\,dx$$

Now we choose $(x+1)^{3/2}\,dx = dv$, $x^2 = u$.

$$I = \tfrac{2}{3} x^3 (x+1)^{3/2} - 2\left[\tfrac{2}{5} x^2 (x+1)^{5/2} - \tfrac{4}{5} \int (x+1)^{5/2} x\,dx \right]$$

Finally, we choose $(x+1)^{5/2}\,dx = dv$, $x = u$.

$$I = \tfrac{2}{3} x^3 (x+1)^{3/2} - \tfrac{4}{5} x^2 (x+1)^{5/2}$$
$$+ \tfrac{8}{5}\left[\tfrac{2}{7}(x+1)^{7/2} x - \tfrac{2}{7} \int (x+1)^{7/2}\,dx \right]$$
$$= \tfrac{2}{3} x^3 (x+1)^{3/2} - \tfrac{4}{5} x^2 (x+1)^{5/2} + \tfrac{16}{35} x(x+1)^{7/2}$$
$$- \tfrac{32}{315}(x+1)^{9/2} + c$$

Notice that our object in each of these examples is to differentiate the part that is the power of x repeatedly until it disappears. The use of brackets enables us to keep our signs straight. Reduction formulas can often be developed, as in 10.3.7.

10.3.1. $I = \int xe^x \, dx$

Choose $u = x$, $dv = e^x \, dx$.

$$I = xe^x - \int e^x \, dx = xe^x - e^x + c$$

10.3.2. $I = \int \ln x \, dx$

Choose $dv = dx$, $\ln x = u$.

$$I = x \ln x - \int x \cdot \frac{1}{x} \, dx = x \ln x - x + c$$

10.3.3. $I = \int \text{Arc} \tan x \, dx$

Choose $dv = dx$, $\text{Arc} \tan x = u$.

$$I = x \, \text{Arc} \tan x - \int \frac{x \, dx}{1 + x^2} = x \, \text{Arc} \tan x - \frac{1}{2} \ln |1 + x^2| + c$$

10.3.4. $I = \int x \ln x \, dx$

Choose $x \, dx = dv$, $\ln x = u$.

$$I = \frac{1}{2} x^2 \ln x - \frac{1}{2} \int x \, dx = \frac{1}{2} x^2 \ln x - \frac{1}{4} x^2 + c$$

Notice that here we do not differentiate the power of x.

10.3.5. $I = \int \ln^2 x \, dx$

Choose $dv = dx$, $\ln^2 x = u$.

$$I = x \ln^2 x - 2 \int \ln x \, dx = x \ln^2 x - 2x \ln x + 2 \int dx$$
$$= x \ln^2 x - 2x \ln x + 2x + c$$

10.3.6. $I = \int e^{2x} \sin 3x \, dx$.

Choose $dv = \sin 3x\, dx$, $u = e^{2x}$.

$$I = -\frac{1}{3} e^{2x} \cos 3x + \frac{2}{3} \int e^{2x} \cos 3x\, dx$$

Now we choose $dv = \cos 3x\, dx$, $u = e^{2x}$.

$$I = -\frac{1}{3} e^{2x} \cos 3x + \frac{2}{9} e^{2x} \sin 3x - \frac{4}{9} \int e^{2x} \sin 3x\, dx$$

so that

$$\frac{13}{9} \int e^{2x} \sin 3x\, dx = -\frac{1}{3} e^{2x} \cos 3x + \frac{2}{9} e^{2x} \sin 3x + c$$

$$\int e^{2x} \sin 3x\, dx = -\frac{3}{13} e^{2x} \cos 3x + \frac{2}{13} e^{2x} \sin 3x + c$$

10.3.7. $I = \int \sec^3 x\, dx$

Choose $dv = \sec^2 x\, dx$, $\sec x = u$.

$$I = \sec x \tan x - \int \tan x \sec x \tan x\, dx$$

$$= \sec x \tan x - \int (\sec^2 x - 1) \sec x\, dx$$

$$= \sec x \tan x - \int \sec^3 x\, dx + \int \sec x\, dx$$

$$2 \int \sec^3 x\, dx = \sec x \tan x + \ln |\sec x + \tan x| + c$$

$$\int \sec^3 x\, dx = \frac{1}{2} \sec x \tan x + \frac{1}{2} \ln |\sec x + \tan x| + c$$

Using these same methods, the student can prove the reduction formula

$$\int \sec^n x\, dx = \frac{\sec^{n-2} x \tan x}{n-1} + \frac{n-2}{n-1} \int \sec^{n-2} x\, dx$$

Note in particular that this formula is true for $n > 2$ but not for $n = 1$.

10.3.8. $I = \int x^3 \sqrt{1 + x^2}\, dx$

Choose $x\sqrt{1+x^2}\, dx = dv$, $x^2 = u$

$$I = \frac{1}{3} x^2 (1 + x^2)^{3/2} - \frac{2}{3} \int x(1 + x^2)^{3/2}\, dx$$

$$= \frac{1}{3} x^2 (1 + x^2)^{3/2} - \frac{2}{15}(1 + x^2)^{5/2} + c$$

10.3.9. $I = \int \dfrac{x^3\, dx}{(4 + x^2)^2}$

Choose $dv = \dfrac{x\, dx}{(4 + x^2)^2}$, $u = x^2$

$$= \frac{-\frac{1}{2}x^2}{4 + x^2} + \frac{1}{2} \int \frac{2x\, dx}{4 + x^2} = -\frac{1}{2} \frac{x^2}{4 + x^2} + \frac{1}{2} \ln|4 + x^2| + c$$

10.4. Trigonometric integrals. If an integral is of the form $\int \sin^2 x \cos^3 x\, dx$, we solve in the following way:

$$\int \sin^2 x \cos^3 x\, dx = \int \sin^2 x \cos^2 x \cos x\, dx$$

$$= \int \sin^2 x (1 - \sin^2 x) \cos x\, dx$$

$$= \int \sin^2 x \cos x\, dx - \int \sin^4 x \cos x\, dx$$

$$= \tfrac{1}{3} \sin^3 x - \tfrac{1}{5} \sin^5 x + c$$

Notice that we have factored out the differential of $\sin x$, ($\cos x\, dx$), and have expressed the rest of the integrand in terms of $\sin x$. This is convenient because $\cos x$ appears in an odd power. If neither $\sin x$ nor $\cos x$ appears in an odd power, the double-angle formulas

$$\cos^2 x = \frac{1 + \cos 2x}{2}, \qquad \sin^2 x = \frac{1 - \cos 2x}{2}$$

can be used as in the following example:

$$\int \sin^2 x \cos^2 x \, dx = \int \left(\frac{1 - \cos 2x}{2}\right)\left(\frac{1 + \cos 2x}{2}\right) dx$$

$$= \frac{1}{4} \int (1 - \cos^2 2x) \, dx = \frac{1}{4} \int \left(1 - \frac{\cos 4x + 1}{2}\right) dx$$

$$= \frac{1}{8} \int dx - \frac{1}{8} \int \cos 4x \, dx = \frac{x}{8} - \frac{1}{32} \sin 4x + c$$

Illustrated below are two examples involving tan x and sec x. The same principle of separating the differential of one function and converting the rest of the integrand to that function applies

$$\int \tan^3 x \sec^4 x \, dx = \int \tan^3 x \sec^2 x \sec^2 x \, dx$$

$$= \int \tan^3 x (1 + \tan^2 x) \sec^2 x \, dx$$

$$= \int \tan^3 x \sec^2 x \, dx + \int \tan^5 x \sec^2 x \, dx$$

$$= \tfrac{1}{4} \tan^4 x + \tfrac{1}{6} \tan^6 x + c$$

or $\quad \int \tan^3 x \sec^4 x \, dx = \int \tan^2 x \sec^3 x (\sec x \tan x) \, dx$

$$= \int (\sec^2 x - 1) \sec^3 x (\sec x \tan x) \, dx$$

$$= \int \sec^5 x \sec x \tan x \, dx - \int \sec^3 x \sec x \tan x \, dx$$

$$= \tfrac{1}{6} \sec^6 x - \tfrac{1}{4} \sec^4 x + c$$

If, however, tan x appears in an even power and sec x in an odd power, as in the integral $\int \tan^4 x \sec^3 x \, dx$, the substitution $\tan^2 x = \sec^2 x - 1$ produces an integral in odd powers of sec x, and the integral must be treated as in 10.3.7. The following formulas can be useful in solving trigonometric integrals:

$$\sin Ax \cos Bx = \tfrac{1}{2}[\sin (A+B)x + \sin (A-B)x]$$
$$\sin Ax \sin Bx = -\tfrac{1}{2}[\cos (A+B)x - \cos (A-B)x]$$
$$\cos Ax \cos Bx = \tfrac{1}{2}[\cos (A+B)x + \cos (A-B)x]$$

Example: $\displaystyle\int \sin 7x \cos 9x \, dx = \frac{1}{2}\int \left[\sin 16x + \sin (-2x)\right] dx$

$\displaystyle = \frac{1}{2}\left[-\frac{1}{16}\cos 16x + \frac{1}{2}\cos 2x\right] + c$

10.4.1. $\displaystyle\int \tan^3 x \, dx = \int \tan x \tan^2 x \, dx$

$\displaystyle = \int \tan x (\sec^2 x - 1) \, dx = \int \tan x \sec^2 x \, dx - \int \tan x \, dx$

$\displaystyle = \tfrac{1}{2} \tan^2 x + \ln |\cos x| + C$

10.4.2. $\displaystyle I = \int \sin \left(\frac{5-x}{2}\right) dx$

Let $\dfrac{5-x}{2} = u$, $dx = -2 \, du$.

$\displaystyle I = -2 \int \sin u \, du = 2 \cos u + c = 2 \cos \left(\frac{5-x}{2}\right) + C$

10.4.3. $\displaystyle\int \cos^{2/3} 2\theta \sin 2\theta \, d\theta$

Let $\cos 2\theta = u$, $\sin 2\theta \, d\theta = -\dfrac{du}{2}$.

$\displaystyle I = -\frac{1}{2}\int u^{2/3} \, du = -\tfrac{3}{10} u^{5/3} + c = -\tfrac{3}{10} \cos^{5/3} 2\theta + C$

10.4.4. $\displaystyle\int \tan^6 \theta \, d\theta = \int \tan^4 \theta \tan^2 \theta \, d\theta$

$\displaystyle = \int \tan^4 \theta \sec^2 \theta \, d\theta - \int \tan^4 \theta \, d\theta = \tfrac{1}{5}\tan^5 \theta - \int \tan^4 \theta \, d\theta$

$\displaystyle = \tfrac{1}{5} \tan^5 \theta - \int \tan^2 \theta \sec^2 \theta \, d\theta + \int \tan^2 \theta \, d\theta$

$$= \tfrac{1}{5} \tan^5 \theta - \tfrac{1}{3} \tan^3 \theta + \int (\sec^2 \theta - 1) \, d\theta$$

$$= \tfrac{1}{5} \tan^5 \theta - \tfrac{1}{3} \tan^3 \theta + \tan \theta - \theta + C$$

10.4.5. $\displaystyle\int \frac{\cos^5 \theta \, d\theta}{\sin^{1/3} \theta} = \int \frac{(1 - \sin^2 \theta)^2}{\sin^{1/3} \theta} \cos \theta \, d\theta$

$$= \int [\sin^{-1/3} \theta - 2 \sin^{5/3} \theta + \sin^{11/3} \theta] \cos \theta \, d\theta$$

$$= \tfrac{3}{2} \sin^{2/3} \theta - \tfrac{6}{8} \sin^{8/3} \theta + \tfrac{3}{14} \sin^{14/3} \theta + C$$

10.4.6. $\displaystyle\int \tan^7 \alpha \sec^4 \alpha \, d\alpha = \int \tan^7 \alpha (1 + \tan^2 \alpha) \sec^2 \alpha \, d\alpha$

$$= \int \tan^7 \alpha \sec^2 \alpha \, d\alpha + \int \tan^9 \alpha \sec^2 \alpha \, d\alpha$$

$$= \tfrac{1}{8} \tan^8 \alpha + \tfrac{1}{10} \tan^{10} \alpha + C$$

10.4.7. $\displaystyle\int \sin^2 x \cos^2 x \, dx = \frac{1}{4} \int \sin^2 2x = \frac{1}{4} \int \frac{1 - \cos 4x}{2} \, dx$

$$= \tfrac{1}{8} x - \tfrac{1}{32} \sin 4x + c$$

This is shorter than the methods used above, but this method does not have so wide an application.

10.4.8. $\displaystyle\int \frac{\tan x \, dx}{\sec^2 x} = \int \frac{\dfrac{\sin x}{\cos x}}{\dfrac{1}{\cos^2 x}} \, dx$

$$= \int \sin x \cos x \, dx = \tfrac{1}{2} \sin^2 x + C$$

Here we changed over to sines and cosines, since $\tan x \, dx$ does not lead to a practicable du.

10.4.9. $\displaystyle\int \sec^4 x \, dx = \int \sec^2 x (1 + \tan^2 x) \, dx$

$$= \tan x + \tfrac{1}{3} \tan^3 x + C$$

10.4.10. $\displaystyle\int \cos mx \cos nx \, dx$

$$= \frac{1}{2} \int \cos(m+n)x\, dx + \frac{1}{2} \int \cos(m-n)x\, dx$$

$$= \frac{1}{2(m+n)} \sin(m+n)x + \frac{1}{2(m-n)} \sin(m-n)x + C$$

$$(m \neq n)$$

$m = n$ in the above,

$$\cos^2 mx = \frac{1}{2} \int (\cos 2mx + 1)\, dx = \frac{1}{2}x + \frac{1}{4m} \sin 2mx + C$$

10.4.11. $\displaystyle\int \tan^3 \theta \sec \theta\, d\theta = \int \tan^2 \theta \sec \theta \tan \theta\, d\theta$

$$= \int (\sec^2 \theta - 1) \sec \theta \tan \theta\, d\theta = \tfrac{1}{3} \sec^3 \theta - \sec \theta + C$$

10.4.12. $I = \displaystyle\int \sin^n x\, dx = \int \sin^{n-1} x \sin x\, dx.$

Choose $dv = \sin x\, dx.$

$$= -\sin^{n-1} x \cos x + (n-1) \int \sin^{n-2} x \cos^2 x\, dx$$

$$= -\sin^{n-1} x \cos x + (n-1) \int \sin^{n-2} x(1 - \sin^2 x)\, dx$$

$$= -\sin^{n-1} x \cos x + (n-1) \int \sin^{n-2} x\, dx$$

$$\quad\quad\quad\quad\quad\quad\quad\quad\quad - (n-1) \cdot \int \sin^n x\, dx$$

$$= -\frac{1}{n} \sin^{n-1} x \cos x + \frac{n-1}{n} \int \sin^{n-2} x\, dx$$

10.4.13. $\displaystyle\int \cos^4 x\, dx = \int \left(\frac{\cos 2x + 1}{2}\right)^2 dx$

$$= \frac{1}{4} \int \cos^2 2x\, dx + \frac{1}{2} \int \cos 2x\, dx + \frac{1}{4} \int dx$$

$$= \frac{1}{4} \int \frac{\cos 4x + 1}{2}\, dx + \tfrac{1}{4} \sin 2x + \tfrac{1}{4}x$$

$$= \tfrac{1}{32} \sin 4x + \tfrac{1}{4} \sin 2x + \tfrac{3}{8}x + C$$

10.5. Trigonometric substitutions

10.5.1. $I = \displaystyle\int \dfrac{dx}{\sqrt{4+x^2}}$

Let $x = 2\tan\theta$, $\quad dx = 2\sec^2\theta\, d\theta$.

$$I = 2\int \dfrac{\sec^2\theta\, d\theta}{\sqrt{4+4\tan^2\theta}} = \int \sec\theta\, d\theta = \ln|\sec\theta + \tan\theta| + C$$

To convert to our original system, we draw the accompanying triangle and find $\sec\theta = \dfrac{\sqrt{4+x^2}}{2}$ and $\tan\theta = \dfrac{x}{2}$.

10.5.1.

$$\int \dfrac{dx}{\sqrt{4+x^2}} = \ln\left|\dfrac{\sqrt{4+x^2}}{2} + \dfrac{x}{2}\right|$$

$$+ C = \ln|\sqrt{4+x^2} + x| + C_1$$

10.5.2. $I = \displaystyle\int \sqrt{a^2 - x^2}\, dx$

Let $\quad x = a\sin\theta, \quad dx = a\cos\theta\, d\theta$.

$$I = a\int \sqrt{a^2 - a^2\sin^2\theta}\, \cos\theta\, d\theta = a^2\int \cos^2\theta\, d\theta$$

$$= a^2 \int \dfrac{1 + \cos 2\theta}{2}\, d\theta = a^2(\tfrac{1}{4}\sin 2\theta + \tfrac{1}{2}\theta) + C$$

$$= a^2(\tfrac{1}{2}\sin\theta\cos\theta + \tfrac{1}{2}\theta) + C$$

10.5.2.

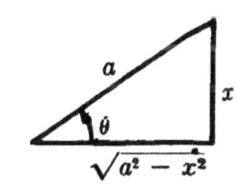

Now that we have reduced $\sin 2\theta$ to functions of θ, we can use our triangle to see that $\sin\theta = \dfrac{x}{a}$, $\cos\theta = \dfrac{\sqrt{a^2-x^2}}{a}$, $\theta = \text{Arc}\sin\dfrac{x}{a}$. We could use Arc $\cos\dfrac{\sqrt{a^2-x^2}}{a}$, or Arc $\tan\dfrac{x}{\sqrt{a^2-x^2}}$, but prefer this expression for θ, and

$$\int \sqrt{a^2 - x^2}\, dx = a^2 \left(\frac{1}{2} \frac{x\sqrt{a^2 - x^2}}{a^2} + \frac{1}{2} \text{Arc sin}\, \frac{x}{a} \right) + C$$

$$= \frac{1}{2} x \sqrt{a^2 - x^2} + \frac{a^2}{2} \text{Arc sin}\, \frac{x}{a} + C$$

10.5.3. $I = \int \sqrt{x^2 - a^2}\, dx$

Let $x = a \sec \theta, \quad dx = a \sec \theta \tan \theta\, d\theta.$

$$= a \int \sqrt{a^2 \sec^2 \theta - a^2}\, \sec \theta \tan \theta\, d\theta = \int a^2 \tan^2 \theta \sec \theta\, d\theta$$

$$= a^2 \int (\sec^2 \theta - 1) \sec \theta\, d\theta$$

$$= a^2 \int \sec^3 \theta\, d\theta - a^2 \ln |\sec \theta + \tan \theta|$$

Referring to **10.3.7**,

$$I = \frac{a^2}{2} \sec \theta \tan \theta - \frac{a^2}{2} \ln |\sec \theta + \tan \theta| + C.$$

From our triangle $\sec \theta = \frac{x}{a}$, and $\tan \theta = \frac{\sqrt{x^2 - a^2}}{a}$

10.5.3.

$$\int \sqrt{x^2 - a^2}\, dx = \frac{1}{2} x \sqrt{x^2 - a^2} - \frac{a^2}{2} \ln \left| \frac{x}{a} + \frac{\sqrt{x^2 - a^2}}{a} \right| + C$$

10.5.4. $\int \frac{(2x^2 + 7)^{3/2}}{x}\, dx$

Let $\sqrt{2} x = \sqrt{7} \tan \theta, \quad dx = \frac{\sqrt{7}}{\sqrt{2}} \sec^2 \theta\, d\theta.$

$$I = 7^{3/2} \int \frac{\sec^3 \theta\, (\tan^2 \theta + 1)}{\tan \theta}\, d\theta$$

$$= 7^{3/2} \int \sec^3 \theta \tan \theta\, d\theta + 7^{3/2} \int \frac{\sec^3 \theta}{\tan \theta}\, d\theta$$

$$= \frac{7^{3/2}}{3} \sec^3 \theta + 7^{3/2} \int \frac{\sec \theta(\tan^2 \theta + 1)}{\tan \theta}\, d\theta$$

$$= \frac{7^{3/2}}{3}\sec^3\theta + 7^{3/2}\int \sec\theta\tan\theta\,d\theta + 7^{3/2}\int \frac{\sec\theta}{\tan\theta}\,d\theta$$

$$= \frac{7^{3/2}}{3}\sec^3\theta + 7^{3/2}\sec\theta + 7^{3/2}\int \frac{d\theta}{\sin\theta}$$

$$= \frac{7^{3/2}}{3}\sec^3\theta + 7^{3/2}\sec\theta + 7^{3/2}\ln|\csc\theta - \cot\theta| + C$$

From our triangle $\sec\theta = \dfrac{\sqrt{2x^2+7}}{\sqrt{7}}$, $\csc\theta = \dfrac{\sqrt{2x^2+7}}{\sqrt{2}x}$,

and $\cot\theta = \dfrac{\sqrt{7}}{\sqrt{2}x}$

10.5.4.

$$\int \frac{(2x^2+7)^{3/2}}{x}\,dx = 7^{3/2}\left[\frac{1}{3}\left(\frac{\sqrt{2x^2+7}}{\sqrt{7}}\right)^3 + \frac{\sqrt{2x^2+7}}{\sqrt{7}}\right.$$

$$\left. + \ln\left|\frac{\sqrt{2x^2+7}}{\sqrt{2}x} - \frac{\sqrt{7}}{\sqrt{2}x}\right|\right] + C$$

10.5.5. $I = \displaystyle\int \frac{dx}{(1+\sqrt{x})^{3/2}}$.

Let $\sqrt{x} = \tan^2\theta$, $\quad x = \tan^4\theta$, $\quad dx = 4\tan^3\theta\sec^2\theta\,d\theta$.

$$I = \int \frac{4\tan^3\theta\sec^2\theta\,d\theta}{(1+\tan^2\theta)^{3/2}} = 4\int \frac{\tan^3\theta}{\sec\theta}\,d\theta$$

$$= 4\int \frac{(\sec^2\theta - 1)\tan\theta}{\sec\theta}\,d\theta$$

$$= 4\int \sec\theta\tan\theta\,d\theta - 4\int \sin\theta\,d\theta$$

$$= 4\sec\theta + 4\cos\theta + C$$

From our triangle,

10.5.5.

$$\int \frac{dx}{(1+\sqrt{x})^{3/2}} = 4\left(\sqrt{1+\sqrt{x}} + \frac{1}{\sqrt{1+\sqrt{x}}}\right) + c$$

10.5.6. $I = \displaystyle\int \frac{x\,dx}{\sqrt{1+6x-3x^2}} = \int \frac{x\,dx}{\sqrt{4-3(x-1)^2}}$

Let $\sqrt{3}(x-1) = 2\sin\theta$.

$$I = \frac{2}{\sqrt{3}} \int \frac{\frac{2}{\sqrt{3}}\sin\theta + 1}{\sqrt{4-4\sin^2\theta}} \cos\theta\,d\theta$$

$$= \frac{2}{\sqrt{3}} \int \left(\frac{1}{\sqrt{3}}\sin\theta + \frac{1}{2}\right) d\theta = -\frac{2}{3}\cos\theta + \frac{1}{\sqrt{3}}\theta + c$$

From our triangle $\cos\theta = \tfrac{1}{2}\sqrt{1+6x-3x^2}$ and $\theta = \text{Arc sin } \tfrac{\sqrt{3}}{2}(x-1)$.

10.5.6.

$$\int \frac{x\,dx}{\sqrt{1+6x-3x^2}} = -\frac{1}{3}\sqrt{1+6x-3x^2} + \frac{1}{\sqrt{3}}\text{Arc sin } \frac{\sqrt{3}}{2}(x-1) + c$$

10.5.7. $I = \displaystyle\int \frac{x\,dx}{(7+4x+x^2)^{3/2}} = \int \frac{x\,dx}{[3+(x+2)^2]^{3/2}}$

Let $x + 2 = \sqrt{3}\tan\theta$.

$$I = \sqrt{3}\int \frac{\sqrt{3}\tan\theta - 2}{(3+3\tan^2\theta)^{3/2}} \sec^2\theta\,d\theta = \frac{1}{3}\int \frac{\sqrt{3}\tan\theta - 2}{\sec\theta}\,d\theta$$

$$= \frac{1}{3}\int (\sqrt{3}\sin\theta - 2\cos\theta)\,d\theta = -\frac{\sqrt{3}}{3}\cos\theta - \frac{2}{3}\sin\theta + c$$

From our triangle $\cos\theta = \dfrac{\sqrt{3}}{\sqrt{7+4x+x^2}}$ and

$$\sin\theta = \frac{x+2}{\sqrt{7+4x+x^2}}$$

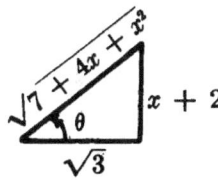

10.5.7

$$\int \frac{x\,dx}{(7+4x+x^2)^{3/2}} = -\frac{1}{3}\cdot\frac{2x+7}{\sqrt{7+4x+x^2}} + c$$

10.6. Algebraic substitutions

10.6.1. $I = \int x^5 \sqrt{2+x^3}\,dx$

Let $z^2 = 2 + x^3$, $2z\,dz = 3x^2\,dx$, and $x^5\,dx = \frac{2}{3}z(z^2 - 2)\,dz$.

$$I = \frac{2}{3}\int z^2(z^2-2)\,dz = \frac{2}{15}z^5 - \frac{4}{9}z^3 + c$$
$$= \tfrac{2}{15}(2+x^3)^{5/2} - \tfrac{4}{9}(2+x^3)^{3/2} + c$$

10.6.2. $I = \int \dfrac{dx}{3+\sqrt{1+2x}}$

Let $1 + 2x = z^2$, $2\,dx = 2z\,dz$.

$$I = \int \frac{z\,dz}{3+z} = \int \frac{(z+3)}{z+3}\,dz - \int \frac{3\,dz}{z+3}$$
$$= z - 3\ln|z+3| + c = \sqrt{1+2x}$$
$$\quad - 3\ln|\sqrt{1+2x}+3| + c$$

10.6.3. $I = \int \dfrac{e^{2x}\,dx}{\sqrt{1+e^x}}$

Let $e^x = z$, $e^x\,dx = dz$.

$$I = \int \frac{z\,dz}{\sqrt{1+z}} = \int \frac{(z+1)\,dz}{\sqrt{z+1}} - \int \frac{dz}{\sqrt{z+1}}$$
$$= \int \sqrt{z+1}\,dz - 2(z+1)^{1/2}$$
$$= \frac{2}{3}(z+1)^{3/2} - 2(z+1)^{1/2} + c$$
$$= \frac{2}{3}(e^x+1)^{3/2} - 2(e^x+1)^{1/2} + c$$

10.6.4. $I = \int \dfrac{x\, dx}{(2 + x)^{1/4}}$

Let $2 + x = z^4$, $dx = 4z^3\, dz$.

$$= \int \dfrac{(z^4 - 2)4z^3\, dz}{z} = 4\int (z^6 - 2z^2)\, dz$$

$$= \tfrac{4}{7}z^7 - \tfrac{8}{3}z^3 + c = \tfrac{4}{7}(2 + x)^{7/4} - \tfrac{8}{3}(2 + x)^{3/4} + c$$

10.6.5. $I = \int \dfrac{x^3\, dx}{(2 + 3x)^{7/2}}$

Let $2 + 3x = z^2$, $dx = \tfrac{2}{3}z\, dz$, $x^3 = \left(\dfrac{z^2 - 2}{3}\right)^3$.

$$= \dfrac{2}{3}\int \dfrac{\left(\dfrac{z^2 - 2}{3}\right)^3 z\, dz}{z^7} = \dfrac{2}{81}\int \left(1 - \dfrac{6}{z^2} + \dfrac{12}{z^4} - \dfrac{8}{z^6}\right) dz$$

$$= \dfrac{2}{81}\left(z + \dfrac{6}{z} - \dfrac{4}{z^3} + \dfrac{8/5}{z^5}\right) + c$$

$$= \dfrac{2}{81}\left(\sqrt{2 + 3x} + \dfrac{6}{\sqrt{2 + 3x}}\right.$$

$$\left. - \dfrac{4}{(2 + 3x)^{3/2}} + \dfrac{8/5}{(2 + 3x)^{5/2}}\right) + c$$

10.6.6. $I = \int \dfrac{dx}{x\sqrt{1 + x^2}}$

Let $x = \dfrac{1}{z}$, $dx = -\dfrac{dz}{z^2}$.

$$I = \int \dfrac{-\dfrac{dz}{z^2}}{\dfrac{1}{z}\sqrt{1 + \dfrac{1}{z^2}}} = -\int \dfrac{dz}{\sqrt{z^2 + 1}}$$

Now let $z = \tan\theta$, $dz = \sec^2\theta\, d\theta$.

10.6.6.

$$I = -\int \dfrac{\sec^2\theta\, d\theta}{\sec\theta} = -\int \sec\theta\, d\theta = -\ln|\sec\theta + \tan\theta| + C$$

$$= -\ln|\sqrt{z^2 + 1} + z| + C = -\ln\left|\sqrt{\dfrac{1}{x^2} + 1} + \dfrac{1}{x}\right| + C$$

$$= -\ln\left|\frac{\sqrt{1+x^2}+1}{x}\right| + C$$

10.6.7. $I = \int \dfrac{dx}{(1+x^2)^{3/2}}$

Let $x = \dfrac{1}{z}$, $dx = -\dfrac{dz}{z^2}$.

$$I = -\int \frac{dz}{z^2\left(1+\dfrac{1}{z^2}\right)^{3/2}} = -\int \frac{z\,dz}{(1+z^2)^{3/2}} = (1+z^2)^{-1/2} + C$$

$$= \left(1 + \frac{1}{x^2}\right)^{-1/2} + C = \frac{x}{\sqrt{1+x^2}} + C$$

10.7. Integration of rational fractions. If the integrand consists of the quotient of two polynomials such that the one in the numerator is of degree equal or greater than the one in the denominator, the fraction must be reduced by division.

Example

$$\int \frac{x^3 + 7x}{x^2 - 2x + 5}\, dx = \int \left(x + 2 + \frac{6x - 10}{x^2 - 2x + 5}\right) dx$$

10.7.1. Distinct, linear factors:

$$I = \int \frac{2x + 7}{(x-1)(x+2)(x-3)}\, dx$$

Assume

$$\frac{2x + 7}{(x-1)(x+2)(x-3)} = \frac{A}{x-1} + \frac{B}{x+2} + \frac{C}{x-3}$$

$$2x + 7 = A(x+2)(x-3) + B(x-1)(x-3) + C(x-1)(x+2)$$

Since the identity must be true for all values of x,

$x = 1$,	$2 + 7 = A(3)(-2)$,	$A = -\tfrac{3}{2}$
$x = -2$,	$-4 + 7 = B(-3)(-5)$,	$B = \tfrac{1}{5}$
$x = 3$,	$6 + 7 = C(2)(5)$,	$C = \tfrac{13}{10}$

$$\int \frac{2x+7}{(x-1)(x+2)(x-3)} dx$$

$$= \int \left(\frac{-\frac{3}{2}}{x-1} + \frac{\frac{1}{5}}{x+2} + \frac{\frac{13}{10}}{x-3} \right) dx$$

$$= -\frac{3}{2} \int \frac{dx}{x-1} + \frac{1}{5} \int \frac{dx}{x+2} + \frac{13}{10} \int \frac{dx}{x-3}$$

$$= -\tfrac{3}{2} \ln|x-1| + \tfrac{1}{5} \ln|x+2| + \tfrac{13}{10} \ln|x-3| + c$$

These same A, B, and C could be obtained in another way. We have

$$2x + 7 = A(x^2 - x - 6) + B(x^2 - 4x + 3) + C(x^2 + x - 2)$$

Equating the powers of x on one side with the same powers on the other: (1) $A + B + C = 0$; (2) $-A - 4B + C = 2$; (3) $-6A + 3B - 2C = 7$. By adding (1), (2), and (3), we have $-6A = 9$, or $A = -\tfrac{3}{2}$. Then (4) is $-3B + 2C = 2$, by adding (1) and (2); and (5) is $9B + 4C = 7$, by adding 6 times (1) to (3). By adding 3 times (4) to (5), we have $10C = 13$, or $C = \tfrac{13}{10}$; and $B = \tfrac{1}{5}$.

Note the coefficient of x^2 on the left-hand side is zero; hence $A + B + C = 0$

10.7.2. Factors linear but repeated:

$$I = \int \frac{(x^2 + 2x + 3)\, dx}{x^3(x-1)(x+3)^2}$$

Corresponding to $(x+3)^2$ we introduce fractions with $(x+3)^2$ and all lower powers as denominators—similarly with x^3.

$$\frac{x^2 + 2x + 3}{x^3(x-1)(x+3)^2} = \frac{A}{x^3} + \frac{B}{x^2} + \frac{C}{x} + \frac{D}{x-1} + \frac{E}{(x+3)} + \frac{F}{(x+3)^2}$$

$$x^2 + 2x + 3 = A(x-1)(x+3)^2 + Bx(x-1)(x+3)^2$$
$$+ Cx^2(x-1)(x+3)^2 + Dx^3(x+3)^2 + Ex^3(x-1)(x+3)$$
$$+ Fx^3(x-1) = A(x^3 + 5x^2 + 3x - 9)$$
$$+ B(x^4 + 5x^3 + 3x^2 - 9x) + C(x^5 + 5x^4 + 3x^3 - 9x^2)$$
$$+ D(x^5 + 6x^4 + 9x^3) + E(x^5 + 2x^4 - 3x^3) + F(x^4 - x^3)$$

(1) $\qquad C + D + E = 0$
(2) $\qquad B + 5C + 6D + 2E + F = 0$
(3) $\qquad A + 5B + 3C + 9D - 3E - F = 0$
(4) $\qquad 5A + 3B - 9C = 1$
(5) $\qquad 3A - 9B = 2$
(6) $\qquad -9A = 3$

From (6), $A = -\frac{1}{3}$; substituting this value in (5), $B = -\frac{1}{3}$. From these values and (4), $C = -\frac{11}{27}$; (1), (2), and (3) then become

$D + E = \frac{11}{27}$; $\qquad 6D + 2E + F = \frac{64}{27}$; $\qquad 9D - 3E - F = \frac{22}{9}$

$\qquad\qquad 15D - E = \frac{151}{27}$; $\qquad\qquad 16D = \frac{162}{27}$;

$\qquad\qquad D = \frac{3}{8}$; $E = \frac{7}{216}$; $\qquad F = \frac{1}{18}$

$$\int \frac{(x^2 + 2x + 3)\,dx}{x^3(x-1)(x+3)^2}$$

$$= -\frac{1}{3}\int \frac{dx}{x^3} - \frac{1}{3}\int \frac{dx}{x^2} - \frac{11}{27}\int \frac{dx}{x} + \frac{3}{8}\int \frac{dx}{x-1} + \frac{7}{216}\int \frac{dx}{x+3}$$

$$+ \frac{1}{18}\int \frac{dx}{(x+3)^2}$$

$$= \frac{1}{6x^2} + \frac{1}{3x} - \frac{1}{18(x+3)} - \frac{11}{27}\ln|x| + \frac{3}{8}\ln|x-1|$$

$$+ \frac{7}{216}\ln|x+3| + c$$

10.7.3. Distinct quadratic factors: $\int \frac{x\,dx}{(x+1)(x^2 + 2x + 5)}$

$$\frac{x}{(x+1)(x^2+2x+5)} = \frac{A}{x+1} + \frac{Cx+D}{x^2+2x+5}$$

Notice that we choose the linear function $Cx + D$ for the numerator corresponding to the quadratic factor.

$$x = A(x^2 + 2x + 5) + (Cx + D)(x + 1)$$
$$A + C = 0; \quad A = -C; \quad D = 5C;$$

$2A + C + D = 1$; $-2C + C + 5C = 1$;

$5A + D = 0$; $C = \frac{1}{4}$; $A = -\frac{1}{4}$; $D = \frac{5}{4}$

$$\int \frac{x\,dx}{(x+1)(x^2+2x+5)} = -\frac{1}{4}\int \frac{dx}{x+1} + \frac{1}{4}\int \frac{x+5}{x^2+2x+5}\,dx$$

$$= -\frac{1}{4}\ln|x+1| + \frac{1}{4}\int \frac{x+1}{x^2+2x+5}\,dx + \int \frac{dx}{x^2+2x+5}$$

$$= -\frac{1}{4}\ln|x+1| + \frac{1}{8}\int \frac{2x+2}{x^2+2x+5}\,dx + \int \frac{dx}{(x+1)^2+4}$$

$$= -\frac{1}{4}\ln|x+1| + \frac{1}{8}\ln|x^2+2x+5|$$

$$+ \frac{1}{2}\text{Arc tan}\frac{x+1}{2} + C$$

Now $\dfrac{x+5}{x^2+2x+5}$ is split so that the first fraction has a numerator which is a constant multiple of the derivative of the denominator, while the second has a constant as numerator.

10.7.4. Factors of the second degree and repeated:

$$I = \int \frac{x^2 - x + 1}{(x-1)(x^2+2x+5)^2}\,dx$$

$$\frac{x^2 - x + 1}{(x-1)(x^2+2x+5)^2}$$

$$= \frac{A}{x-1} + \frac{Bx+C}{x^2+2x+5} + \frac{Dx+E}{(x^2+2x+5)^2}$$

(1) $x^2 - x + 1 = A(x^2+2x+5)^2 + (Bx+C)(x-1)$

$$(x^2+2x+5) + (Dx+E)(x-1)$$

$$= A(x^4 + 4x^3 + 14x^2 + 20x + 25)$$

$$+ (Bx+C)(x^3 + x^2 + 3x - 5)$$

$$+ (Dx+E)(x-1)$$

Equating the coefficients of the powers of x:

(2) $\qquad A + B = 0$
(3) $\qquad 4A + B + C = 0$
(4) $\qquad 14A + 3B + C + D = 1$
(5) $\qquad 20A - 5B + 3C - D + E = -1$
(6) $\qquad 25A - 5C - E = 1$

From (2), $A = -B$, and from (3), $C = -3A$. From (6), $E = 40A - 1$, and from (4), $D = 1 - 8A$. Substituting in (5),

$$20A + 5A - 9A + 8A - 1 + 40A - 1 = -1; \quad A = \tfrac{1}{64};$$
$$B = -\tfrac{1}{64}; \quad C = -\tfrac{3}{64}; \quad D = \tfrac{7}{8}; \quad E = -\tfrac{3}{8}$$

Notice, by setting $x = 1$ in (1), we obtain $A = \tfrac{1}{64}$, and we could have used this value in (2), (3), (4), (5), and (6), shortening our work.

$$\int \frac{x^2 - x + 1}{(x-1)(x^2 + 2x + 5)^2}\, dx$$

$$= \frac{1}{64}\int \frac{dx}{x-1} + \int \frac{(-\tfrac{1}{64}x - \tfrac{3}{64})\, dx}{x^2 + 2x + 5} + \int \frac{(\tfrac{7}{8}x - \tfrac{3}{8})\, dx}{(x^2 + 2x + 5)^2}$$

$$= \frac{1}{64}\ln|x-1| - \frac{1}{128}\int \frac{(2x+2)\, dx}{x^2 + 2x + 5} - \frac{2}{64}\int \frac{dx}{(x+1)^2 + 4}$$

$$\quad + \frac{7}{16}\int \frac{(2x+2)\, dx}{(x^2 + 2x + 5)^2} - \frac{10}{8}\int \frac{dx}{(x^2 + 2x + 5)^2}$$

$$= \frac{1}{64}\ln|x-1| - \frac{1}{128}\ln|x^2 + 2x + 5| - \frac{1}{64}\text{Arc tan}\,\frac{x+1}{2}$$

$$\quad - \frac{7}{16}\cdot\frac{1}{x^2 + 2x + 5} - \frac{10}{8}\int \frac{dx}{[(x+1)^2 + 4]^2}$$

Only the last integral is troublesome. Let $x + 1 = 2\tan\theta$, $dx = 2\sec^2\theta\, d\theta$.

10.7.4.

$$\int \frac{dx}{[(x+1)^2+4]^2} = 2\int \frac{\sec^2\theta\, d\theta}{(4\tan^2\theta+4)^2} = \frac{1}{8}\int \frac{d\theta}{\sec^2\theta}$$

$$= \frac{1}{8}\int \cos^2\theta\, d\theta = \frac{1}{8}\int \frac{\cos 2\theta + 1}{2}\, d\theta = \frac{1}{32}\sin 2\theta + \frac{\theta}{16} + C$$

$$= \frac{1}{16}\sin\theta\cos\theta + \frac{\theta}{16} + C$$

$$= \frac{1}{8}\cdot\frac{x+1}{x^2+2x+5} + \frac{1}{16}\text{Arc tan}\frac{x+1}{2} + C$$

$$I = \frac{1}{64}\ln|x-1| - \frac{1}{128}\ln|x^2+2x+5|$$

$$- \frac{3}{32}\text{Arc tan}\frac{x+1}{2} - \frac{1}{32}\cdot\frac{5x+19}{x^2+2x+5} + C$$

10.7.5. $I = \int \dfrac{dx}{x^2 - 9}$

$$\frac{1}{x^2-9} = \frac{A}{x-3} + \frac{B}{x+3}; \quad A = \frac{1}{6}; \quad B = -\frac{1}{6}$$

$$I = \frac{1}{6}\int \frac{dx}{x-3} - \frac{1}{6}\int \frac{dx}{x+3} = \frac{1}{6}\ln|x-3| - \frac{1}{6}\ln|x+3| + C$$

$$= \frac{1}{6}\ln\left|\frac{x-3}{x+3}\right| + C$$

Compare this with a trigonometric substitution.

10.7.6. $I = \int \dfrac{(x^2+1)\, dx}{(x+3)^3}$

$$\frac{x^2+1}{(x+3)^3} = \frac{A}{(x+3)^3} + \frac{B}{(x+3)^2} + \frac{D}{x+3}$$

$$x^2 + 1 = A + B(x+3) + D(x^2+6x+9)$$

$$A = 10; \quad D = 1; \quad B = -6$$

$$\int \frac{(x^2+1)\, dx}{(x+3)^3} = 10\int \frac{dx}{(x+3)^3} - 6\int \frac{dx}{(x+3)^2} + \int \frac{dx}{x+3}$$

$$= \frac{-5}{(x+3)^2} + \frac{6}{x+3} + \ln|x+3| + C$$

Compare this with the substitution $x + 3 = z$:

$$\int \frac{(x^2 + 1)\, dx}{(x + 3)^3} = \int \frac{z^2 - 6z + 10}{z^3}\, dz = \int \left(\frac{1}{z} - \frac{6}{z^2} + \frac{10}{z^3}\right) dz$$

$$= \ln|z| + \frac{6}{z} - \frac{5}{z^2} + C = \ln|x + 3| + \frac{6}{x + 3}$$

$$- \frac{5}{(x + 3)^2} + C$$

10.7.7. $I = \displaystyle\int \frac{x^2 + 8}{(x - 1)(x + 2)^2}\, dx$

$$\frac{x^2 + 8}{(x - 1)(x + 2)^2} = \frac{A}{x - 1} + \frac{B}{(x + 2)^2} + \frac{C}{x + 2}$$

$x^2 + 8 = A(x^2 + 4x + 4) + B(x - 1) + C(x^2 + x - 2)$

$$\begin{cases} A + C = 1; \\ 4A + B + C = 0; \\ 4A - B - 2C = 8; \end{cases} \quad \begin{cases} A + C = 1 \\ 8A - C = 8 \end{cases}$$

$A = 1; \quad C = 0; \quad B = -4$

$$\int \frac{x^2 + 8}{(x - 1)(x + 2)^2}\, dx = \int \frac{dx}{x - 1} - 4\int \frac{dx}{(x + 2)^2}$$

$$= \ln|x - 1| + \frac{4}{x + 2} + C$$

10.8. Other substitutions. As shown in the examples below, many integrals can be simplified by the substitution $\tan \dfrac{x}{2} = z$, or, as we can see from the triangle, with this substitution

$$\sin \frac{x}{2} = \frac{z}{\sqrt{z^2 + 1}} \text{ and } \cos \frac{x}{2} = \frac{1}{\sqrt{z^2 + 1}}; \text{ and hence}$$

$$\sin x = \frac{2z}{1 + z^2}; \quad \cos x = \frac{1 - z^2}{1 + z^2}; \quad \tan x = \frac{2z}{1 - z^2};$$

$$dx = \frac{2\, dz}{1 + z^2} \text{ from } x = 2 \text{ Arc tan } z$$

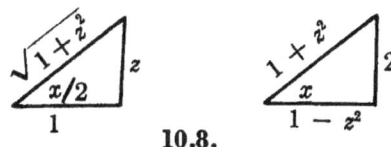

10.8.

Example:

$$\int \frac{dx}{1+\sin x + \cos x} = \int \frac{\frac{2dz}{1+z^2}}{1 + \frac{2z}{1+z^2} + \frac{1-z^2}{1+z^2}} = \int \frac{2dz}{2z+2}$$

$$= \ln|z+1| + C = \ln\left|\tan\frac{x}{2} + 1\right| + C$$

10.8.1. $I = \int \dfrac{dx}{4 + 2\cos 3x}$.   First let $3x = u$, then let $z = \tan\dfrac{u}{2}$.

$$I = \frac{1}{3}\int \frac{du}{4+2\cos u} = \frac{1}{3}\int \frac{\frac{2dz}{1+z^2}}{4 + 2\frac{1-z^2}{1+z^2}} = \frac{2}{3}\int \frac{dz}{6+2z^2}$$

$$= \frac{1}{3\sqrt{3}} \operatorname{Arc\,tan}\frac{z}{\sqrt{3}} + C = \frac{1}{3\sqrt{3}} \operatorname{Arc\,tan}\frac{1}{\sqrt{3}}\tan\frac{u}{2}$$

$$= \frac{1}{3\sqrt{3}} \operatorname{Arc\,tan}\left[\frac{1}{\sqrt{3}}\tan\frac{3x}{2}\right] + C$$

10.8.2. $I = \int \dfrac{dx}{12 + 5\tan x}$

Setting $\tan\dfrac{x}{2} = z$,

$$I = \int \frac{\frac{2dz}{z^2+1}}{12 + \frac{10z}{1-z^2}} = \int \frac{(z^2-1)\,dz}{(z^2+1)(6z^2-5z-6)}$$

$$= \int \frac{(z^2-1)\,dz}{(z^2+1)(3z+2)(2z-3)}$$

$$\frac{z^2-1}{(z^2+1)(3z+2)(2z-3)} = \frac{Az+B}{z^2+1} + \frac{C}{3z+2} + \frac{D}{2z-3}$$

$$z^2 - 1 = (Az+B)(6z^2 - 5z - 6) + C(z^2+1)(2z-3)$$
$$\quad + D(z^2+1)(3z+2)$$

Setting $z = -\frac{2}{3}$, $C = \dfrac{-5/9}{\dfrac{13}{9} \cdot \left(\dfrac{-13}{3}\right)} = \dfrac{15}{169}$

From $z = \frac{3}{2}$, $D = \dfrac{5/4}{\dfrac{13}{4} \cdot \dfrac{13}{2}} = \dfrac{10}{169}$

From constant terms, $-6B - 3C + 2D = -1$; $B = \frac{24}{169}$. Also from z^2 terms, $6A + 2C + 3D = 0$; $A = -\frac{10}{169}$.

$$I = \frac{1}{169} \int \frac{-10z + 24}{z^2 + 1}\, dz + \frac{15}{169} \int \frac{dz}{3z + 2} + \frac{10}{169} \int \frac{dz}{2z - 3}$$

$$= -\tfrac{5}{169} \ln |z^2 + 1| + \tfrac{24}{169} \operatorname{Arc\,tan} z + \tfrac{5}{169} \ln |3z + 2|$$

$$+ \tfrac{5}{169} \ln |2z - 3| + C$$

$$= -\frac{5}{169} \ln \sec^2 \frac{x}{2} + \frac{12x}{169} + \frac{5}{169} \ln \left| 3 \tan \frac{x}{2} + 2 \right|$$

$$+ \frac{5}{169} \ln \left| 2 \tan \frac{x}{2} - 3 \right| + C$$

10.8.3. $I = \displaystyle\int \frac{\cot x\, dx}{1 - \cos x}$

Let $z = \tan \dfrac{x}{2}$.

$$I = \int \frac{\dfrac{1 - z^2}{2z} \cdot \dfrac{2\, dz}{1 + z^2}}{1 - \dfrac{1 - z^2}{1 + z^2}} = \int \frac{(1 - z^2)\, dz}{z(2z^2)} = \frac{1}{2} \int \frac{dz}{z^3} - \frac{1}{2} \int \frac{dz}{z}$$

$$= -\frac{1}{4} \cdot \frac{1}{z^2} - \tfrac{1}{2} \ln |z| + C$$

$$= -\frac{1}{4} \cdot \frac{1}{\tan^2 \dfrac{x}{2}} - \tfrac{1}{2} \ln \left| \tan \frac{x}{2} \right| + C$$

Chapter 11

DEFINITE INTEGRALS

11.1. Definite integrals and substitution. If a substitution is used in evaluating a definite integral, there are two procedures that can be adopted. The limits of the integral can be changed in a corresponding manner, and the new limits used when the integration has been performed in terms of the new variable. The alternative is to substitute back when the integration has been performed and use the old limits. When the integration leads to inverse trigonometric functions, care should be taken in choosing the principal values to evaluate the integral.

Example:
$$\int_{-3}^{3} \frac{dx}{x^2 + 9} = \frac{1}{3} \text{Arc tan} \frac{x}{3}\Big|_{-3}^{3} = \frac{\pi}{12} - \left(-\frac{\pi}{12}\right) = \frac{\pi}{6}$$

Similarly, care must be exercised in changing limits when making a trigonometric substitution. One might otherwise be tempted to the following.

Example: $I = \int_{0}^{-\frac{1}{2}} \frac{dx}{\sqrt{1 - x^2}}$

Let $x = \sin \theta$; $dx = \cos \theta \, d\theta$. When $x = 0$, $\theta = 0$; when $x = -\frac{1}{2}$, $\theta = \frac{7\pi}{6}$.

$$I = \int_{0}^{7\pi/6} \frac{\cos \theta \, d\theta}{\sqrt{1 - \sin^2 \theta}} = \int_{0}^{7\pi/6} d\theta = \theta \Big|_{0}^{7\pi/6} = \frac{7\pi}{6}$$

On the other hand,
$$\int_{0}^{-\frac{1}{2}} \frac{dx}{\sqrt{1 - x^2}} = \text{Arc sin } x \Big|_{0}^{-\frac{1}{2}} = \text{Arc sin}(-\frac{1}{2}) = \frac{-\pi}{6},$$
since the principal value of Arc sin$(-\frac{1}{2}) = \frac{-\pi}{6}$.

This example illustrates the importance of principal values when choosing new limits under a trigonometric substitution. If we had stated $x = 0$, $\theta = 0$, $x = -\frac{1}{2}$, $\theta = \frac{-\pi}{6}$ above, our solution would have agreed with the second and correct solution.

Finally, attention must be paid to the branch of the function in the integrand.

Example: $I = \int_0^2 \sqrt{x+1}\, dx$

Let $x + 1 = z^2$, $dx = 2z\, dz$. If $x = 0$, $z = 1$; $x = 2$, $z = \sqrt{3}$.

$$I = \int_1^{\sqrt{3}} \sqrt{z^2} \cdot 2z\, dz = \int_1^{\sqrt{3}} 2z^2\, dz$$

We do not choose $z = -1$ and $-\sqrt{3}$, since we are using the positive root of z^2 in the integration. Of course, we could also write

$$\sqrt{x+1} = -z; \quad I = -\int_{-1}^{-\sqrt{3}} 2z^2\, dz$$

Notice, in changing a variable, the change affects three things: the integrand, limits, and differential.

11.1.1. $I = \int_1^9 \dfrac{dx}{3 + 5x}$

Let $3 + 5x = w$, $dx = \tfrac{1}{5} dw$. If $x = 1$, $w = 8$; $x = 9$, $w = 48$.

$$I = \frac{1}{5}\int_8^{48} \frac{dw}{w} = \frac{1}{5} \ln w \Big|_8^{48} = \frac{1}{5}(\ln 48 - \ln 8) = \frac{1}{5}\ln 6$$

11.1.2. $\int_{-4}^{4} \dfrac{dx}{16 + x^2} = \dfrac{1}{4} \operatorname{Arc\,tan} \dfrac{x}{4}\Big|_{-4}^{4}$

$$= \frac{1}{4}(\operatorname{Arc\,tan} 1 - \operatorname{Arc\,tan}(-1)) = \frac{1}{4}\left[\frac{\pi}{4} - \left(-\frac{\pi}{4}\right)\right] = \frac{\pi}{8}$$

11.1.3. $I = \int_0^{-5} \sqrt{25 - x^2}\, dx$

Let $x = 5 \sin \theta$, $dx = 5 \cos \theta\, d\theta$. If $x = 0$, $\theta = 0$, $x = -5$, $\theta = -\dfrac{\pi}{2}$.

$$I = \int_0^{-\pi/2} \sqrt{25 - 25 \sin^2 \theta} \cdot 5 \cos \theta\, d\theta = 25 \int_0^{-\pi/2} \cos^2 \theta\, d\theta$$

$$= 25 \int_0^{-\pi/2} \left(\frac{\cos 2\theta + 1}{2}\right) d\theta = \frac{25}{2}\left[\frac{1}{2}\sin 2\theta + \theta\right]_0^{-\pi/2} = -\frac{25\pi}{4}$$

11.1.4. $\displaystyle\int_{-\sqrt{2}}^{-2} \frac{dx}{x\sqrt{x^2-1}} = \text{Arc sec } x \Big|_{-\sqrt{2}}^{-2}$

$$= \text{Arc sec }(-2) - \text{Arc sec }(-\sqrt{2}) = \frac{2\pi}{3} - \frac{3\pi}{4} = -\frac{\pi}{12}$$

11.1.5. $I = \displaystyle\int_0^{2\pi/3} \frac{\tan x \, dx}{1 + \sin x}$

Let $\tan \dfrac{x}{2} = z$, $\sin x = \dfrac{2z}{1+z^2}$, $\tan x = \dfrac{2z}{1-z^2}$;

$dx = \dfrac{2dz}{1+z^2}$. If $x = 0$, $z = 0$; $x = \dfrac{2\pi}{3}$, $z = \sqrt{3}$.

$$I = \int_0^{\sqrt{3}} \frac{\dfrac{2z}{1-z^2} \cdot \dfrac{2dz}{1+z^2}}{1 + \dfrac{2z}{1+z^2}} = 4\int_0^{\sqrt{3}} \frac{z\,dz}{(1-z^2)(1+2z+z^2)}$$

$$= 4\int_0^{\sqrt{3}} \frac{z\,dz}{(1-z)(1+z)^3}$$

$$\frac{z}{(1-z)(1+z)^3} = \frac{A}{1-z} + \frac{B}{(1+z)^3} + \frac{C}{(1+z)^2} + \frac{D}{1+z};$$

$$A = \frac{1}{8}; \quad B = -\frac{1}{2}; \quad C = \frac{1}{4}; \quad D = \frac{1}{8}$$

$$I = \frac{1}{8}\int_0^{\sqrt{3}}\left(\frac{1}{1-z} + \frac{1}{1+z}\right)dz - \frac{1}{2}\int_0^{\sqrt{3}} \frac{1}{(1+z)^3}\,dz$$

$$+ \frac{1}{4}\int_0^{\sqrt{3}} \frac{1}{(1+z)^2}\,dz = \left(\frac{1}{8}\ln\left|\frac{1+z}{1-z}\right| + \frac{1}{4}\frac{1}{(1+z)^2} - \frac{1}{4}\frac{1}{1+z}\right)_0^{\sqrt{3}}$$

$$= \frac{1}{8}\ln\frac{\sqrt{3}+1}{\sqrt{3}-1} + \frac{1}{4}\left(\frac{1}{(1+\sqrt{3})^2} - \frac{1}{1+\sqrt{3}}\right)$$

11.2. Improper integrals. If the integrand has an infinite discontinuity in the range of integration at point $x = \xi$, we define our integral thus:

$$\int_a^b f(x)\,dx = \lim_{x_1 \to \xi^-} \int_a^{x_1} f(x)\,dx + \lim_{x_2 \to \xi^+} \int_{x_2}^b f(x)\,dx$$

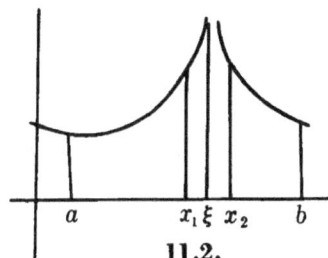

11.2.

If $\xi = b$ or $\xi = a$, the definition becomes

$$\int_a^b f(x)\,dx = \lim_{x_1 \to b^-} \int_a^{x_1} f(x)\,dx \quad \text{or} \quad \int_a^b f(x)\,dx = \lim_{x_2 \to a^+} \int_{x_2}^b f(x)\,dx.$$

Notice that x_1 and x_2 are independent of each other, else

$$\int_{-1}^1 \frac{dx}{x^3} = \lim_{x_1 \to 0}\left(\int_{-1}^{-x_1}\frac{dx}{x^3} + \int_{x_1}^1 \frac{dx}{x^3}\right) = \lim_{x_1 \to 0}\left(-\frac{1}{2x^2}\bigg|_{-1}^{-x_1} - \frac{1}{2x^2}\bigg|_{x_1}^1\right)$$

$$= \lim_{x_1 \to 0}\left(-\frac{1}{2x_1^2} + \frac{1}{2} - \frac{1}{2} + \frac{1}{2x_1^2}\right) = \lim_{x_1 \to 0} 0 = 0$$

However,

$$\int_{-1}^1 \frac{dx}{x^3} = \lim_{x_1 \to 0^-}\int_{-1}^{+x_1}\frac{dx}{x^3} + \lim_{x_2 \to 0^+}\int_{x_2}^1 \frac{dx}{x^3} = \lim_{x_1 \to 0^-} -\frac{1}{2x^2}\bigg|_{-1}^{x_1}$$

$$+ \lim_{x_2 \to 0^+} -\frac{1}{2x^2}\bigg|_{x_2}^1 = \lim_{x_1 \to 0^-}\left(-\frac{1}{2x_1^2} + \frac{1}{2}\right) + \lim_{x_2 \to 0^+}\left(-\frac{1}{2} + \frac{1}{2x_2^2}\right)$$

Neither limit exists and hence the integral does not exist.

11.2.1. $\int_0^1 \frac{dx}{\sqrt{1-x^2}} = \lim_{x_1 \to 1^-}\int_0^{x_1}\frac{dx}{\sqrt{1-x^2}}$

Here the discontinuity is at $x = 1$.

$$\int_0^1 \frac{dx}{\sqrt{1-x^2}} = \lim_{x_1 \to 1^-} \text{Arc sin } x\bigg|_0^{x_1} = \lim_{x_1 \to 1^-} \text{Arc sin } x_1 = \frac{\pi}{2}$$

11.2.2. $\int_0^1 \frac{x\,dx}{\sqrt{1-x^2}} = \lim_{x_1 \to 1}\int_0^{x_1}\frac{x\,dx}{\sqrt{1-x^2}} = \lim_{x_1 \to 1} -\sqrt{1-x^2}\bigg|_0^{x_1}$

$$= \lim_{x_1 \to 1}(-\sqrt{1-x_1^2} + 1) = 1$$

11.2.3. $\displaystyle\int_3^4 \frac{dx}{x-3} = \lim_{x_1 \to 3^+} \int_{x_1}^4 \frac{dx}{x-3} = \lim_{x_1 \to 3^+} \ln|x-3|\Big|_{x_1}^4$

$\displaystyle\qquad\qquad = \lim_{x_1 \to 3^+} (-\ln|x_1 - 3|)$

Neither limit nor integral exists.

1.2.4. $\displaystyle\int_0^1 \ln x\, dx = \lim_{x_1 \to 0^+} \int_{x_1}^1 \ln x\, dx = \lim_{x_1 \to 0^+}\left(x \ln x\Big|_{x_1}^1 - \int_{x_1}^1 dx\right)$

$\displaystyle\qquad = \lim_{x_1 \to 0^+}(-x_1 \ln x_1 + x_1 - 1) = \lim_{x_1 \to 0^+}(-x_1 \ln x_1) - 1$

But $\displaystyle\lim_{x_1 \to 0^+} \frac{\ln x_1}{\dfrac{1}{x_1}} = \lim_{x_1 \to 0^+} \frac{\dfrac{1}{x_1}}{\dfrac{-1}{x_1^2}} = 0,$ and $\displaystyle\int_0^1 \ln x\, dx = -1$

11.2.5. $\displaystyle\int_1^2 \frac{x\, dx}{x-1} = \lim_{x_1 \to 1^+} \int_{x_1}^2 \frac{x\, dx}{x-1}$

$\displaystyle\qquad = \lim_{x_1 \to 1^+}\left(\int_{x_1}^2 dx + \int_{x_1}^2 \frac{dx}{x-1}\right) = \lim_{x_1 \to 1^+}(x + \ln|x-1|)\Big|_{x_1}^2$

$\displaystyle\qquad = \lim_{x_1 \to 1^+}(2 - x_1 - \ln|x_1 - 1|)$

Neither limit nor integral exists.

11.2.6. $\displaystyle\int_1^2 \frac{dx}{x\sqrt{x^2-1}} = \lim_{x_1 \to 1^+} \int_{x_1}^2 \frac{dx}{x\sqrt{x^2-1}}$

$\displaystyle\qquad = \lim_{x_1 \to 1^+} \text{Arc sec } x\Big|_{x_1}^2 = \lim_{x_1 \to 1^+}\left(\frac{\pi}{6} - \text{Arc sec } x_1\right) = \frac{\pi}{6}$

11.2.7. $\displaystyle\int_1^2 \frac{dx}{(x-1)^{4/3}} = \lim_{x_1 \to 1^+} \int_{x_1}^2 \frac{dx}{(x-1)^{4/3}}$

$\displaystyle\qquad = \lim_{x_1 \to 1^+} \frac{-3}{(x-1)^{1/3}}\Big|_{x_1}^2 = \lim_{x_1 \to 1^+} \frac{3}{(x_1-1)^{1/3}} - 3$

Neither limit nor integral exists.

11.2.8. $\displaystyle\int_1^2 \frac{dx}{(x-1)^{1/3}} = \lim_{x_1 \to 1^+} \int_{x_1}^2 \frac{dx}{(x-1)^{1/3}}$

$\displaystyle\qquad = \lim_{x \to 1^+} \frac{3}{2}(x-1)^{2/3}\Big|_{x_1}^2 = \lim_{x_1 \to 1^+} -\frac{3}{2}(x_1-1)^{2/3} + \frac{3}{2} = \frac{3}{2}$

11.2.9. $\int_0^{\pi/2} \tan x \, dx = \lim_{x_1 \to \pi/2^-} \int_0^{x_1} \tan x \, dx = \lim_{x_1 \to \pi/2^-} -\ln \cos x \Big|_0^{x_1}$

$$= \lim_{x_1 \to \pi/2^-} (-\ln \cos x_1)$$

Neither limit nor integral exists.

11.3. Infinite limits $\int_a^{+\infty} f(x) \, dx$ is defined as $\lim_{A \to +\infty} \int_a^A f(x) \, dx$, $\int_{-\infty}^a f(x) \, dx$ is similarly defined as $\lim_{B \to -\infty} \int_B^a f(x) \, dx$, and

$$\int_{-\infty}^{+\infty} f(x) \, dx = \int_{-\infty}^a f(x) \, dx + \int_a^{+\infty} f(x) \, dx$$

$$= \lim_{B \to -\infty} \int_B^a f(x) \, dx + \lim_{A \to +\infty} \int_a^A f(x) \, dx$$

11.3.1. $\int_2^{+\infty} \frac{dx}{(x-1)^{4/3}} = \lim_{A \to +\infty} \int_2^A \frac{dx}{(x-1)^{4/3}}$

$$= \lim_{A \to +\infty} \frac{-3}{(x-1)^{1/3}} \Big|_2^A = \lim_{A \to +\infty} \frac{-3}{(A-1)^{1/3}} + 3 = 3$$

11.3.2. $\int_2^{+\infty} \frac{dx}{(x-1)^{1/3}} = \lim_{A \to +\infty} \int_2^A \frac{dx}{(x-1)^{1/3}}$

$$= \lim_{A \to +\infty} \frac{3}{2} (x-1)^{2/3} \Big|_2^A = \lim_{A \to +\infty} \frac{3}{2} (A-1)^{2/3} - \frac{3}{2}$$

Neither limit nor integral exists.

11.3.3. $\int_2^{+\infty} \frac{dx}{1-x^2} = \lim_{A \to +\infty} \int_2^A \frac{dx}{1-x^2}$

$$= \lim_{A \to +\infty} \frac{1}{2} \int_2^A \left(\frac{1}{1-x} + \frac{1}{1+x} \right) dx$$

$$= \frac{1}{2} \lim_{A \to +\infty} \left(-\ln|1-x| + \ln|1+x| \Big|_2^A \right)$$

$$= \frac{1}{2} \lim_{A \to +\infty} (-\ln|1-A| + \ln|1+A| - \ln 3)$$

$$= \frac{1}{2} \lim_{A \to +\infty} \left(\ln \left| \frac{1+A}{1-A} \right| - \ln 3 \right) = -\frac{1}{2} \ln 3$$

Here we used L'Hospital's rule.

11.3.4. $\int_{-\infty}^{+\infty} \frac{dx}{1+x^2} = \lim_{A \to +\infty} \int_{0}^{A} \frac{dx}{1+x^2} + \lim_{B \to -\infty} \int_{B}^{0} \frac{dx}{1+x^2}$

$= \lim_{A \to +\infty} \text{Arc tan } x \Big|_0^A + \lim_{B \to -\infty} \text{Arc tan } x \Big|_B^0$

$= \lim_{A \to +\infty} \text{Arc tan } A - \lim_{B \to -\infty} \text{Arc tan } B = \frac{\pi}{2} - \left(-\frac{\pi}{2}\right) = \pi$

11.3.5. $\int_{1}^{+\infty} \frac{dx}{x} = \lim_{A \to +\infty} \int_{1}^{A} \frac{dx}{x} = \lim_{A \to +\infty} \ln x \Big|_1^A = \lim_{A \to +\infty} \ln A$

Neither limit nor integral exists.

11.3.6. $\int_{-\infty}^{+\infty} \frac{dx}{e^x + e^{-x}} = \lim_{A \to +\infty} \int_{0}^{A} \frac{e^x\, dx}{e^{2x} + 1} + \lim_{B \to -\infty} \int_{B}^{0} \frac{e^x\, dx}{e^{2x} + 1}$

$= \lim_{A \to +\infty} \left(\text{Arc tan } e^A - \frac{\pi}{4}\right) + \lim_{B \to -\infty} \left(\frac{\pi}{4} - \text{Arc tan } e^B\right)$

$= \frac{\pi}{2} - \frac{\pi}{4} + \frac{\pi}{4} - 0 = \frac{\pi}{2}$

CHAPTER 12

GEOMETRIC APPLICATIONS OF THE DEFINITE INTEGRAL

12.1. Area in Rectangular Coordinates

12.1.1. Find the area inside the ellipse $\dfrac{x^2}{a^2} + \dfrac{y^2}{b^2} = 1$.

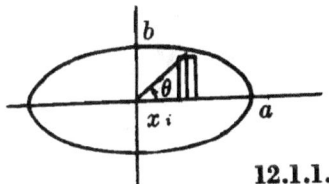

12.1.1.

Solution. Here $y = \pm b\sqrt{1 - \dfrac{x^2}{a^2}}$, where $y = +b\sqrt{1 - \dfrac{x^2}{a^2}}$ represents the branch above the x axis and $y = -b\sqrt{1 - \dfrac{x^2}{a^2}}$ represents the branch below the x-axis. The area above the x axis will be given by $b\int_{-a}^{a} \sqrt{1 - \dfrac{x^2}{a^2}}\, dx$ and the area below the x axis by $-b\int_{-a}^{a} \sqrt{1 - \dfrac{x^2}{a^2}}\, dx$. However, by symmetry about the x axis, the two may be combined and $A = 2b\int_{-a}^{a} \sqrt{1 - \dfrac{x^2}{a^2}}\, dx$. By symmetry about the y axis, $A = 4b\int_{0}^{a} \sqrt{1 - \dfrac{x^2}{a^2}}\, dx$. Substitute $x = a \sin\theta$; then $dx = a\cos\theta\, d\theta$; when $x = 0, \theta = 0$; $x = a, \theta = \dfrac{\pi}{2}$.

$$A = 4ba\int_0^{\pi/2} \sqrt{1 - \sin^2\theta}\,\cos\theta\, d\theta = 4ab\int_0^{\pi/2} \cos^2\theta\, d\theta$$

$$= 4ab\int_0^{\pi/2} \frac{\cos 2\theta + 1}{2}\, d\theta = 4ab\left(\frac{\sin 2\theta}{4} + \frac{\theta}{2}\right)\Big|_0^{\pi/2} = \pi ab$$

We could also have proceeded directly from the parametric equations of the ellipse $x = a\cos\theta$, $y = b\sin\theta$:

$$A = \int_{2\pi}^{0} y\,dx = -ab\int_{2\pi}^{0}\sin^2\theta\,d\theta$$

$$= -ab\int_{2\pi}^{0}\frac{1-\cos 2\theta}{2}\,d\theta = \pi ab$$

12.1.2. Find the area bounded by the x axis and one arch of the cycloid $x = a(\theta - \sin\theta)$, $y = a(1 - \cos\theta)$.

12.1.2.
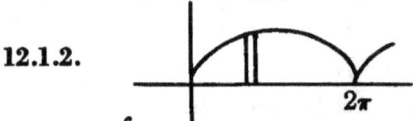

Solution. $A = \int y\,dx$. When θ goes from 0 to 2π, the point (x, y) traces out one arch of the cycloid.

$$A = a^2\int_0^{2\pi}(1-\cos\theta)(1-\cos\theta)\,d\theta$$

$$= a^2\int_0^{2\pi}(1 - 2\cos\theta + \cos^2\theta)\,d\theta$$

$$= a^2(\theta - 2\sin\theta)\Big|_0^{2\pi} + a^2\int_0^{2\pi}\left(\frac{\cos 2\theta + 1}{2}\right)d\theta$$

$$= 2\pi a^2 + a^2\left(\frac{\sin 2\theta}{4} + \frac{\theta}{2}\right)\Big|_0^{2\pi} = 3\pi a^2$$

12.1.3. Find the area in the first quadrant under the curve $y = e^{-x}$.

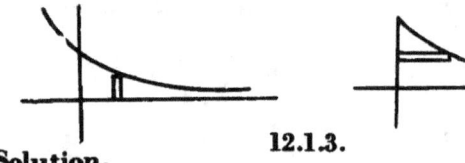

12.1.3.

Solution.

$$A = \int_0^{+\infty} y\,dx = \int_0^{+\infty} e^{-x}\,dx = \lim_{B\to+\infty}\int_0^{B} e^{-x}\,dx$$

$$= \lim_{B\to+\infty} -e^{-x}\Big|_0^{B} = \lim_{B\to+\infty}(-e^{-B} + 1) = 1$$

Notice also

$$A = \int_0^1 x \, dy = -\int_0^1 \ln y \, dy = \lim_{\xi \to 0} -\int_\xi^1 \ln y \, dy$$

$$= \lim_{\xi \to 0} (-y \ln y + y)\Big|_\xi^1 = \lim_{\xi \to 0} (\xi \ln \xi - \xi + 1) = 1$$

12.1.4. Find the area between $y = x^2$ and the witch $y = \dfrac{2}{x^2 + 1}$.

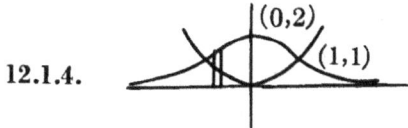

12.1.4.

Solution

$$A = \int_{-1}^1 (y_2 - y_1) \, dx = \int_{-1}^1 \left(\frac{2}{x^2 + 1} - x^2\right) dx$$

$$= \left(2 \operatorname{Arc tan} x - \frac{1}{3} x^3\right)\Big|_{-1}^1 = \pi - \frac{2}{3}$$

We could also have chosen our strips horizontally. As y varies from 0 to 1, the length of the strip is found from the two branches of $y = x^2$, $x = \sqrt{y}$, $x = -\sqrt{y}$. From 1 to 2 we use $y = \dfrac{2}{x^2 + 1}$, $x = \pm\sqrt{-1 + \dfrac{2}{y}}$.

$$A = \int_0^1 [\sqrt{y} - (-\sqrt{y})] \, dy$$

$$+ \int_1^2 \left[\sqrt{-1 + \frac{2}{y}} - \left(-\sqrt{-1 + \frac{2}{y}}\right)\right] dy$$

$$= 2 \int_0^1 y^{1/2} \, dy + 2 \int_1^2 \sqrt{-1 + \frac{2}{y}} \, dy$$

$$= \frac{4}{3} y^{3/2}\Big|_0^1 + 2 \int_1^2 \sqrt{\frac{2}{y} - 1} \, dy$$

Let $\dfrac{2}{y} - 1 = z^2$, $y = \dfrac{2}{z^2 + 1}$, $dy = \dfrac{-4z\,dz}{(z^2 + 1)^2}$.

$$I = \int_1^2 \sqrt{\dfrac{2}{y} - 1}\,dy = \int_1^0 \dfrac{-4z^2\,dz}{(z^2 + 1)^2}$$

$$= -4\int_1^0 \dfrac{dz}{z^2 + 1} + 4\int_1^0 \dfrac{dz}{(z^2 + 1)^2}$$

In the second integral, let $z = \tan\theta$.

$$I = 4 \times \text{Arc tan } 1 + 4\int_{\pi/4}^0 \dfrac{\sec^2\theta\,d\theta}{\sec^4\theta}$$

$$= \pi + 4\int_{\pi/4}^0 \left(\dfrac{\cos 2\theta + 1}{2}\right)d\theta$$

$$= \pi + 2\left(\dfrac{\sin 2\theta}{2} + \theta\right)\Big|_{\pi/4}^0 = \dfrac{\pi}{2} - 1$$

$$A = \pi - \dfrac{2}{3}$$

12.1.5. Find the area between the witch $y = \dfrac{8a^3}{x^2 + 4a^2}$ and the x axis.

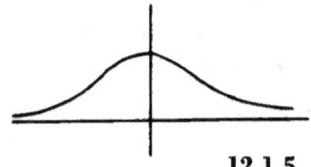

12.1.5.

Solution. Since the curve is symmetric with respect to the y axis, we may write

$$A = 2\int_0^{+\infty} y\,dx = 2\int_0^{+\infty} \dfrac{8a^3\,dx}{x^2 + 4a^2} = 2\lim_{B\to+\infty} 8a^3 \int_0^B \dfrac{dx}{x^2 + 4a^2}$$

$$= 8a^2 \lim_{B\to+\infty} \text{Arc tan}\dfrac{x}{2a}\Big|_0^B = 8a^2 \lim_{B\to+\infty} \text{Arc tan}\dfrac{B}{2a} = 4\pi a^2$$

Again, the integration could have been taken along the y axis, leading to an improper integral.

$$A = 2\int_0^{2a} \sqrt{-4a^2 + \dfrac{8a^3}{y}}\,dy = 2\lim_{\xi\to 0}\int_\xi^{2a} \sqrt{-4a^2 + \dfrac{8a^3}{y}}\,dy$$

12.1.6. Find the area bounded by the x axis, the y axis, and the curve $y = a \sin^4 \theta$, $x = a \cos^4 \theta$. (This is the parabola $x^{1/2} + y^{1/2} = a^{1/2}$.)

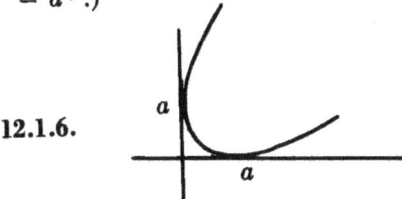

12.1.6.

Solution. The parametric equations trace out only that portion cut off by the axes. When $x = 0$, $\theta = \dfrac{\pi}{2}$; $x = a$, $\theta = 0$.

$$A = \int_0^a y\, dx = -4a^2 \int_{\pi/2}^0 \sin^4\theta \cos^3\theta \sin\theta\, d\theta$$

$$= -4a^2 \int_{\pi/2}^0 (\sin^5\theta - \sin^7\theta) \cos\theta\, d\theta$$

$$= -4a^2 \left(\dfrac{\sin^6\theta}{6} - \dfrac{\sin^8\theta}{8}\right)\bigg|_{\pi/2}^0 = -4a^2\left(-\dfrac{1}{6} + \dfrac{1}{8}\right) = \dfrac{a^2}{6}$$

12.1.7. Find the area between two intersections of $y = \sin x$ and $y = \cos x$.

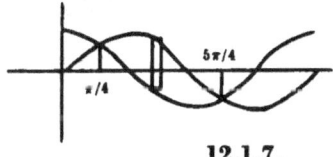

12.1.7.

Solution

$$A = \int_{\pi/4}^{5\pi/4} (y_2 - y_1)\, dx$$

$$= \int_{\pi/4}^{5\pi/4} (\sin x - \cos x)\, dx = \left(-\cos x - \sin x\right)\bigg|_{\pi/4}^{5\pi/4}$$

$$= \dfrac{\sqrt{2}}{2} + \dfrac{\sqrt{2}}{2} + \dfrac{\sqrt{2}}{2} + \dfrac{\sqrt{2}}{2} = 2\sqrt{2}$$

12.1.8. Find the area between -2 and 2 of $y = \ln(x^2 - 1)$.

Solution

$$\int_{-2}^{2} \ln(x^2 - 1)\, dx = x \ln(x^2 - 1) \Big|_{-2}^{2} - 2 \int_{-2}^{2} \frac{x^2\, dx}{x^2 - 1}$$

$$= \left[x \ln(x^2 - 1) - 2x - \ln|x - 1| + \ln|x + 1| \right]_{-2}^{2}$$

$$= 4 \ln 3 - 8 + 2 \ln 3 = 6 \ln 3 - 8$$

However, our curve does not exist for values of x between -1 and 1 so that our answer has no meaning.

12.2. Volumes of revolution. When a plane area is revolved about an axis, we can, by considering our volume as made up by infinitesimal disks, obtain the formula

$$V = \lim_{n \to +\infty} \pi \sum_{i=1}^{n} r_i^2 \Delta h_i = \pi \int_a^b r^2\, dh$$

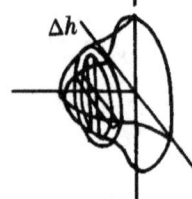
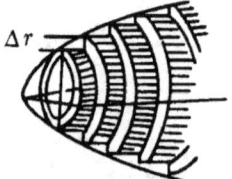

12.2.

We may also consider the volume as made up of cylindrical shells of infinitesimal thickness and obtain the formula

$$V = 2\pi \int_a^b rh\, dr$$

12.2.1. Find the volume when the area between $y = \dfrac{1}{x}$ and the x axis for $x \geqslant 1$ is rotated about the x axis.

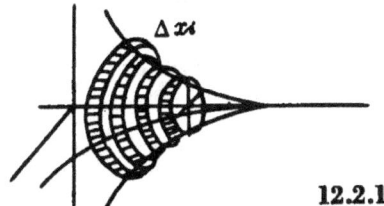
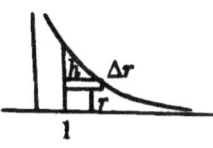

12.2.1.

Solution. Choosing the disk generated by the element of area shown, $r = y$ and $dh = dx$.

$$V = \pi \int_1^\infty \frac{dx}{x^2} = \pi\left(-\frac{1}{x}\right)\Big|_1^\infty = \pi$$

This volume could also have been obtained using shells, $h = x - 1, r = y$.

$$V = 2\pi \int_0^1 y\left(\frac{1}{y} - 1\right) dy = \pi$$

Notice that the area does not exist under this curve since $\int_1^\infty \frac{dx}{x}$ is not defined.

12.2.2. Find the volume obtained by rotating the circle $(x - a)^2 + y^2 = c^2$ about the y axis.

Solution. Here we assume $a > c$ and use the cylindrical shells generated by the elements of area shown.

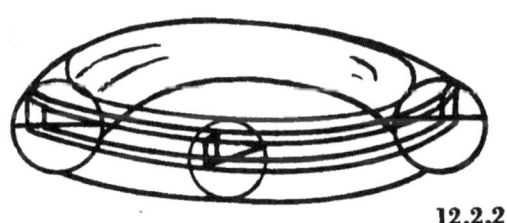

12.2.2.

$$V = 2\pi \int_{a-c}^{a+c} rh\, dr = 2\pi \int_{a-c}^{a+c} 2x \sqrt{c^2 - (x - a)^2}\, dx$$

Let $x - a = c \sin \theta$; $dx = c \cos \theta\, d\theta$; $x = a - c$; $\theta = -\frac{\pi}{2}$;

$$x = a + c, \theta = \frac{\pi}{2}.$$

$$V = 4\pi \int_{-\pi/2}^{\pi/2} (a + c \sin \theta) \sqrt{c^2 - c^2 \sin^2 \theta} \, c \cos \theta \, d\theta$$

$$= 4\pi c^2 \int_{-\pi/2}^{\pi/2} (a + c \sin \theta) \cos^2 \theta \, d\theta$$

$$= \frac{-4\pi c^3 \cos^3 \theta}{3} \bigg|_{-\pi/2}^{\pi/2} + 4\pi ac^2 \int_{-\pi/2}^{\pi/2} \frac{1 + \cos 2\theta}{2} \, d\theta$$

$$= 4\pi ac^2 \left(\frac{\theta}{2} + \frac{1}{4} \sin 2\theta \right) \bigg|_{-\pi/2}^{\pi/2} = 2\pi^2 ac^2$$

This problem could have been worked in the following way: The volume generated by the element of area shown revolving about the y axis is

$$\Delta V = \pi x_2^2 \, \Delta y - \pi x_1^2 \, \Delta y; \qquad x_2 = a + \sqrt{c^2 - y^2};$$

$$x_1 = a - \sqrt{c^2 - y^2};$$

$$= \pi (a^2 + 2a \sqrt{c^2 - y^2} + c^2 - y^2 - a^2 + 2a \sqrt{c^2 - y^2}$$

$$- (c^2 - y^2)) \Delta y;$$

$$V = \pi \int_{-c}^{c} 4a \sqrt{c^2 - y^2} \, dy$$

12.2.3. Find the volume generated by revolving about the x axis one arch of the cycloid $y = a(1 - \cos \theta)$, $x = a(\theta - \sin \theta)$.

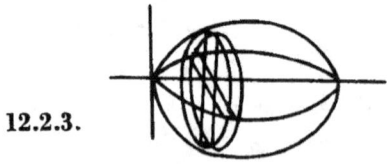

12.2.3.

Solution. Here we use the disk generated by the element of area shown: $dx = a(1 - \cos\theta)\,d\theta$.

$$V = \pi\int y^2\,dx = \pi a^3 \int_0^{2\pi}(1 - \cos\theta)^3\,d\theta =$$

$$\pi a^3 \int_0^{2\pi}\left[1 - 3\cos\theta + \frac{3}{2}(\cos 2\theta + 1) - \cos\theta + \cos\theta\sin^2\theta\right]d\theta$$

$$= \pi a^3\left(\frac{5}{2}\theta - 4\sin\theta + \frac{3}{4}\sin 2\theta + \frac{1}{3}\sin^3\theta\right)\Big|_0^{2\pi} = 5\pi^2 a^3$$

12.2.4. Find the volume when the witch $y = \dfrac{8a^3}{x^2 + 4a^2}$ is revolved about the x axis.

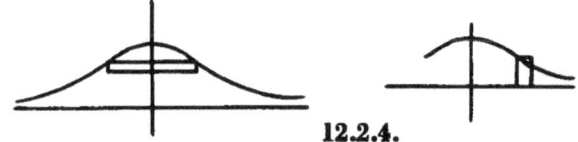

12.2.4.

Solution. Choosing the ring-shaped element of volume generated by the element of area shown, $\gamma = y$, $h = 2x$.

$$V = 4\pi\int_0^{2a} y\sqrt{-4a^2 + \frac{8a^3}{y}}\,dy$$

The substitution $-4a^2 + \dfrac{8a^3}{y} = z^2$ changes this to the integral used below. Choosing our element as the volume of the disk generated by the element of area $y\,dx$,

$$V = \pi\int_{-\infty}^{\infty}\left(\frac{8a^3}{x^2 + 4a^2}\right)^2 dx = 2\pi\int_0^{\infty}\left(\frac{8a^3}{x^2 + 4a^2}\right)^2 dx$$

Let $x = 2a\tan\theta$. When $x = 0$, $\theta = 0$; when $x = \infty$, $\theta = \dfrac{\pi}{2}$.

$$V = 4a\pi\int_0^{\pi/2}\left(\frac{2a}{1 + \tan^2\theta}\right)^2 \sec^2\theta\,d\theta$$

$$= 16a^3\pi\int_0^{\pi/2}\cos^2\theta\,d\theta = 16\pi a^3\left(\frac{\sin 2\theta}{4} + \frac{\theta}{2}\right)\Big|_0^{\pi/2} = 4\pi^2 a^3$$

12.2.5. The ellipse $x = a \cos \theta$, $y = b \sin \theta$ is rotated about the x axis; find the volume generated.

Solution. Using the disks generated by the elements of area shown,

$$V = \pi \int_{-a}^{a} y^2 \, dx = \pi b^2 \int_{\pi}^{0} \sin^2 \theta (-a \sin \theta) \, d\theta$$

$$= -\pi a b^2 \int_{\pi}^{0} (1 - \cos^2 \theta) \sin \theta \, d\theta$$

$$= \pi a b^2 \left(\cos \theta - \frac{\cos^3 \theta}{3} \right) \Big|_{\pi}^{0} = \frac{4\pi a b^2}{3}$$

See **4.4.8** for another solution to this problem.

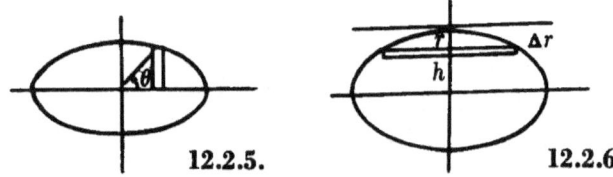

12.2.5.   12.2.6.

12.2.6. Find the volume when the ellipse $\dfrac{x^2}{a^2} + \dfrac{y^2}{b^2} = 1$ is revolved about the line $y = b$.

Solution. Here we can use the shell-shaped volume generated by the element of area shown: $h = 2x$, $r = b - y$.

$$V = 2\pi \int_{b}^{-b} 2x(b-y)(-dy) = -4\pi \int_{b}^{-b} (b-y) a \sqrt{1 - \frac{y^2}{b^2}} \, dy$$

$$= -4\pi a b \int_{b}^{-b} \sqrt{1 - \frac{y^2}{b^2}} \, dy - \frac{4\pi a b^2}{3} \left(1 - \frac{y^2}{b^2} \right)^{3/2} \Big|_{b}^{-b}$$

Let $y = b \sin \theta$. Then

$$V = -4\pi a b^2 \int_{\pi/2}^{-\pi/2} \cos^2 \theta \, d\theta$$

$$= -2\pi a b^2 \left(\frac{\sin 2\theta}{2} + \theta \right) \Big|_{+\pi/2}^{-\pi/2} = 2\pi^2 a b^2$$

12.2.7. Find the volume generated when the area bounded by x axis and $y = \sin x$ from 0 to π is revolved about $x = \pi$.

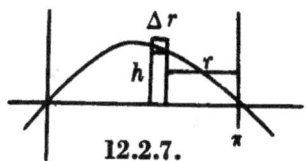

12.2.7.

Solution. Here we shall use the shells generated by the element of area shown: $h = y = \sin x; r = \pi - x$.

$$V = 2\pi \int rh \, dr = -2\pi \int_\pi^0 (\pi - x) \sin x \, dx$$

$$= +2\pi^2 \cos x \Big|_\pi^0 + 2\pi \int_\pi^0 x \sin x \, dx$$

$$= 4\pi^2 + 2\pi(-x \cos x + \sin x) \Big|_\pi^0 = 4\pi^2 - 2\pi^2 = 2\pi^2$$

12.3. Length of arc. Length of arc may be found from any of the three forms:

$$S = \int_a^b \sqrt{1 + \left(\frac{dy}{dx}\right)^2} \, dx, \qquad S = \int_c^d \sqrt{\left(\frac{dx}{dy}\right)^2 + 1} \, dy,$$

$$S = \int_e^f \sqrt{\left(\frac{dx}{dt}\right)^2 + \left(\frac{dx}{dt}\right)^2} \, dt$$

12.3.1. Find the length of one arch of the cycloid $y = a(1 - \cos \theta)$, $x = a(\theta - \sin \theta)$.

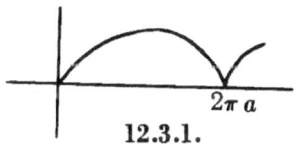

12.3.1.

Solution

$$\frac{dx}{d\theta} = a(1 - \cos \theta), \qquad \frac{dy}{d\theta} = a \sin \theta$$

$$S = \int_0^{2\pi} \sqrt{a^2(1 - \cos \theta)^2 + a^2 \sin^2 \theta} \, d\theta$$

$$= a \int_0^{2\pi} \sqrt{2 - 2\cos \theta} \, d\theta = 2a \int_0^{2\pi} \sin \frac{\theta}{2} \, d\theta$$

$$= -4a \cos \frac{\theta}{2} \Big|_0^{2\pi} = 8a$$

Notice that, if we wanted to find the length of two arches, i.e., limits 0 and 4π, we should have to be careful to give in $\frac{\theta}{2}$ a negative sign between 2π and 4π. In this way we make $\sqrt{\left(\frac{dx}{d\theta}\right)^2 + \left(\frac{dy}{d\theta}\right)^2}$ positive over the range of integration.

$$\int_0^{4\pi} \sqrt{2 - 2\cos\theta}\, d\theta = 2\int_0^{2\pi} \sin\frac{\theta}{2}\, d\theta + 2\int_{2\pi}^{4\pi} \left(-\sin\frac{\theta}{2}\right) d\theta$$

12.3.2. Find the length of $y = x^2$ from $x = 0$ to $x = 1$.

12.3.2.

Solution

$$S = \int_0^1 \sqrt{1 + y'^2}\, dx = \int_0^1 \sqrt{1 + 4x^2}\, dx$$

$$x = \frac{1}{2}\tan\theta$$

$$S = \frac{1}{2}\int_0^{\text{arc tan 2}} \sqrt{1 + \tan^2\theta}\, \sec^2\theta\, d\theta$$

$$= \frac{1}{2}\int_0^{\text{arc tan 2}} \sec^3\theta\, d\theta \qquad \text{Arc tan } 2 = \text{Arc sec } \sqrt{5}$$

$$= \left(\frac{1}{4}\sec\theta\tan\theta + \frac{1}{4}\ln|\sec\theta + \tan\theta|\right)\Big|_0^{\text{arc tan 2}}$$

$$= \frac{1}{4}\sqrt{5}\cdot 2 + \frac{1}{4}\ln|\sqrt{5} + 2| = \frac{\sqrt{5}}{2} + \frac{1}{4}\ln|\sqrt{5} + 2|$$

12.3.3. Find the length of the curve $y = \ln\sec x$ from $x = 0$ to $x = \frac{\pi}{6}$.

Solution

$$y' = \tan x$$

$$S = \int_0^{\pi/6} \sqrt{1 + \tan^2 x}\, dx = \int_0^{\pi/6} \sec x\, dx$$

$$= \ln|\sec x + \tan x|\Big|_0^{\pi/6} = \ln\left|\frac{2\sqrt{3}}{3} + \frac{\sqrt{3}}{3}\right| = \ln\sqrt{3}$$

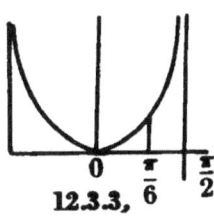
12.3.3,

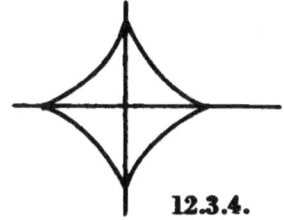
12.3.4.

12.3.4. Find the entire length of the curve $x^{2/3} + y^{2/3} = a^{2/3}$ using the parametric equations $x = a\cos^3\theta$, $y = a\sin^3\theta$.

Solution

$$\frac{dx}{d\theta} = -3a\cos^2\theta\sin\theta; \qquad \frac{dy}{d\theta} = 3a\sin^2\theta\cos\theta$$

$$S = 4\int_0^{\pi/2} \sqrt{9a^2\cos^4\theta\sin^2\theta + 9a^2\sin^4\theta\cos^2\theta}\, d\theta$$

$$= 12a\int_0^{\pi/2} \sin\theta\cos\theta\, d\theta = 6a\sin^2\theta\Big|_0^{\pi/2} = 6a$$

Here we use the symmetry of the figure in order to integrate from 0 to $\frac{\pi}{2}$. We could integrate from 0 to 2π, but we would have to consider the signs of $\sin\theta$ and $\cos\theta$ in all four quadrants and in each case to replace the positive root of $\sin^2\cos^2\theta$ by $|\sin\theta\cos\theta|$. In problems of arc length, the student must always remember the tacit assumption that we have taken the positive root of $\sqrt{1 + y'^2}\, dx$ everywhere.

12.3.5. Find the length of the curve $x^{1/2} + y^{1/2} = a^{1/2}$ cut off by the coordinate axes.

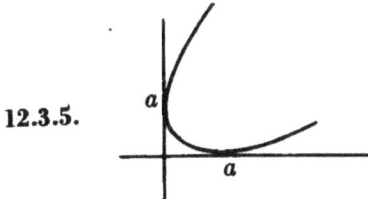
12.3.5.

Solution

$$y' = -\frac{a^{1/2} - x^{1/2}}{x^{1/2}}; \qquad S = \int_0^a \sqrt{1 + \left(\frac{a^{1/2} - x^{1/2}}{x^{1/2}}\right)^2}\, dx$$

Let $x = a\cos^4\theta$.

$$S = -\int_{+\pi/2}^{0} \sqrt{1 + \frac{a\sin^4\theta}{a\cos^4\theta}} \cdot 4a\cos^3\theta \sin\theta\, d\theta$$

$$= -4a\int_{\pi/2}^{0} \sqrt{\cos^4\theta + \sin^4\theta}\, \cos\theta \sin\theta\, d\theta$$

$$= -2a\int_{\pi/2}^{0} \sqrt{\left(\frac{1+\cos 2\theta}{2}\right)^2 + \left(\frac{1-\cos 2\theta}{2}\right)^2}\, \sin 2\theta\, d\theta$$

$$= -a\int_{\pi/2}^{0} \sqrt{2 + 2\cos^2 2\theta}\, \sin 2\theta\, d\theta$$

Let $\cos 2\theta = \tan z$; $-2\sin 2\theta\, d\theta = \sec^2 z\, dz$; $\theta = 0, z = \dfrac{\pi}{4}$;

$$\theta = \frac{\pi}{2}, z = -\frac{\pi}{4}.$$

$$= \frac{\sqrt{2}}{2} a \int_{-\pi/4}^{\pi/4} \sqrt{1 + \tan^2 z}\, \sec^2 z\, dz = \frac{\sqrt{2}}{2} a \int_{-\pi/4}^{\pi/4} \sec^3 z\, dz$$

$$= \frac{\sqrt{2}a}{4} (\sec z \tan z + \ln |\sec z + \tan z|) \Big|_{-\pi/4}^{\pi/4}$$

$$= \frac{\sqrt{2}a}{4} \left(\ln \left| \frac{\sqrt{2}+1}{\sqrt{2}-1} \right| + 2\sqrt{2} \right)$$

$$= \frac{\sqrt{2}a}{4} (\ln [\sqrt{2}+1]^2 + 2\sqrt{2})$$

$$= \frac{\sqrt{2}a}{2} (\ln [\sqrt{2}+1] + \sqrt{2})$$

Note that we chose $x^{1/2}$ as the positive root, else we would be integrating along the upper branch. Our first substitution $x = a\cos^4\theta$ is equivalent to using the parametric form $x = a\cos^4\theta$, $y = a\sin^4\theta$.

2.4. Area of a surface of revolution. The area of a surface generated by revolution about the x axis is $S = 2\pi \int_a^b y\, ds$, or a more general form is $2\pi \int_a^b r\, ds$, where r is the radius of revolution and ds is the differential of arc length.

12.4.1. One arch of the cycloid $x = a(\theta - \sin\theta)$, $y = a(1 - \cos\theta)$ is revolved about the x axis, find the area generated

Solution

$$\frac{dx}{d\theta} = (1 - \cos\theta); \qquad \frac{dy}{d\theta} = a\sin\theta$$

$$ds = a\sqrt{(1-\cos\theta)^2 + \sin^2\theta}\, d\theta = a\sqrt{2 - 2\cos\theta}\, d\theta$$

$$S = 2\pi \int y\, ds = 2\pi \int_0^{2\pi} a(1 - \cos\theta)a\sqrt{2 - 2\cos\theta}\, d\theta$$

$$= 2\sqrt{2}\pi a^2 \int_0^{2\pi} (1 - \cos\theta)^{3/2}\, d\theta = 8\pi a^2 \int_0^{2\pi} \sin^3\frac{\theta}{2}\, d\theta$$

$$= 8\pi a^2 \int_0^{2\pi} \left(1 - \cos^2\frac{\theta}{2}\right) \sin\frac{\theta}{2}\, d\theta$$

$$= 8\pi a^2 \left(-2\cos\frac{\theta}{2} + \frac{2}{3}\cos^3\frac{\theta}{2}\right)\bigg|_0^{2\pi} = \frac{64\pi a^2}{3}$$

12.4.1.

12.4.2. Find the area when $\dfrac{x^2}{a^2} + \dfrac{y^2}{b^2} = 1$ is rotated around the y axis. Use the parametric form $x = a\cos\theta$, $y = b\sin\theta$.

Solution

$$\frac{dx}{d\theta} = -a\sin\theta; \qquad \frac{dy}{d\theta} = b\cos\theta;$$

$$ds = \sqrt{a^2\sin^2\theta + b^2\cos^2\theta}\, d\theta; \qquad r = x = a\cos\theta$$

$$S = 2\pi \int_{-\pi/2}^{\pi/2} a\cos\theta \sqrt{(a^2 - b^2)\sin^2\theta + b^2}\, d\theta$$

Let $\sqrt{a^2 - b^2}\sin\theta = b\tan z$; $\quad \sqrt{a^2 - b^2}\cos\theta\, d\theta = b\sec^2 z\, dz$. When $\theta = -\dfrac{\pi}{2}$; $\quad z = \text{Arc}\tan - \dfrac{\sqrt{a^2 - b^2}}{b}$; $\theta = \dfrac{\pi}{2}$;

$z = \text{Arc}\tan \dfrac{\sqrt{a^2 - b^2}}{b}$

$$= \frac{2\pi ab}{\sqrt{a^2 - b^2}} \int_{\text{Arc tan}-\sqrt{a^2-b^2}/b}^{\text{Arc tan}\sqrt{a^2-b^2}/b} \sec^2 z \sqrt{b^2 \tan^2 z + b^2} \, dz$$

$$= \frac{2\pi ab^2}{\sqrt{a^2 - b^2}} \int_{\text{Arc tan}-\sqrt{a^2-b^2}/b}^{\text{Arc tan}\sqrt{a^2-b^2}/b} \sec^3 z \, dz$$

$$\frac{2\pi ab^2}{\sqrt{a^2 - b^2}} \left(\frac{1}{2} \sec z \tan z + \frac{1}{2} \ln |\sec z + \tan z| \right)_{\text{Arc tan}-\sqrt{a^2-b^2}/b}^{\text{Arc tan}\sqrt{a^2-b^2}/b}$$

From the triangles Arc tan $\pm \dfrac{\sqrt{a^2 - b^2}}{b} = $ Arc sec $\dfrac{a}{b}$,

12.4.2.

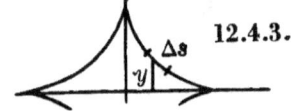

$$= \frac{\pi ab^2}{\sqrt{a^2 + b^2}} \left(\frac{2a}{b} \cdot \frac{\sqrt{a^2 - b^2}}{b} + \ln \left| \frac{\frac{a}{b} + \frac{\sqrt{a^2 - b^2}}{b}}{\frac{a}{b} - \frac{\sqrt{a^2 - b^2}}{b}} \right| \right)$$

$$= 2\pi a^2 + \frac{\pi ab^2}{\sqrt{a^2 - b^2}} \ln \left| \frac{a + \sqrt{a^2 - b^2}}{a - \sqrt{a^2 - b^2}} \right|$$

12.4.3. Find the area generated when $x^{2/3} + y^{2/3} = a^{2/3}$ is revolved about the x axis.

Solution

$$y' = -\frac{y^{1/3}}{x^{1/3}}; \quad 1 + y'^2 = 1 + \frac{y^{2/3}}{x^{2/3}} = \frac{x^{2/3} + y^{2/3}}{x^{2/3}} = \frac{a^{2/3}}{x^{2/3}}$$

$$S = 4\pi \int_0^a y \sqrt{1 + y'^2} \, dx = 4\pi \int_0^a (a^{2/3} - x^{2/3})^{3/2} \cdot \frac{a^{1/3}}{x^{1/3}} \, dx$$

$$= -\frac{12\pi a^{1/3}}{5} (a^{2/3} - x^{2/3})^{5/2} \Big|_0^a = \frac{12\pi a^2}{5}$$

Here again we chose to take advantage of symmetry and take twice the integral from 0 to a; otherwise, since $x^{-1/3}$ is negative in the second quadrant, we would have to evaluate in each quadrant separately and add absolute values.

12.4.3.

12.4.4.

12.4.4. Revolve $y = \dfrac{a}{2}(e^{x/a} + e^{-x/a})$ about the x axis from $x = -a$ to $x = a$ and find the area generated.

Solution

$$y' = \frac{1}{2}(e^{x/a} - e^{-x/a}); \quad 1 + y'^2 = 1 + \frac{e^{2x/a}}{4} - \frac{2}{4} + \frac{e^{-2x/a}}{4}$$

$$= \left[\frac{1}{2}(e^{x/a} + e^{-x/a})\right]^2$$

$$S = 2\pi \int_{-a}^{a} \frac{a}{2}(e^{x/a} + e^{-x/a}) \frac{1}{2}(e^{x/a} + e^{-x/a}) \, dx$$

$$= \frac{\pi a}{2} \int_{-a}^{a} (e^{2x/a} + 2 + e^{-2x/a}) \, dx$$

$$= \frac{\pi a}{2} \left(\frac{a}{2} e^{2x/a} + 2x - \frac{a}{2} e^{-2x/a}\right)\bigg|_{-a}^{a}$$

$$= \frac{\pi a}{2}(ae^2 - ae^{-2} + 4a) = \frac{\pi a^2}{2}(e^2 - e^{-2} + 4$$

12.4.5. Find the area generated when the arc of the parabola $y^2 = x$ between $y = 1$ and -1 is revolved about the line $x = $

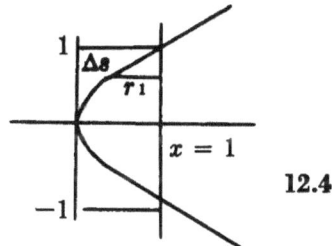

12.4.5

Solution

$$r = 1 - x = 1 - y^2; \quad \frac{dx}{dy} = 2y$$

$$S = 2\pi \int_{-1}^{1} (1 - y^2)\sqrt{1 + 4y^2} \, dy$$

Let $y = \frac{1}{2}\tan\theta$.

$$S = 2\pi \int_{\text{Arc tan}-2}^{\text{Arc tan}2} \left(1 - \frac{1}{4}\tan^2\theta\right)\sqrt{1+\tan^2\theta}\,\frac{1}{2}\sec^2\theta\,d\theta$$

$$= \pi \int_{\text{Arc tan}-2}^{\text{Arc tan}2} \left(1 - \frac{1}{4}\tan^2\theta\right)\sec^3\theta\,d\theta$$

$$= \pi \int_{\text{Arc tan}-2}^{\text{Arc tan}2} \left(\frac{5}{4}\sec^3\theta - \frac{1}{4}\sec^5\theta\right)d\theta$$

Using the results of 10.3.7,

$$\int \sec^5\theta\,d\theta = \frac{\sec^3\theta\tan\theta}{4} + \frac{3}{4}\int \sec^3\theta\,d\theta$$

$$= \pi\left(\frac{\sec^3\theta\tan\theta}{-16}\bigg|_{\text{Arc tan}-2}^{\text{Arc tan}2} + \frac{17}{16}\int_{\text{Arc tan}-2}^{\text{Arc tan}2}\sec^3\theta\,d\theta\right)$$

$$= \pi\left(\frac{\sec^3\theta\tan\theta}{-16} + \frac{17}{32}\sec\theta\tan\theta + \frac{17}{32}\ln|\sec\theta + \tan\theta|\right)\bigg|_{\text{Arc tan}-2}^{\text{Arc tan}2}$$

From our triangle, $\sec(\text{Arc tan}\pm 2) = \sqrt{5}$.

$$S = \pi\left(\frac{-10\sqrt{5}}{16} + \frac{17\sqrt{5}}{16} + \frac{17}{32}\ln|\sqrt{5}+2|\right.$$

$$\left. - \frac{10\sqrt{5}}{16} + \frac{17\sqrt{5}}{16} - \frac{17}{32}\ln|\sqrt{5}-2|\right)$$

$$= \frac{17\pi}{32}\ln\frac{\sqrt{5}+2}{\sqrt{5}-2} + \frac{7\sqrt{5}\pi}{8}$$

12.4.6. One arch of $y = \cos x$ is revolved about the x axis; find the area generated.
Solution

$$S = 2\pi \int_{-\pi/2}^{\pi/2} y\sqrt{1+(y')^2}\,dx = 2\pi \int_{-\pi/2}^{\pi/2} \cos x\sqrt{1+\sin^2 x}\,dx$$

Let $\sin x = \tan \theta$; $\cos x\, dx = \sec^2 \theta\, d\theta$; $x = -\frac{\pi}{2}$, $\theta = -\frac{\pi}{4}$;
$$y = \frac{\pi}{2}; \quad \theta = \frac{\pi}{4}.$$

12.4.6.

$$S = 2\pi \int_{-\pi/4}^{\pi/4} \sqrt{1 + \tan^2 \theta}\, \sec^2 \theta\, d\theta = \int_{-\pi/4}^{\pi/4} \sec^3 \theta\, d\theta$$

$$= \pi(\sec\theta \tan\theta + \ln|\sec\theta + \tan\theta|)\Big|_{-\pi/4}^{\pi/4}$$

$$= \pi\left(2\sqrt{2} + \ln\frac{\sqrt{2}+1}{\sqrt{2}-1}\right) = 2\pi\sqrt{2} + 2\pi \ln(\sqrt{2}+1)$$

12.5. Miscellaneous volumes and areas. In this section are found volumes of solids having various cross sectional areas.

12.5.1. Find the volume of the ellipsoid $\dfrac{x^2}{a^2} + \dfrac{y^2}{b^2} + \dfrac{z^2}{c^2} = 1$.

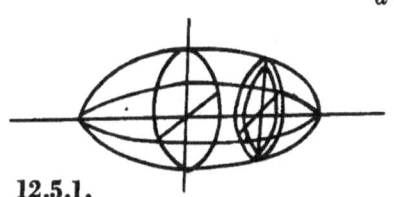

12.5.1.

Solution. Take as an element of volume a disk perpendicular to the x axis at a distance $x = k$. Then

$$\frac{y^2}{b^2} + \frac{z^2}{c^2} = 1 - \frac{k^2}{a^2}, \quad \text{or} \quad \frac{y^2}{b^2\left(1-\dfrac{k^2}{a^2}\right)} + \frac{z^2}{c^2\left(1-\dfrac{k^2}{a^2}\right)} = 1,$$

is the equation satisfied by a cross section at distance k, and using the results of **12.1.1**, we have the cross-sectional area as π times the product of the semiaxes, and the elementary volume

$$\Delta V = \pi \sqrt{b^2\left(1-\frac{k^2}{a^2}\right)} \sqrt{c^2\left(1-\frac{k^2}{a^2}\right)}\, \Delta x = \pi bc\left(1-\frac{k^2}{a^2}\right)\Delta x$$

Using the fundamental theorem,
$$V = \pi bc \int_{-a}^{a}\left(1 - \frac{x^2}{a^2}\right) dx = \pi bc \left(x - \frac{x^3}{3a^2}\right)\Big|_{-a}^{a} = \frac{4}{3}\pi abc$$

12.5.2. Find the volume cut from the cylinder $y^2 + z^2 = a^2$ by the cylinder $x^2 + z^2 = a^2$; that is, find the volume common to the two cylinders.

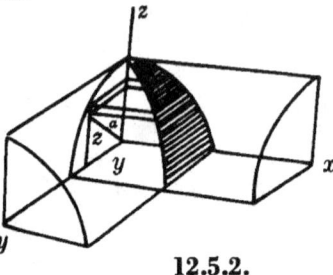

12.5.2.

In the figure is shown that portion of this volume lying in the first octant. We choose for our element of volume the rectangular parallelepiped of height Δz whose bases are parallel to the xy plane.

$$\Delta V = xy\Delta z \qquad \Delta V = \sqrt{a^2 - z^2}\sqrt{a^2 - z^2}\,\Delta z$$

$$V = 8\int_0^a (a^2 - z^2)\,dz = 8\left(a^2 z - \frac{z^3}{3}\right)\Big|_0^a = \frac{16}{3}a^3$$

12.5.3. Find the area cut from the cylinder $x^2 + z^2 = a^2$ by the cylinder $y^2 + z^2 = a^2$.

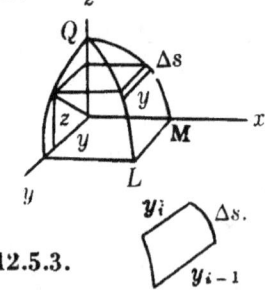

12.5.3.

Solution. An element of this area in the first octant is shown. The area of this element is $y\,\Delta s$ where Δs is an element of arc length on the circle $x^2 + z^2 = a^2$; $y = 0$. Notice that we do not choose Δs on the arc QL, since the sides of our elementary area are not normal to this arc. Our elementary area is not a rectangle but a cylinder whose area is $y\,\Delta s$. Using our fundamental theorem,

$$A = 8\int_0^a y\,ds;$$

$$ds = \sqrt{1+\left(\frac{dx}{dz}\right)^2}\,dz = \sqrt{1+\frac{z^2}{x^2}}\,dz = \frac{a\,dz}{\sqrt{a^2-z^2}}$$

$$A = 8\int_0^a \sqrt{a^2-z^2}\,\frac{a\,dz}{\sqrt{a^2-z^2}} = 8a^2$$

The total area cut out of the two cylinders is $16\,a^2$.

12.5.4. Find the volume above the xy plane and bounded by the cylinder $x^2 + y^2 = a^2$ and the plane $x = z$.

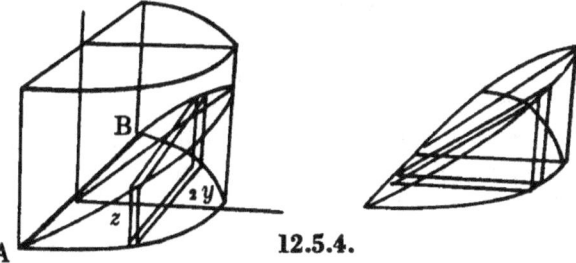

12.5.4.

Solution. An element of this volume is shown, a rectangular parallelepiped with faces parallel to the yz plane. Its volume is $2yz\,\Delta x$.

$$V = 2\int_0^a yz\,dx = 2\int_0^a \sqrt{a^2-x^2}\,x\,dx$$

$$= -\frac{2}{3}(a^2-x^2)^{3/2}\Big|_0^a = \frac{2a^3}{3}$$

his volume could also be calculated using the triangular prism shown as an element of volume. The area of the triangle is $\frac{1}{2}xz$, and the volume of the prism is $\frac{1}{2}xz\,\Delta y$.

$$V = \frac{1}{2}\int_{-a}^{a} xz\,dy = \frac{1}{2}\int_{-a}^{a} x^2\,dy = \frac{1}{2}\int_{-a}^{a} (a^2 - y^2)\,dy$$

$$= \frac{1}{2}\left(a^2 y - \frac{y^3}{3}\right)_{-a}^{a} = \frac{2a^3}{3}$$

12.5.5. Find the area cut from the cylinder $x^2 + y^2 - ax = 0$ by the sphere $x^2 + y^2 + z^2 = 4a^2$.

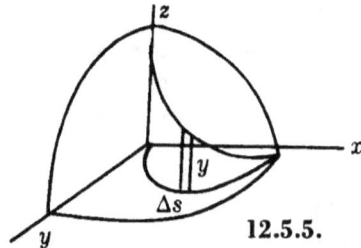

12.5.5.

Solution. The part of this area lying in the first octant is shown. Our element has area $z\,\Delta s$, where Δs is an element of arc of the circle which is a cross section of the cylinder. From the cylinder

$$\frac{dy}{dx} = \frac{a - x}{\sqrt{2ax - x^2}};$$

$$ds = \sqrt{1 + \left(\frac{a - x}{\sqrt{2ax - x^2}}\right)^2}\,dx = \frac{a\,dx}{\sqrt{2ax - x^2}}$$

$$z^2 = 4a^2 - (x^2 + y^2) = 4a^2 - 2ax$$

$$A = 4\int_0^{2a} z\,ds = 4\int_0^{2a} z \cdot \frac{a}{\sqrt{2ax - x^2}}\,dx$$

$$= 4\int_0^{2a} \sqrt{4a^2 - 2ax}\,\frac{a}{\sqrt{2ax - x^2}}\,dx = 4a\int_0^{2a} \sqrt{\frac{2a}{x}}\,dx$$

$$= 2(2a)^{3/2} \lim_{\xi \to 0} \int_\xi^{2a} \frac{dx}{\sqrt{x}} = 4(2a)^{3/2} \lim_{\xi \to 0} \sqrt{x}\,\Big|_\xi^{2a} = 16a^2$$

12.6. Area in polar coordinates. Area in polar coordinates is given by $\frac{1}{2}\int_\alpha^\beta r^2\, d\theta$.

12.6.1. Find the area of the cardioid $r = 1 + \cos\theta$.

Solution

$$A = \frac{1}{2}\int_0^{2\pi} r^2\, d\theta = \frac{1}{2}\int_0^{2\pi} (1+\cos\theta)^2\, d\theta$$

$$= \frac{1}{2}\int_0^{2\pi}\left(1 + 2\cos\theta + \frac{1+\cos 2\theta}{2}\right) d\theta$$

$$= \frac{1}{2}\left(\frac{3}{2}\theta + 2\sin\theta + \frac{1}{4}\sin 2\theta\right)_0^{2\pi} = \frac{3\pi}{2}$$

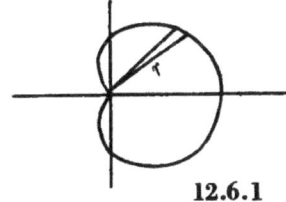
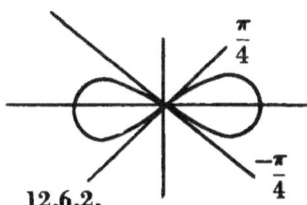

12.6.1 12.6.2.

12.6.2. Find the area bounded by the curve $r^2 = \cos 2\theta$.

Solution. It would be tempting to proceed as follows:

$$A = \frac{1}{2}\int_0^{2\pi} r^2\, d\theta = \frac{1}{2}\int_0^{2\pi}\cos 2\theta\, d\theta = 0$$

However, it must be remembered that the integral we have evaluated is simply $\lim_{n\to\infty} \frac{1}{2}\sum_{i=1}^{n}\cos 2\theta_i\, \Delta\theta_i$. Those θ_i between $\frac{\pi}{4}$ and $\frac{3\pi}{4}$, for example, contribute terms to this series (if we divide the interval 0 to 2π in n parts) even though the curve is not existent $\frac{\pi}{4} < \theta < \frac{3\pi}{4}$ (r^2 is negative when $\frac{\pi}{4} < \theta < \frac{3\pi}{4}$).

Hence

$$A = 2\times\frac{1}{2}\int_{-\pi/4}^{\pi/4}\cos 2\theta\, d\theta = \frac{1}{2}\sin 2\theta\,\Big|_{-\pi/4}^{\pi/4} = 1$$

12.6.3. Find the area of one leaf of $r = a \cos n\theta$.

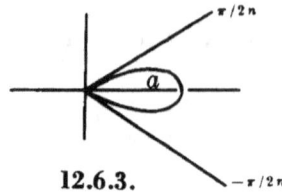

Solution

12.6.3.

$$A = \frac{1}{2} a^2 \int_{-\pi/2n}^{\pi/2n} \cos^2 n\theta \, d\theta = \frac{1}{2} a^2 \int_{-\pi/2n}^{\pi/2n} \frac{\cos 2n\theta + 1}{2} \, d\theta$$

$$= \frac{a^2}{2} \left(\frac{\sin 2n\theta}{4n} + \frac{\theta}{2} \right)_{-\pi/2n}^{\pi/2n} = \frac{\pi a^2}{4n}$$

Notice that the area of any of the "roses" is $\dfrac{\pi a^2}{4}$ if n is odd and $\dfrac{\pi a^2}{2}$ if n is even.

12.6.4. Find the area bounded by the curve $r = \sin \tfrac{1}{2}\theta$.

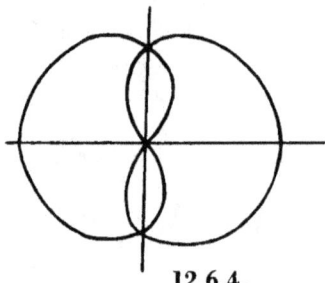

12.6.4.

Solution. By examining values of θ between 0 and 4π, we trace out the curve graphed in the figure. If we integrate from 0 to 4π, we shall cover the area of the inner loops twice. The upper one between 0 and π and 3π and 4π is swept over twice while the rest of the area above the polar axis is swept over once.

$$A = \frac{1}{2}\int_0^{4\pi} \sin^2\frac{\theta}{2}\,d\theta - 4 \times \frac{1}{2}\int_0^{\pi/2} \sin^2\frac{\theta}{2}\,d\theta$$

$$= \frac{1}{4}\int_0^{4\pi}(1-\cos\theta)\,d\theta - \int_0^{\pi/2}(1-\cos\theta)\,d\theta$$

$$= \frac{1}{4}(\theta - \sin\theta)_0^{4\pi} - (\theta - \sin\theta)_0^{\pi/2} = \pi - \frac{\pi}{2} + 1 = \frac{\pi}{2} +$$

12.6.5. Find the area between

$$r = \frac{4}{1-\cos\theta} \quad \text{and} \quad r = \frac{4}{1+\cos\theta}$$

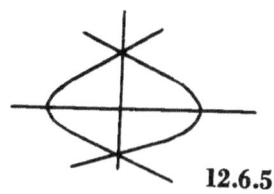

12.6.5.

Solution

$$A = \frac{1}{2}\int_{-\pi/2}^{\pi/2} \frac{16\,d\theta}{(1+\cos\theta)^2} + \frac{1}{2}\int_{\pi/2}^{3\pi/2} \frac{16\,d\theta}{(1-\cos\theta)^2}$$

$$= 2\int_{-\pi/2}^{\pi/2} \frac{d\theta}{\cos^4\frac{\theta}{2}} + 2\int_{+\pi/2}^{3\pi/2} \frac{d\theta}{\sin^4\frac{\theta}{2}}$$

$$= 2\int_{-\pi/2}^{\pi/2} \sec^4\frac{\theta}{2}\,d\theta + 2\int_{\pi/2}^{3\pi/2} \csc^4\frac{\theta}{2}\,d\theta$$

$$= 2\int_{-\pi/2}^{\pi/2}\left(1+\tan^2\frac{\theta}{2}\right)\sec^2\frac{\theta}{2}\,d\theta$$

$$\qquad + 2\int_{\pi/2}^{3\pi/2}\left(1+\cot^2\frac{\theta}{2}\right)\csc^2\frac{\theta}{2}\,d\theta$$

$$= 4\left(\tan\frac{\theta}{2} + \frac{1}{3}\tan^3\frac{\theta}{2}\right)\Big|_{-\pi/2}^{\pi/2} - 4\left(\cot\frac{\theta}{2} + \frac{1}{3}\cot^3\frac{\theta}{2}\right)\Big|_{\pi/2}^{3\pi/2}$$

$$= 4\left(2+\frac{2}{3}\right) - 4\left(-2-\frac{2}{3}\right) = \frac{64}{3}$$

12.6.6. Find the area inside the circle $r = \cos\theta$ and outside the cardioid $r = 1 - \cos\theta$.

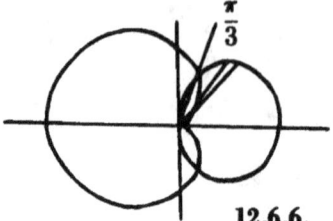

12.6.6.

Solution

$$dA = \tfrac{1}{2}(r_1^2 - r_2^2)\,d\theta; \quad 1 - \cos\theta = \cos\theta \quad \text{when}$$

$$\theta = \frac{\pi}{3} \text{ or } -\frac{\pi}{3}$$

$$A = \frac{1}{2}\int_{-\pi/3}^{\pi/3} [\cos^2\theta - (1 - \cos\theta)^2]\,d\theta$$

$$= \frac{1}{2}\int_{-\pi/3}^{\pi/3} (2\cos\theta - 1)\,d\theta$$

$$= \frac{1}{2}\left(2\sin\theta - \theta\right)\Big|_{-\pi/3}^{\pi/3} = \sqrt{3} - \frac{\pi}{3}$$

12.7. Arc length in polar coordinates

$$S = \int_\alpha^\beta \sqrt{r^2 + \left(\frac{dr}{d\theta}\right)^2}\,d\theta$$

12.7.1. Find the length of the cardioid $r = a(1 + \cos\theta)$.
Solution

$$\frac{dr}{d\theta} = -a\sin\theta$$

$$S = 2a\int_0^\pi \sqrt{(1 + \cos\theta)^2 + \sin^2\theta}\,d\theta$$

$$= 2a\int_0^\pi \sqrt{2 + 2\cos\theta}\,d\theta$$

$$= 4a\int_0^\pi \cos\tfrac{1}{2}\theta\,d\theta = 8a\sin\frac{\theta}{2}\Big|_0^\pi = 8a$$

We take advantage of the symmetry and take twice the integral from 0 to π. If we had chosen the limits 0 and 2π, $\cos\tfrac{1}{2}\theta$ would be $-\sqrt{r^2 + \left(\dfrac{dr}{d\theta}\right)^2}$ from π to 2π, and we would have to split our integral into two parts.

12.7.2. Find the length of the spiral $r = a\theta$ around one turn 0 to 2π.

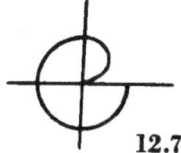

12.7.2.

Solution $\quad S = a\displaystyle\int_0^{2\pi} \sqrt{\theta^2 + 1}\, d\theta$

Let $\theta = \tan z$.

$$S = a\int_0^{\text{Arc tan}2\pi} \sqrt{\tan^2 z + 1}\, \sec^2 z\, dz = a\int_0^{\text{Arc tan}2\pi} \sec^3 z\, dz$$

$$= \frac{a}{2}\left(\sec z \tan z + \ln|\sec z + \tan z|\right)\Big|^{\text{Arc tan}2\pi}$$

From our triangle, Arc tan 2π = Arc sec $\sqrt{4\pi^2 + 1}$.

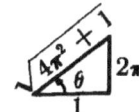

$$S = \frac{a}{2}[2\pi(4\pi^2 + 1)^{1/2} + \ln|\sqrt{4\pi^2 + 1} + 2\pi|]$$

12.7.3. Find the length of the cissoid $r = 2a \tan\theta \sin\theta$ from $\theta = 0$ to $\theta = \dfrac{\pi}{3}$.

12.7.3.

Solution

$$\frac{dr}{d\theta} = 2a(\sin\theta \sec^2\theta + \tan\theta \cos\theta) = 2a\sin\theta(\sec^2\theta + 1)$$

$$S = 2a \int_0^{\pi/3} \sqrt{\sin^2\theta \tan^2\theta + \sin^2\theta(\sec^2\theta + 1)^2}\, d\theta$$

$$= 2a \int_0^{\pi/3} \sin\theta \sqrt{\tan^2\theta + \sec^4\theta + 2\sec^2\theta + 1}\, d\theta$$

$$= 2a \int_0^{\pi/3} \sin\theta \sqrt{\sec^4\theta + 3\sec^2\theta}\, d\theta$$

$$= 2a \int_0^{\pi/3} \sin\theta \sec\theta \sqrt{\sec^2\theta + 3}\, d\theta$$

$$= 2a \int_0^{\pi/3} \sqrt{\sec^2\theta + 3}\, \tan\theta\, d\theta$$

Let $\sec^2\theta + 3 = u^2$, $2\sec^2\theta \tan\theta\, d\theta = 2u\, du$,

$$\tan\theta\, d\theta = \frac{u\, du}{u^2 - 3}; \quad \theta = 0, u = 2; \quad \theta = \frac{\pi}{3}, u = \sqrt{7}.$$

$$S = 2a \int_2^{\sqrt{7}} \frac{u^2\, du}{u^2 - 3} = 2a \int_2^{\sqrt{7}} du + 6a \int_2^{\sqrt{7}} \frac{du}{u^2 - 3}$$

$$= 2a(\sqrt{7} - 2) + \sqrt{3}\,a \left(\ln\left|\frac{u - \sqrt{3}}{u + \sqrt{3}}\right|\right)_2^{\sqrt{7}}$$

$$= 2a(\sqrt{7} - 2) + \sqrt{3}\,a \ln\left|\frac{\sqrt{7} - \sqrt{3}}{\sqrt{7} + \sqrt{3}} \cdot \frac{2 + \sqrt{3}}{2 - \sqrt{3}}\right|$$

$$= 2a(\sqrt{7} - 2) + \sqrt{3}\,a$$

$$\ln\left|\frac{(7 + 3 - 2\sqrt{21})(4 + 3 + 4\sqrt{3})}{4 \cdot 1}\right|$$

$$= 2a(\sqrt{7} - 2) + \sqrt{3}\,a \ln\left[\frac{1}{2}(5 - \sqrt{21})(7 + 4\sqrt{3})\right]$$

12.7.4. Find the length of the curve $r = a\sin^3\frac{\theta}{3}$.

Solution. As θ varies from 0 to $\frac{3\pi}{2}$, half of the curve is traced out.

$$S = 2a \int_0^{3\pi/2} \sqrt{\sin^6 \frac{\theta}{3} + \sin^4 \frac{\theta}{3} \cos^2 \frac{\theta}{3}}\, d\theta = 2a \int_0^{3\pi/2} \sin^2 \frac{\theta}{3}\, d\theta$$

$$= a \int_0^{3\pi/2} \left(1 - \cos \frac{2\theta}{3}\right) d\theta = a\left(\theta - \frac{3}{2}\sin\frac{2\theta}{3}\right)\Big|_0^{3\pi/2} = \frac{3\pi a}{2}$$

12.7.4.   12.8.

12.8. Area of a surface of revolution in polar coordinates.

In the following problems, an element of arc is rotated about the initial line, the radius of revolution is $r \sin \theta$, and the area is

$$A = 2\pi \int_\alpha^\beta r \sin \theta\, ds$$

About the y axis

$$A = 2\pi \int_\alpha^\beta r \cos \theta\, ds$$

12.8.1. Find the area generated when the cardiod $r = a(1 + \cos \theta)$ is revolved about the initial line.

Solution

$$\sqrt{r^2 + \left(\frac{dr}{d\theta}\right)^2} = a\sqrt{1 + 2\cos\theta + \cos^2\theta + \sin^2\theta}$$
$$= a\sqrt{2 + 2\cos\theta}$$

$$A = 2\pi a^2 \int_0^\pi \sqrt{2 + 2\cos\theta}\,(1 + \cos\theta) \sin\theta\, d\theta$$

$$= 2\sqrt{2}\,\pi a^2 \int_0^\pi (1 + \cos\theta)^{3/2} \sin\theta\, d\theta$$

$$= 2\sqrt{2}\pi a^2 \left(-\frac{2}{5}\right)(1 + \cos\theta)^{5/2}\Big|_0^\pi = \frac{32\pi a^2}{5}$$

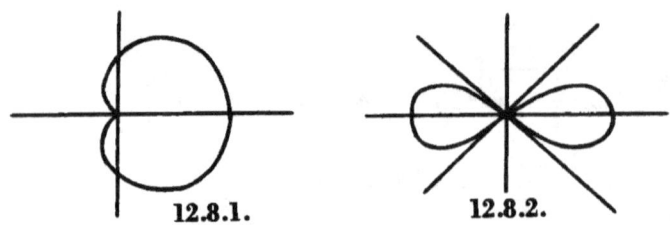

12.8.1.   12.8.2.

12.8.2. Find the area generated when $r^2 = \cos 2\theta$ is rotated out the initial line.

Solution

$$\frac{r'}{\theta} = \frac{-\sin 2\theta}{\sqrt{\cos 2\theta}}; \quad ds = \sqrt{\cos 2\theta + \frac{\sin^2 2\theta}{\cos 2\theta}}\, d\theta = \frac{d\theta}{\sqrt{\cos 2\theta}}$$

$$A = 2 \cdot 2\pi \int_0^{\pi/4} r \sin\theta \, ds$$

$$= 4\pi \int_0^{\pi/4} \sin\theta \, d\theta = -4\pi \cos\theta \Big|_0^{\pi/4}$$

$$= 4\pi\left(1 - \frac{\sqrt{2}}{2}\right) = 2\sqrt{2}\pi(\sqrt{2} - 1)$$

12.8.3. Find the area generated when the curve $r = a\cos\theta$ is evolved about the y axis.

12.8.3.

Solution. Here the area passed through by an elementary arc is $2\pi r \cos\theta \, ds$.

$$A = 2\pi \int_0^\pi r \cos\theta \, ds = 2\pi \int_0^\pi a \cos^2\theta \sqrt{a^2\cos^2\theta + a^2\sin^2\theta} \, d\theta$$

$$= 2\pi a^2 \int_0^\pi \cos^2\theta \, d\theta = 2\pi a^2 \int_0^\pi \left(\frac{1 + \cos 2\theta}{2}\right) d\theta$$

$$= \pi a^2 \left(\theta + \frac{1}{2}\sin 2\theta\right)\Big|_0^\pi = \pi^2 a^2$$

Notice we choose the limits 0 and π since the entire length of the curve is traced out between these limits.

CHAPTER 13

PHYSICAL APPLICATIONS OF THE DEFINITE INTEGRAL

13.1. Work. If a force $F(x)$ is applied to a point that move along the x-axis from a to b, the work done is $W = \int_a^b F(x)\,d$

Our first problems will be concerned with Hooke's law, which states that the force required to stretch a helical spring proportional to the distance stretched. Notice that, if we ca the elongation t, then $F(t) = kt$. Also, we shall use the trivia fact that the change in elongation is the change of length $\Delta t = \Delta x$.

13.1.1. A spring of natural length 10 in. is stretched to 12 in by a force of 10 lb. Find the work done in stretching it from length of 14 in. to 18 in.

Solution

$$F = kt; \quad 10 = k2; \quad k = 5 \text{ lb/in.}$$

$$W = 5\int_4^8 t\,dt = \frac{5}{2} t^2 \Big|_4^8 = 120 \text{ in.-lb} = 10 \text{ ft-lb}$$

We choose limits 4 and 8 since the extension is changed from 4 in. to 8 in.

13.1.2. The spring of **13.1.1** is compressed from 8 in. to 6 in. Find the work done.

Solution. Here our extension changes from -2 in. to -4 in.

$$W = 5\int_{-2}^{-4} t\,dt = \frac{5}{2} t^2 \Big|_{-2}^{-4} = 30 \text{ in.-lb} = 2\tfrac{1}{2} \text{ ft-lb}$$

If we had asked for the work done in changing from a length of 6 in. to 8 in.,

$$W = 5\int_{-4}^{-2} t\,dt = -2\tfrac{1}{2} \text{ ft-lb}$$

The minus sign indicates that the work done on the spring is negative, i.e., the spring does work.

13.1.3. The weight of a particle is $\frac{k}{R^2}$ lb, where R is the distance from the center of the earth. Find the work done in ving the particle from the surface of the earth ($R = 4000$ les) to a point 2000 miles from the surface.

Solution

$$W = \int F\,ds = \int_{4000}^{6000} \frac{k}{R^2}\,dR = -\frac{k}{R}\Big|_{4000}^{6000}$$

$$= -k\left(\frac{1}{6000} - \frac{1}{4000}\right) \text{mile-lb} = \frac{1}{12000} k \text{ mile-lb}$$

$$= \frac{5280}{12000} k \text{ ft-lb} = 0.44k \text{ ft-lb}$$

d we changed the distance from 6000 to 4000 miles, and so ersed the limits of integration, $W = -0.44$ ft-lb, again licating that the particle does work in the process.

13.1.4. A gas is confined in a cylinder. Assuming Boyle's v is satisfied, i.e., the gas is compressed isothermally and = pressure, v = volume, $pv = k$, find the work done in mpressing the gas by a movable piston in a cylinder from 200 in. to 20 cu in. if the original volume is at atmospheric essure (14.7 lb/sq in.).

Solution. Since $v = hA$, where A is a constant cross-sectional ea, $F = A \cdot p = A \cdot \frac{k}{Ah} = \frac{k}{h}$; h varies from $\frac{200}{A}$ to $\frac{20}{A}$; $= 14.7 \times 200$.

$$W = \int F\,ds = \int_{20/A}^{200/A} F\,dh = k \int_{20/A}^{200/A} \frac{dh}{h} = k \ln h \Big|_{20/A}^{200/A}$$

$$= k \ln 10 = 2940 \ln 10 \text{ in.-lb} = 564^+ \text{ ft-lb}$$

13.1.5. A cylindrical cistern of radius 10 ft and height 20 ft is ll of water; find the work done in raising the water to its top.

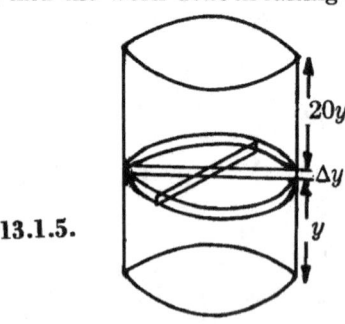

13.1.5.

Solution. In this problem we consider the water to b removed by infinitesimal layers. The volume of such a layer shown in the figure is $\pi \times 100 \times \Delta y$, and its weight is $62.4 \times \pi \times 100 \times \Delta y$.

$$W = \int_0^{20} (20 - y) \times 62.4 \times \pi \times 100 \, dy$$

$$= 6240\pi \left(20y - \frac{y^2}{2}\right)\Big|_0^{20} = 6240\pi \times 200 \text{ ft-lb}$$

$$= 1{,}248{,}000\pi \text{ ft-lb}$$

13.1.6. A cylindrical tank 10 ft long and 4 ft in radius i lying on its side. If it is full of oil (density 40 lb per cu ft), fin the work required to pump it over the top.

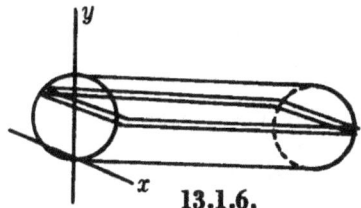

13.1.6.

Solution. We regard the oil as being removed in infinitesima layers of weight, $40 \times 10 \times 2x \times \Delta y$; $x^2 + y^2 - 8y = 0$.

$$W = 800 \int_0^8 \sqrt{8y - y^2} \, dy = 800 \int_0^8 \sqrt{16 - (y-4)^2} \, dy$$

Let $y - 4 = 4 \sin \theta$.

$$W = 800 \int_{-\pi/2}^{\pi/2} 16 \cos^2 \theta \, d\theta = 12{,}800 \left(\frac{\theta}{2} + \frac{\sin 2\theta}{4}\right)\Big|_{-\pi/2}^{\pi/2}$$

$$= 6400\pi \text{ ft-lb}$$

13.1.7. If an old-fashioned well has a bucket holding 1 cu ft or 62.4 lb of water, find the work done in winding 20 ft of rope on a windlass. The rope is 2 lb per foot.

Solution. If x is feet of rope out, $F = 62.4 + x \cdot 2$; $\Delta W = Fx(-\Delta x)$.

$$W = -\int_{20}^0 (62.4 + 2x) \, dx = -(62.4x + x^2)\Big|_{20}^0 = 1648 \text{ ft-lb}$$

13.2. Fluid pressure. The force acting on a vertical submerged plane in terms of density w of the fluid, depth h and breadth l is $F = \int_{h_1}^{h_2} whl\,dh$.

13.2.1. Find the force exerted on a triangle submerged in water as shown with vertex at the surface and base horizontal if its height is 4 ft and base 9 ft; $w = 62.4$.

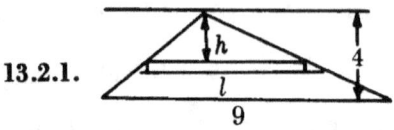

13.2.1.

Solution. Here $\dfrac{h}{l} = \dfrac{4}{9}$ or $l = \dfrac{9}{4} h$.

$$F = 62.4 \int_0^4 h \cdot \frac{9}{4} h\,dh = 62.4 \times \frac{3}{4} h^3 \Big|_0^4 = 2995.2 \text{ lbs}$$

All triangles of these dimensions have this force on them, regardless of shape.

13.2.2. Find the force exerted on a triangle submerged in water as shown with its vertex 5 ft below the surface, height 4 ft and base 9 ft.

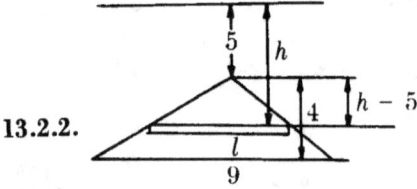

13.2.2.

Solution. Here $\dfrac{h-5}{l} = \dfrac{4}{9}$; $l = \dfrac{9}{4}(h-5)$.

$$F = 62.4 \int_5^9 h \cdot \frac{9}{4}(h-5)\,dh = 62.4 \times \frac{9}{4}\left(\frac{h^3}{3} - \frac{5}{2}h^2\right)\Big|_5^9$$

$$= 140.4(243 - 202.5 - 41.67 + 62.5) = 8610.7^+ \text{ lb}$$

Limits 5 and 9 are chosen since h varies from 5 ft to 9 ft.

13.2.3. Find the force on a triangle submerged in water as shown, height 4 ft, base 9 ft, depth of vertex 6 ft.

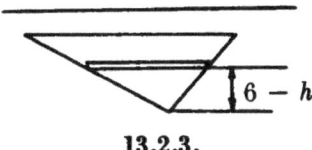

Solution

13.2.3.

$$\frac{6-h}{l} = \frac{4}{9}; \quad l = \frac{9}{4}(6-h)$$

$$F = 62.4 \int_2^6 h \cdot \frac{9}{4}(6-h)\, dh = 140.4\left(3h^2 - \frac{h^3}{3}\right)\Big|_2^6$$

$$= 140.4\left(108 - 72 - 12 + \frac{8}{3}\right) = 3744 \text{ lb}$$

13.2.4. Find the pressure on the end of an oil tank filled with oil of density 40 lb per cu ft if its axis is horizontal and the oil is under a pressure of 2 lb per sq ft at the top. The radius of the tank is 5 ft.

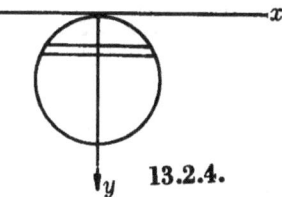

13.2.4.

Solution. We have taken the downward direction as positive, and the equation of our circle is $x^2 + y^2 - 10y = 0$. This has been done that the direction of increasing depth may be considered positive; $l = 2\sqrt{10h - h^2}$.

Here $\Delta F = (40h + 2)\Delta A$, i.e., the force due to the column of oil of height h, added to the force due to the pressure at the top.

$$F = 2\int_0^{10}(40h + 2)\sqrt{10h - h^2}\, dh$$

$$= 2\int_0^{10}(40h + 2)\sqrt{25 - (h-5)^2}\, dh$$

Let $h - 5 = 5\sin\theta$.

$$F = 2\int_{-\pi/2}^{\pi/2}(200\sin\theta + 202)\sqrt{25 - 25\sin^2\theta}\,5\cos\theta\,d\theta$$

$$= 50\int_{-\pi/2}^{\pi/2} 200\sin\theta\cos^2\theta\,d\theta + 50\int_{-\pi/2}^{\pi/2} 202\left(\frac{1+\cos 2\theta}{2}\right)d\theta$$

$$= -\frac{10000}{3}\cos^3\theta\Big|_{-\pi/2}^{\pi/2} + 5050\left(\theta + \frac{\sin 2\theta}{2}\right)_{-\pi/2}^{\pi/2} = 5050\pi \text{ lb}$$

13.2.5. The ends of a trough are equilateral triangles of edge 2 ft. If the trough is 10 ft long and filled with water, find the force acting on one of the sides.

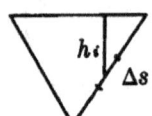

13.2.5.
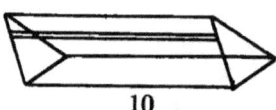

Solution. Our strips are all 10 ft long, and $\Delta s = \frac{2\,\Delta h}{\sqrt{3}}$;

$$\Delta F = wh \times 10 \cdot \frac{2\,\Delta h}{\sqrt{3}}.$$

$$F = \frac{20}{\sqrt{3}} w \int_0^{\sqrt{3}} h\,dh = \frac{20}{\sqrt{3}} w \frac{h^2}{2}\Big|_0^{\sqrt{3}}$$

$$= \frac{10\sqrt{3}}{3} w \cdot 3 = 10\sqrt{3} \times 62.4 = 1080.8 \text{ lb}$$

13.2.6. Find the total force on a hemispherical bowl full of water if the radius is 2 ft.

13.2.6.
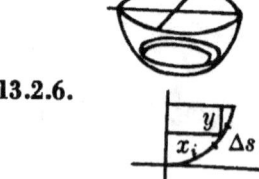

Solution. The force on a ring-shaped area traced out by the arc Δs is

$$\Delta F = 2\pi x_i \, \Delta s \times w y_i; \quad x^2 + y^2 = 4; \quad ds = \sqrt{1 + \left(\frac{dx}{dy}\right)^2}\, dy$$

$$F = 2w\pi \int_0^2 \sqrt{4 - y^2} \sqrt{1 + \left(\frac{y}{\sqrt{4 - y^2}}\right)^2}\, y\, dy$$

$$= 2w\pi \int_0^2 \sqrt{4 - y^2} \times \frac{2}{\sqrt{4 - y^2}}\, y\, dy = 2\pi w y^2 \Big|_0^2 = 8\pi w \text{ lb}$$

13.2.7. A vertical dam is in the form of a trapezoid, 300 ft across the top, 100 ft across the bottom, and 60 ft deep. Find the force acting on it.

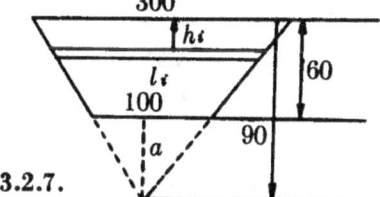

13.2.7.

Solution. In the figure $\dfrac{a}{100} = \dfrac{a + 60}{300}$; $a = 30$. Now using similar triangles, $\dfrac{90}{300} = \dfrac{90 - h}{l}$.

$$F = 62.4 \int_0^{60} lh\, dh = 62.4 \int_0^{60} \frac{10}{3}(90 - h)h\, dh$$

$$= \frac{624}{3}\left(45h^2 - \frac{h^3}{3}\right)\Big|_0^{60} = \frac{624}{3} \times 3600 \times 25 = 18{,}720{,}000 \text{ lb}$$

$$= 9360 \text{ tons}$$

The trapezoid has this force acting on it, regardless of its shape.

13.3. Centroids of plane areas. See **4.5** and **6.3** for formulas.

13.3.1. Find the centroid of one quadrant of the area of an ellipse $x = a \cos \theta$, $y = b \sin \theta$.

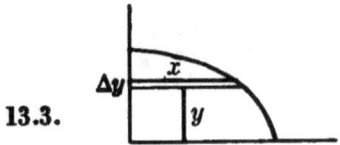

13.3.

Solution. By a previous example $A = \dfrac{\pi ab}{4}$.

$$M_x = \int_0^b xy\, dy = \int_0^{\pi/2} a\cos\theta\, b\sin\theta\, b\cos\theta\, d\theta$$

$$= -\dfrac{ab^2}{3}\cos^3\theta\Big|_0^{\pi/2} = \dfrac{ab^2}{3}$$

$$M_y = \int_0^a xy\, dx = \int_{\pi/2}^0 a\cos\theta\, b\sin\theta(-a\sin\theta)\, d\theta$$

$$= \dfrac{-a^2 b}{3}\sin^3\theta\Big|_{\pi/2}^0 = \dfrac{a^2 b}{3}$$

$$\bar{x} = \dfrac{\dfrac{a^2 b}{3}}{\dfrac{\pi ab}{4}} = \dfrac{4a}{3\pi}; \qquad \bar{y} = \dfrac{\dfrac{ab^2}{3}}{\dfrac{\pi ab}{4}} = \dfrac{4b}{3\pi}$$

Of course, we could also have performed our calculation this way:

$$M_x = \dfrac{1}{2}\int_0^a y^2\, dx = \dfrac{1}{2}\int_{\pi/2}^0 b^2 \sin^2\theta(-a\sin\theta)\, d\theta$$

$$= \dfrac{ab^2}{2}\left(\cos\theta - \dfrac{\cos^3\theta}{3}\right)\Big|_{\pi/2}^0 = \dfrac{ab^2}{3}$$

13.3.2. Find the centroid of the area in the first quadrant under the curve $y = xe^{-x}$.

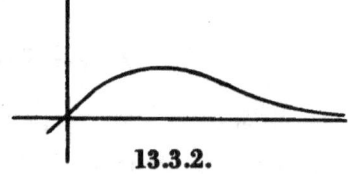

13.3.2.

Solution

$$A = \int_0^{+\infty} xe^{-x}\,dx = -xe^{-x} - e^{-x}\Big|_0^\infty = 1$$

$$M_x = \frac{1}{2}\int_0^\infty y^2\,dx = \frac{1}{2}\int_0^\infty x^2 e^{-2x}\,dx$$

$$= \frac{1}{2}\left(-\frac{1}{2}x^2 e^{-2x} - \frac{1}{2}xe^{-2x} - \frac{1}{4}e^{-2x}\right)\Big|_0^\infty = \frac{1}{8}$$

$$M_y = \int_0^\infty xy\,dx = \int_0^\infty x^2 e^{-x}\,dx = (-x^2 e^{-x} - 2xe^{-x} - 2e^{-x})\Big|_0^\infty = 2$$

$$\bar{x} = 2; \quad \bar{y} = \tfrac{1}{8}$$

13.3.3. Find the centroid of the area enclosed by one arch of the cycloid $x = a(\theta - \sin\theta)$, $y = a(1 - \cos\theta)$.

Solution

$$A = a^2 \int_0^{2\pi} (1 - \cos\theta)^2\,d\theta = a^2 \int_0^{2\pi}\left(1 - 2\cos\theta + \frac{1 + \cos 2\theta}{2}\right)d\theta$$

$$= a^2\left(\frac{3}{2}\theta - 2\sin\theta + \frac{\sin 2\theta}{4}\right)\Big|_0^{2\pi} = 3\pi a^2$$

$$M_x = \frac{1}{2}\int_0^{2\pi} y^2\,dx = \frac{1}{2}a^3 \int_0^{2\pi}(1 - \cos\theta)^3\,d\theta$$

$$= \frac{a^3}{2}\int_0^{2\pi}\left[1 - 3\cos\theta + 3\left(\frac{1 + \cos 2\theta}{2}\right) - (1 - \sin^2\theta)\cos\theta\right]d\theta$$

$$= \frac{a^3}{2}\left(\frac{5}{2}\theta - 3\sin\theta + \frac{3}{4}\sin 2\theta - \sin\theta + \frac{\sin^3\theta}{3}\right)\Big|_0^{2\pi} = \frac{5\pi a^3}{2}$$

$$M_y = \int_0^{2\pi} xy\,dx = a^3 \int_0^{2\pi} (\theta - \sin\theta)(1 - \cos\theta)(1 - \cos\theta)\,d\theta$$

$$= a^3 \int_0^{2\pi} (1 - \cos\theta)^2(-\sin\theta)\,d\theta + a^3 \int_0^{2\pi} \frac{3}{2}\theta\,d\theta$$

$$+ a^3 \int_0^{2\pi}\left(-2\cos\theta + \frac{\cos 2\theta}{2}\right)\theta\,d\theta$$

$$= -\frac{a^3}{3}(1-\cos\theta)^3\Big|_0^{2\pi} + a^3\left(-2\sin\theta + \frac{\sin 2\theta}{4}\right)\theta\Big|_0^{2\pi}$$
$$+ \frac{3a^3\theta^2}{4}\Big|_0^{2\pi} - a^3\int_0^{2\pi}\left(-2\sin\theta + \frac{\sin 2\theta}{4}\right)d\theta$$
$$= 3a^3\pi^2 - a^3\left(2\cos\theta - \frac{\cos 2\theta}{8}\right)\Big|_0^{2\pi} = 3a^3\pi^2$$

$$\bar{x} = \frac{3\pi^2 a^3}{3\pi a^2} = \pi a; \qquad \bar{y} = \frac{\frac{5\pi a^3}{2}}{3\pi a^2} = \frac{5a}{6}$$

13.3.3.

We could have obtained $\bar{x}$ by considerations of symmetry. Notice that we could not calculate M_x by means of $\int xy\,dy$. The area of our elementary strip is not $x\,\Delta y$ but $(x\,|_{2\pi-\theta} - x\,|_\theta)\,\Delta y$.
$$dA = a^2[(2\pi - \theta - \sin(2\pi - \theta)) - (\theta - \sin\theta)]\sin\theta\,d\theta$$
$$= a^2(2\pi - 2\theta + 2\sin\theta)\sin\theta\,d\theta$$

Notice also that the limits for θ would be from 0 to π.

13.3.4. Find the centroid of the area bounded by the x axis, the curves $y = \sin x$, and $y = \cos x$ from 0 to $\frac{\pi}{2}$.

Solution

$$A = \int_0^{\pi/4}\sin x\,dx + \int_{\pi/4}^{\pi/2}\cos x\,dx = -\cos x\Big|_0^{\pi/4} + \sin x\Big|_{\pi/4}^{\pi/2}$$
$$= 2 - \sqrt{2}$$

13.3.4.

Choosing strips parallel to the x axis of area $(x_2 - x_1)\,\Delta y$, we have their moment about the x axis,

$$M_x = \int y\,dA = \int_0^{\sqrt{2}/2} y(x_2 - x_1)\,dy$$

$$= \int_0^{\sqrt{2}/2} y(\operatorname{Arc\,cos} y - \operatorname{Arc\,sin} y)\,dy$$

$$= \frac{1}{2}y^2(\operatorname{Arc\,cos} y - \operatorname{Arc\,sin} y)\bigg|_0^{\sqrt{2}/2}$$

$$- \int_0^{\sqrt{2}/2} \frac{1}{2}y^2\left(-\frac{1}{\sqrt{1-y^2}} - \frac{1}{\sqrt{1-y^2}}\right)dy = \int_0^{\sqrt{2}/2} \frac{y^2\,dy}{\sqrt{1-y^2}}$$

Let $y = \sin\theta$.

$$M_x = \int_0^{\pi/4} \frac{\sin^2\theta \cos\theta\,d\theta}{\sqrt{1-\sin^2\theta}} = \int_0^{\pi/4} \frac{1-\cos 2\theta}{2}\,d\theta$$

$$= \left(\frac{\theta}{2} - \frac{\sin 2\theta}{4}\right)\bigg|_0^{\pi/4} = \frac{\pi}{8} - \frac{1}{4} = \frac{1}{4}\left(\frac{\pi}{2} - 1\right)$$

Alternately

$$M_x = \frac{1}{2}\int y^2\,dx = \frac{1}{2}\int_0^{\pi/4} \sin^2 x\,dx + \frac{1}{2}\int_{\pi/4}^{\pi/2} \cos^2 x\,dx$$

$$= \frac{1}{2}\int_0^{\pi/4} \frac{1-\cos 2x}{2}\,dx + \frac{1}{2}\int_{\pi/4}^{\pi/2} \frac{1+\cos 2x}{2}\,dx$$

$$= \frac{1}{4}\left(x - \frac{\sin 2x}{2}\right)\bigg|_0^{\pi/4} + \frac{1}{4}\left(x + \frac{\sin 2x}{2}\right)\bigg|_{\pi/4}^{\pi/2} = \frac{1}{4}\left(\frac{\pi}{2} - 1\right)$$

$$M_y = \int xy\,dx = \int_0^{\pi/4} x\sin x\,dx + \int_{\pi/4}^{\pi/2} x\cos x\,dx$$

$$= (-x\cos x + \sin x)\bigg|_0^{\pi/4} + (x\sin x + \cos x)\bigg|_{\pi/4}^{\pi/2}$$

$$= -\frac{\pi}{4}\cdot\frac{\sqrt{2}}{2} + \frac{\sqrt{2}}{2} + \frac{\pi}{2} - \frac{\pi}{4}\cdot\frac{\sqrt{2}}{2} - \frac{\sqrt{2}}{2} = \frac{\pi}{4}(2 - \sqrt{2})$$

$$\bar{x} = \frac{\frac{\pi}{4}(2-\sqrt{2})}{2-\sqrt{2}} = \frac{\pi}{4}; \quad \bar{y} = \frac{\frac{1}{4}\left(\frac{\pi}{2}-1\right)}{2-\sqrt{2}}$$

$$= \frac{1}{8}\left(\frac{\pi}{2}-1\right)(2+\sqrt{2})$$

$\bar{x}$ could also have been found by symmetry, $x = \frac{\pi}{4}$ being an axis of symmetry for the area.

13.3.5. Find the centroid in the first quadrant of the arc $x^{2/3} + y^{2/3} = a^{2/3}$.

Solution. From a previous example, $s = \frac{3}{2}a$,

$$ds = \sqrt{1 + y'^2}\, dx = \frac{a^{1/3}}{x^{1/3}}\, dx.$$

$$M_x = \int_0^a y\, ds = \int_0^a (a^{2/3} - x^{2/3})^{3/2} \frac{a^{1/3}}{x^{1/3}}\, dx$$

$$= -\frac{3}{5}(a^{2/3} - x^{2/3})^{5/2} a^{1/3}\Big|_0^a = \frac{3}{5}a^2$$

$$M_y = \int_0^a x\, ds = a^{1/3}\int_0^a x^{2/3}\, dx = \frac{3}{5}a^{1/3} x^{5/3}\Big|_0^a = \frac{3}{5}a^2$$

$$\bar{x} = \frac{2}{5}a; \quad \bar{y} = \frac{2}{5}a$$

13.3.6. Find the centroid for the arc that is the first arch of the cycloid $x = a(\theta - \sin\theta)$, $y = a(1 - \cos\theta)$.

Solution

$$ds = \sqrt{a^2(1-\cos\theta)^2 + a^2\sin^2\theta}\, d\theta = a\sqrt{2 - 2\cos\theta}\, d\theta$$

We have previously determined the length of the arch as $8a$.

$$M_x = \int_0^{2\pi} y\, ds = a^2\int_0^{2\pi}(1-\cos\theta)\sqrt{2-2\cos\theta}\, d\theta$$

$$= 4a^2\int_0^{2\pi}\sin^3\frac{\theta}{2}\, d\theta = 4a^2\int_0^{2\pi}\left(1-\cos^2\frac{\theta}{2}\right)\sin\frac{\theta}{2}\, d\theta$$

$$= 4a^2\left(-2\cos\frac{\theta}{2} + \frac{2}{3}\cos^3\frac{\theta}{2}\right)\Big|_0^{2\pi} = \frac{32}{3}a^2$$

$$M_y = \int_0^{2\pi} x\, ds = a^2 \int_0^{2\pi} (\theta - \sin\theta)\sqrt{2 - 2\cos\theta}\, d\theta$$

$$= 2a^2 \int_0^{2\pi} \theta \sin\frac{\theta}{2}\, d\theta - a^2 \sqrt{2} \int_0^{2\pi} (1 - \cos\theta)^{1/2} \sin\theta\, d\theta$$

$$= 2a^2 \left(-2\theta \cos\frac{\theta}{2} + 4\sin\frac{\theta}{2}\right)\Big|_0^{2\pi}$$

$$- a^2\sqrt{2}\,\frac{2}{3}(1-\cos\theta)^{3/2}\Big|_0^{2\pi} = 8\pi a^2$$

$$\bar{x} = \pi a; \qquad \bar{y} = \tfrac{4}{3}a$$

Notice $\bar{x}$ could be obtained from considerations of symmetry.

13.3.7. If the equation of the curve is given in polar coordinates, we procede as follows; first, we note that the triangular element of area has, by similar triangles (see **4.6.6**),

$$\bar{x}_i = \tfrac{2}{3}r_i \cos\theta_i; \qquad \bar{y}_i = \tfrac{2}{3}r_i \sin\theta_i$$

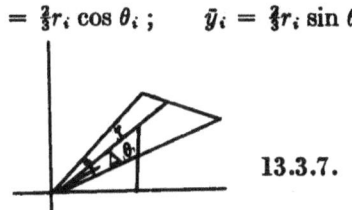

13.3.7.

The moment of this triangle can be given as

$$M_{x_i} = \tfrac{2}{3}r_i \sin\theta_i\, \Delta A_i; \qquad M_{y_i} = \tfrac{2}{3}r_i \cos\theta_i\, \Delta A_i,$$

or since $\Delta A_i = \tfrac{1}{2}r_i^2\, \Delta\theta_i$,

$$M_{x_i} = \tfrac{1}{3}r_i^3 \sin\theta_i\, \Delta\theta_i; \qquad M_{y_i} = \tfrac{1}{3}r_i^3 \cos\theta_i\, \Delta\theta_i$$

$$A\bar{x} = \frac{1}{3}\int_\alpha^\beta r^3 \cos\theta\, d\theta; \qquad A\bar{y} = \frac{1}{3}\int_\alpha^\beta r^3 \sin\theta\, d\theta$$

For example, for the area of a cardioid $r = a(1 + \cos\theta)$:

$$A = \frac{a^2}{2}\int_0^{2\pi}(1+\cos\theta)^2\, d\theta$$

$$= \frac{a^2}{2}\int_0^{2\pi}\left(1 + 2\cos\theta + \frac{1+\cos 2\theta}{2}\right) d\theta$$

$$= \frac{a^2}{2}\left(\frac{3}{2}\theta + 2\sin\theta + \frac{\sin 2\theta}{4}\right)\Big|_0^{2\pi} = \frac{3\pi a^2}{2}$$

$$\frac{3}{2}\pi a^2 \bar{x} = \frac{1}{3} a^3 \int_0^{2\pi} (1 + \cos\theta)^3 \cos\theta \, d\theta$$

$$= \frac{1}{3} a^3 \int_0^{2\pi} \left[\cos\theta + 3\left(\frac{1+\cos 2\theta}{2}\right) \right.$$
$$+ 3(1 - \sin^2\theta) \cos\theta$$
$$\left. + \frac{1}{4}\left(1 + 2\cos 2\theta + \frac{1+\cos 4\theta}{2}\right) \right] d\theta$$

$$= \frac{1}{3} a^3 \left(\frac{15}{8}\theta + 4\sin\theta + \sin 2\theta - \sin^3\theta + \frac{1}{32}\sin 4\theta \right)\Big|_0^{2\pi}$$

$$= \frac{5}{4} \pi a^3$$

$$\frac{3}{2}\pi a^2 \bar{y} = \frac{1}{3} a^3 \int_0^{2\pi} (1 + \cos\theta)^3 \sin\theta \, d\theta$$

$$= -\frac{1}{12} a^3 (1 + \cos\theta)^4 \Big|_0^{2\pi} = 0$$

$$\bar{x} = \frac{5}{6} a; \qquad \bar{y} = 0$$

which also follows from considerations of symmetry.

13.3.8. In order to find the centroid of an arc given in polar coordinates, we must consider the moment of an element of arc length.

13.3.8.

$$M_{x_i} = r_i \sin\theta_i \, \Delta s_i$$

$$M_x = \int_\alpha^\beta r \sin\theta \sqrt{r^2 + \left(\frac{dr}{d\theta}\right)^2} \, d\theta$$

$$M_y = \int_\alpha^\beta r \cos\theta \sqrt{r^2 + \left(\frac{dr}{d\theta}\right)^2} \, d\theta$$

For example, to find the moment of the arc of a cardioid $r = a(1 + \cos\theta)$:

$$ds = a\sqrt{(1+\cos\theta)^2 + \sin^2\theta}\, d\theta = \sqrt{2}a\sqrt{1+\cos\theta}\, d\theta$$

$$S = 2\int_0^\pi \sqrt{2}a(1+\cos\theta)^{1/2}\, d\theta = 4a\int_0^\pi \cos\frac{\theta}{2}\, d\theta$$

$$= 8a\sin\frac{\theta}{2}\Big|_0^\pi = 8a$$

$$4a\bar{x} = \sqrt{2}a^2 \int_0^\pi (1+\cos\theta)\cos\theta(1+\cos\theta)^{1/2}\, d\theta$$

$$= 4a^2 \int_0^\pi \cos^3\frac{\theta}{2}\left(1 - 2\sin^2\frac{\theta}{2}\right) d\theta$$

$$= 4a^2 \int_0^\pi \left(1 - \sin^2\frac{\theta}{2}\right)\left(1 - 2\sin^2\frac{\theta}{2}\right)\cos\frac{\theta}{2}\, d\theta$$

$$= 4a^2 \int_0^\pi \left(2\sin^4\frac{\theta}{2} - 3\sin^2\frac{\theta}{2} + 1\right)\cos\frac{\theta}{2}\, d\theta$$

$$= 4a^2\left(\frac{4}{5}\sin^5\frac{\theta}{2} - 2\sin^3\frac{\theta}{2} + 2\sin\frac{\theta}{2}\right)\Big|_0^\pi = 4a^2\frac{4}{5}$$

$$\bar{x} = \frac{4}{5}a$$

We had to use the symmetry of the figure and write $M_y = 2\int_0^\pi r\cos\theta\, ds$ as $\cos\frac{\theta}{2} = +\sqrt{\frac{1+\cos\theta}{2}}$ is true only for $0 < \theta < \pi$. By symmetry we have $\bar{y} = 0$, or $M_x = \sqrt{2}a^2 \int_0^{2\pi} (1+\cos\theta)^{3/2}\sin\theta\, d\theta = 0$.

13.3.9. Find the centroid of the area bounded by the leaf of the curve $r = a\cos n\theta$ from $\theta = -\frac{\pi}{2n}$ to $\theta = \frac{\pi}{2n}$.

Solution. We have previously shown the area to be $\frac{\pi a^2}{4n}$. By symmetry $\bar{y} = 0$.

$$\frac{\pi a^2}{4n}\bar{x} = \frac{a^3}{3}\int_{-\pi/2n}^{\pi/2n} \cos^3 n\theta \cos\theta\, d\theta$$

$\cos^3 n\theta \cos \theta = \cos^2 n\theta \cos n\theta \cos \theta$

$$= \tfrac{1}{2} \cos^2 n\theta [\cos (n+1)\theta + \cos (n-1)\theta]$$
$$= \tfrac{1}{4} \cos n\theta [\cos (2n+1)\theta + 2 \cos \theta + \cos (2n-1)\theta]$$
$$= \tfrac{1}{8} [\cos (3n+1)\theta + 3 \cos (n+1)\theta$$
$$\qquad + 3 \cos (n-1)\theta + \cos (3n-1)\theta]$$

$$\frac{a^2}{n} \bar{x} = \frac{a^3}{24} \int_{-\pi/2n}^{\pi/2n} [\cos (3n+1)\theta + 3 \cos (n+1)\theta$$
$$\qquad + 3 \cos (n-1)\theta + \cos (3n-1)\theta] \, d\theta$$

$$= \frac{a^3}{24} \left[\frac{\sin (3n+1)\theta}{3n+1} + \frac{3 \sin (n+1)\theta}{n+1} \right.$$
$$\left. + \frac{3 \sin (n-1)\theta}{n-1} + \frac{\sin (3n-1)\theta}{3n-1} \right]_{-\pi/2n}^{\pi/2n}$$

$$= \frac{a^3}{12} \left[\frac{1}{3n+1} \sin \left(\frac{3\pi}{2} + \frac{\pi}{2n} \right) + \frac{3}{n+1} \sin \left(\frac{\pi}{2} + \frac{\pi}{2n} \right) \right.$$
$$\left. + \frac{3}{n-1} \sin \left(\frac{\pi}{2} - \frac{\pi}{2n} \right) + \frac{1}{3n+1} \sin \left(\frac{3\pi}{2} - \frac{\pi}{2n} \right) \right]$$

$$= \frac{a^3}{12} \cos \frac{\pi}{2n} \left(- \frac{1}{3n+1} + \frac{3}{n+1} + \frac{3}{n-1} - \frac{1}{3n-1} \right)$$

$$= \frac{4a^3 n^3}{(n^2-1)(9n^2-1)} \cos \frac{\pi}{2n}$$

$$\bar{x} = \frac{16an^4}{\pi(n^2-1)(9n^2-1)} \cos \frac{\pi}{2n} ; \quad \lim_{n \to \infty} \bar{x} = \frac{16a}{9\pi}$$

13.4. Moments with respect to planes. The moments of a volume with respect to the xy plane is given by $M_{xy} = \int_a^b z A(z) \, dz$, where $A(z)$ is the area of a slab parallel to the xy plane. Similar expressions can be written for moments to the xz and yz planes. For a surface generated by revolving about the x axis, the moment with respect to the yz plane is given by $N_{yz} = 2\pi \int_a^b xy \, ds$. By symmetry, the centroid of this surface will be on the x axis. In the problems on moment and moment of inertia in this chapter, we assume that the density of the object we are speaking of is uniform and equal to one.

13.4.1. Find the moment of the volume in the first octant of the ellipsoid $\dfrac{x^2}{a^2} + \dfrac{y^2}{b^2} + \dfrac{z^2}{c^2} = 1$ with respect to the xy plane. Find the centroid.

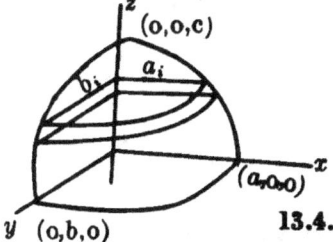

13.4.1.

Solution. An elementary slab parallel to the xy plane has the area $\dfrac{\pi a_i b_i}{4}$.

$$M_{xy} = \int_0^c z A(z)\, dz$$

When $y = 0, x = a_i$,

$$a_i = a\sqrt{1 - \frac{z_i^2}{c^2}};\qquad b_i = b\sqrt{1 - \frac{z_i^2}{c^2}}$$

$$M_{xy} = \frac{\pi ab}{4}\int_0^c \left(1 - \frac{z^2}{c^2}\right) z\, dz = -\frac{\pi ab}{4}\cdot\frac{c^2}{4}\left(1 - \frac{z^2}{c^2}\right)^2\bigg|_0^c = \frac{\pi abc^2}{16}$$

We can similarly find $M_{yz} = \dfrac{\pi a^2 bc}{16}$; $M_{xz} = \dfrac{\pi ab^2 c}{16}$

From **12.5.1**, $V = \dfrac{\pi abc}{6}$; $\bar{x} = \dfrac{3}{8}a$; $\bar{y} = \dfrac{3}{8}b$; $\bar{z} = \dfrac{3}{8}c$

13.4.2. Find the moment with respect to the yz plane of the volume generated by revolving the area in the first quadrant between $y = e^{-x}$ and the x axis about the x axis.

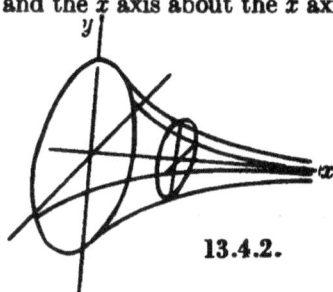

13.4.2.

Solution. Here $A(x) = \pi y^2$.

$$M_{yz} = \pi \int_0^\infty xy^2 \, dx = \pi \int_0^\infty xe^{-2x} \, dx$$

$$= \pi \left(-\frac{1}{2} xe^{-2x} - \frac{1}{4} e^{-2x} \right)\Big|_0^\infty = \frac{\pi}{4}$$

$$M_{zz} = M_{xy} = 0; \quad V = \int_0^\infty \pi e^{-2x} \, dx = \frac{\pi}{2}$$

so $\bar{x} = \frac{1}{2}$, and the centroid is $(\frac{1}{2}, 0, 0)$.

13.4.3. Find the centroid of the solid bounded by

$$\frac{x^2}{a^2} + \frac{y^2}{b^2} - \frac{z^2}{c^2} = 1, \; z = 0, \text{ and } z = c.$$

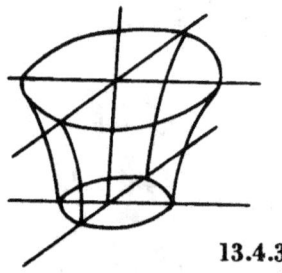

13.4.3.

Solution. By symmetry $M_{xz} = M_{yz} = 0$; $M_{xy} = \int_0^c zA(z)\,dz$. The area of the elliptical cross section is $\pi a_i b_i$, where $a_i = a\sqrt{1 + \frac{z_i^2}{c_i^2}}$; $b_i = b\sqrt{1 + \frac{z_i^2}{c_i^2}}$

$$M_{xy} = \pi ab \int_0^c \left(1 + \frac{z^2}{c^2}\right) z \, dz = \pi ab \left(\frac{z^2}{2} + \frac{z^4}{4c^2}\right)\Big|_0^c = \frac{3}{4} \pi abc^2$$

$$V = \pi ab \int_0^c \left(1 + \frac{z^2}{c^2}\right) dz = \pi ab \left(z + \frac{z^3}{3c^2}\right)\Big|_0^c = \frac{4}{3} \pi abc$$

$$\bar{z} = \frac{9}{16} c$$

13.4.4. Find the centroid of that portion of $x^{2/3} + y^{2/3} + z^{2/3} = c^{2/3}$ that lies above the xy plane.

13.4.4.

Solution. Here $\bar{x} = \bar{y} = 0$; $x^{2/3} + y^{2/3} = c^{2/3} - z_i^{2/3}$ for any z_i. By a previous example $A(z_i) = \frac{3}{8}\pi(c^{2/3} - z_i^{2/3})^3$.

$$M_{xy} = \frac{3\pi}{8}\int_0^c (c^2 - 3c^{4/3}z^{2/3} + 3c^{2/3}z^{4/3} - z^2)z\, dz$$

$$= \frac{3\pi}{8}\left(\frac{c^2 z^2}{2} - \frac{9}{8}c^{4/3}z^{8/3} + \frac{9}{10}c^{2/3}z^{10/3} - \frac{z^4}{4}\right)\Big|_0^c = \frac{3\pi c^4}{320}$$

$$V = \frac{3\pi}{8}\int_0^c (c^2 - 3c^{4/3}z^{2/3} + 3c^{2/3}z^{4/3} - z^2)\, dz$$

$$= \frac{3\pi}{8}\left(c^2 z - \frac{9}{5}c^{4/3}z^{5/3} + \frac{9}{7}c^{2/3}z^{7/3} - \frac{z^3}{3}\right)\Big|_0^c = \frac{6\pi}{105}c^3;$$

$$\bar{z} = \frac{21c}{128}$$

13.4.5. That portion of the sphere $x^2 + y^2 + z^2 = 4a^2$ lying above the xy plane has a cylindrical hole of equation $x^2 + y^2 = a^2$. Find the centroid of the resulting volume.

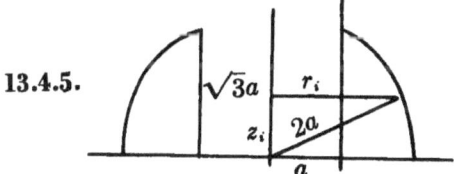

13.4.5.

Solution

$$\bar{x} = \bar{y} = 0$$

$$A(z_i) = \pi r_i^2 - \pi a^2 = \pi(4a^2 - z_i^2) - \pi a^2 = \pi(3a^2 - z_i^2)$$

$$M_{xy} = \pi \int_0^{\sqrt{3}a} (3a^2 - z^2)z\, dz = -\frac{\pi}{4}(3a^2 - z^2)^2\Big|_0^{\sqrt{3}a} = \frac{9}{4}\pi a^4$$

$$V = \pi \int_0^{\sqrt{3}a} (3a^2 - z^2)\, dz = \pi \left(3a^2 z - \frac{z^3}{3}\right)\Big|_0^{\sqrt{3}a} = 2\pi\sqrt{3}a^3$$

$$\bar{z} = \frac{3\sqrt{3}a}{8}$$

13.4.6. Find the centroid of the hemispherical surface generated by revolving about the y axis the half of $x^2 + y^2 = a^2$ lying above the x axis.
Solution

$$\bar{x} = \bar{z} = 0$$

$$N_{xz} = \int_0^a y 2\pi x\, ds = 2\pi \int_0^a y\sqrt{a^2 - y^2}\,\sqrt{\frac{y^2}{a^2 - y^2} + 1}\, dy$$

$$= 2\pi a \int_0^a y\, dy = \pi a y^2\Big|_0^a = \pi a^3$$

$$A = 2\pi a^2; \qquad \bar{z} = \frac{a}{2}$$

13.4.7. Find the centroid of the surface generated when the rst arch of the cycloid $x = a(\theta - \sin\theta)$, $y = a(1 - \cos\theta)$ is evolved about the x axis.
Solution

$$\bar{y} = \bar{z} = 0$$

$$V_{yz} = \int_0^{2\pi} x \cdot 2\pi y\, ds$$

$$= 2\pi a^3 \int_0^{2\pi} (\theta - \sin\theta)(1 - \cos\theta)\sqrt{(1 - \cos\theta)^2 + \sin^2\theta}\, d\theta$$

$$= 2\pi\sqrt{2}a^3 \int_0^{2\pi} \theta(1 - \cos\theta)^{3/2}\, d\theta$$

$$\quad + 2\pi\sqrt{2}a^3 \int_0^{2\pi} (1 - \cos\theta)^{3/2}(-\sin\theta)\, d\theta$$

$$= 8\pi a^3 \int_0^{2\pi} \theta \sin^3\frac{\theta}{2}\, d\theta - 2\pi\sqrt{2}a^3 \frac{2}{5}(1 - \cos\theta)^{5/2}\Big|_0^{2\pi}$$

$$= 8\pi a^3 \int_0^{2\pi} \theta\left(1 - \cos^2\frac{\theta}{2}\right)\sin\frac{\theta}{2}\, d\theta$$

$$= 8\pi a^3 \left[\theta\left(-2\cos\frac{\theta}{2} + \frac{2}{3}\cos^3\frac{\theta}{2}\right)\right]_0^{2\pi}$$

$$- 8\pi a^3 \int_0^{2\pi}\left(-2\cos\frac{\theta}{2} + \frac{2}{3}\cos^3\frac{\theta}{2}\right)d\theta$$

$$= 8\pi\left(2\pi \times \frac{4}{3}\right)a^3$$

$$- 8\pi a^3 \left[-4\sin\frac{\theta}{2} + \frac{2}{3}\left(2\sin\frac{\theta}{2} - \frac{2}{3}\sin^3\frac{\theta}{2}\right)\right]_0^{2\pi}$$

$$= \frac{64\pi^2 a^3}{3}$$

$$A = 2\pi a^2 \int_0^{2\pi}(1 - \cos\theta)\sqrt{2 - 2\cos\theta}\,d\theta = 8\pi a^2 \int_0^{2\pi}\sin^3\frac{\theta}{2}\,d\theta$$

$$= 8\pi a^2\left(-2\cos\frac{\theta}{2} + \frac{2}{3}\cos^3\frac{\theta}{2}\right)\bigg|_0^{2\pi} = \frac{64\pi}{2}a^2$$

$$\bar{x} = \pi a$$

13.5. Moments of inertia of plane areas and arcs. Fo formulas see 4.5 and 6.3.

13.5.1. Find the moment of inertia with respect to each axi of the area in the first quadrant between $y = e^{-x}$ and the x axis.

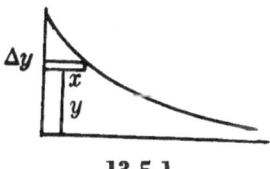

13.5.1.

Solution

$$I_x = \int_0^1 y^2 x\,dy = -\int_0^1 y^2 \ln y\,dy = \left(-\frac{y^3}{3}\ln y + \frac{y^3}{9}\right)_0^1 = \frac{1}{9}$$

$$I_y = \int_0^{+\infty} x^2 y\,dx = \int_0^{+\infty} x^2 e^{-x}\,dx$$

$$= \left(-x^2 e^{-x} - 2xe^{-x} - 2e^{-x}\right)_0^{+\infty} = 2$$

nce $A = \int_0^{+\infty} e^{-x} dx = 1$; $I_x = \frac{1}{9} A$; $I_y = 2A$

13.5.2. Find the moment of inertia with respect to the x axis of the area under the first arch of the cycloid $x = (\theta - \sin \theta)$, $y = a(1 - \cos \theta)$.

Solution. Here our element of area is given by

$$(x_{2\pi - \theta} - x_\theta) \Delta y = [a(2\pi - \theta - \sin(2\pi - \theta)) - a(\theta - \sin \theta)] \Delta y$$
$$= a(2\pi - 2\theta + 2 \sin \theta) \Delta y$$

$$I_x = \int y^2 \, dA = a^4 \int_0^\pi (1 - \cos \theta)^2 (2\pi - 2\theta + 2 \sin \theta) \sin \theta \, d\theta$$

$$= \frac{a^4}{3} (1 - \cos \theta)^3 \left(2\pi - 2\theta + 2 \sin \theta \right) \Big|_0^\pi$$

$$= \frac{-2a^4}{3} \int_0^\pi (1 - \cos \theta)^3 (\cos \theta - 1) \, d\theta$$

$$= \frac{2a^4}{3} \int_0^\pi \left(16 \sin^8 \frac{\theta}{2} \right) d\theta = \frac{64 a^4}{3} \int_0^{\pi/2} \sin^8 u \, du$$

13.5.2.

Using the results of problem **10.4.12**,

$$\int_0^{\pi/2} \sin^8 u \, du = \left(-\frac{\sin^7 u \cos u}{8} - \frac{7}{8 \cdot 6} \sin^5 u \cos u \right.$$
$$\left. - \frac{7 \cdot 5}{8 \cdot 6 \cdot 4} \sin^3 u \cos u - \frac{7 \cdot 5 \cdot 3}{8 \cdot 6 \cdot 4 \cdot 2} \sin u \cos u + \frac{7 \cdot 5 \cdot 3 \cdot 1}{8 \cdot 6 \cdot 4 \cdot 2} u \right)_0^{\pi/2}$$

Now $I_x = \frac{35}{12} \pi a^4$, and since $A = 3\pi a^2$, then $I_x = \frac{35}{36} a^2 A$.

13.5.3. Find the moment of inertia of the area $\frac{x^2}{a^2} + \frac{y^2}{b^2} = 1$ with respect to the x axis.

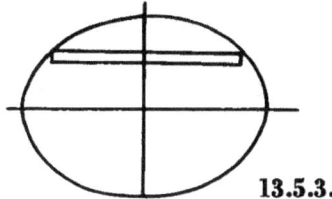

13.5.3.

Solution

$$I_x = \int y^2\, dA = \int_{-b}^{b} y^2 \cdot 2x\, dy = 2a \int_{-b}^{b} y^2 \sqrt{1 - \frac{y^2}{b^2}}\, dy$$

$$y = b \sin \theta$$

$$I_x = 2ab^3 \int_{-\pi/2}^{\pi/2} \sin^2 \theta \sqrt{1 - \sin^2 \theta} \cos \theta\, d\theta$$

$$= 2ab^3 \int_{-\pi/2}^{\pi/2} \sin^2 \theta \cos^2 \theta\, d\theta = \frac{2ab^3}{4} \int_{-\pi/2}^{\pi/2} \left(\frac{1 - \cos 4\theta}{2}\right) d\theta$$

$$= \frac{ab^3}{4}\left(\theta - \frac{\sin 4\theta}{4}\right)\Big|_{-\pi/2}^{\pi/2} = \frac{\pi ab^3}{4} = A\frac{b^2}{4}$$

13.5.4. Find the moment of inertia of the semicircular arc o $x^2 + y^2 = a^2$ above the x axis with respect to the x axis.

Solution

$$I_x = \int_0^a y^2\, ds = \int_0^a y^2 \sqrt{\left(\frac{y}{\sqrt{a^2 - y^2}}\right)^2 + 1}\, dy$$

$$= a \int_0^a \frac{y^2\, dy}{\sqrt{a^2 - y^2}}$$

$$y = a \sin \theta$$

$$I_x = a^3 \int_0^{\pi/2} \frac{\sin^2 \theta \cos \theta\, d\theta}{\sqrt{1 - \sin^2 \theta}} = a^3 \int_0^{\pi/2} \sin^2 \theta\, d\theta$$

$$= \frac{a^3}{2}\left(\theta - \frac{\sin 2\theta}{2}\right)\Big|_0^{\pi/2} = \frac{\pi a^3}{4} = s\frac{a^2}{4}$$

13.5.5. Find the moment of inertia of the arc of one arch of the cycloid $x = a(\theta - \sin \theta)$, $y = a(1 - \cos \theta)$ about the x axis.

Solution

$$= \int y^2 \, ds = a^3 \int_0^{2\pi} (1 - \cos\theta)^2 \sqrt{(1 - \cos\theta)^2 + \sin^2\theta} \, d\theta$$

$$= \sqrt{2}a^3 \int_0^{2\pi} (1 - \cos\theta)^{5/2} \, d\theta = 8a^3 \int_0^{2\pi} \sin^5 \frac{\theta}{2} \, d\theta$$

$$= 8a^3 \int_0^{2\pi} \left(1 - 2\cos^2 \frac{\theta}{2} + \cos^4 \frac{\theta}{2}\right) \sin \frac{\theta}{2}$$

$$= 8a^3 \left(-2\cos\frac{\theta}{2} + \frac{4}{3}\cos^3\frac{\theta}{2} - \frac{2}{5}\cos^5\frac{\theta}{2}\right)\Big|_0^{2\pi} = \frac{256}{15} a^3$$

$$s = 8a; \qquad I_x = \frac{32}{15} sa^2$$

13.5.6. Find the moment of inertia of the area under $y = \sin x$ from $x = 0$ to $x = \pi$ with respect to the y axis.

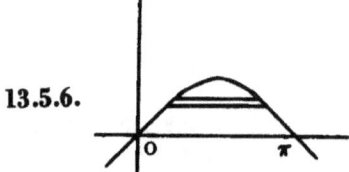

13.5.6.

Solution

$$I_y = \int x^2 \, dA = \int_0^\pi x^2 y \, dx = \int_0^\pi x^2 \sin x \, dx$$

$$= -x^2 \cos x \Big|_0^\pi + 2\int_0^\pi x \cos x \, dx$$

$$= (-x^2 \cos x + 2x \sin x + 2\cos x)_0^\pi = \pi^2 - 4$$

$$I_x = \int y^2 \, dA = \int_0^1 y^2 (\pi - 2x) \, dy$$

$$= \int_0^{\pi/2} (\sin^2 x)(\pi - 2x) \cos x \, dx$$

$$= \frac{(\pi - 2x)}{3} \sin^3 x \Big|_0^{\pi/2} + \frac{2}{3} \int_0^{\pi/2} \sin^3 x \, dx$$

$$= \frac{2}{3}\int_0^{\pi/2}(1-\cos^2 x)\sin x\,dx = \frac{2}{3}\left(-\cos x + \frac{\cos^3 x}{3}\right)\Big|_0^{\pi/2} = \frac{4}{9}$$

$$A = \int_0^\pi \sin x\,dx = 2; \quad I_z = \frac{2}{9}A; \quad I_y = \frac{\pi^2 - 4}{2}A$$

13.6. Moments of inertia of surfaces and solids of revolution. By the cylindrical shell method, we obtain the following formula for the moment of inertia of a solid of revolution about its axis: $I = 2\pi\int_a^b r^3 h\,dr$. If disks of thickness Δh are used, $I = \frac{\pi}{2}\int_a^b r^4\,dh$. For a surface of revolution, we use ring-shaped elements of area generated by an element of arc parallel to the axis of revolution. $I = 2\pi\int_a^b r^3\,ds$.

13.6.1. Find the moment of inertia with respect to the x axis of the volume generated when the area in the first quadrant under $y = e^{-x}$ is revolved about the x axis.

Solution. Using our rings

$$I_z = 2\pi\int_0^1 y^3 x\,dy = -2\pi\int_0^1 y^3 \ln y\,dy$$

$$= \left(-\frac{2\pi}{4}y^4 \ln y + \frac{2\pi}{16}y^4\right)_0^1 = \frac{2\pi}{16}$$

$$V = \pi\int_0^\infty y^2\,dx = \pi\int_0^\infty e^{-2x}\,dx = -\frac{1}{2}\pi e^{-2x}\Big|_0^\infty = \frac{\pi}{2}$$

Solution. Using disks of radius y and altitude Δx, $\Delta I_z = \frac{\pi}{2}y^4 \Delta x$.

$$I_z = \frac{\pi}{2}\int_0^\infty y^4\,dx = \frac{\pi}{2}\int_0^\infty e^{-4x}\,dx = -\frac{\pi}{8}e^{-4x}\Big|_0^\infty = \frac{\pi}{8} = \frac{1}{4}V$$

13.6.2. Find the moment of inertia with respect to the x axis of the volume generated when the cycloid $x = a(\theta - \sin\theta)$, $y = a(1 - \cos\theta)$ is revolved about the x axis, $x = 0$, $x = 2\pi$.

Using disks, we have

$$I_z = \frac{\pi}{2}\int_0^{2\pi} y^4\, dx = \frac{\pi}{2}a^5\int_0^{2\pi}(1-\cos\theta)^4(1-\cos\theta)\,d\theta$$

$$= \frac{\pi}{2}a^5\int_0^{2\pi} 2^5 \sin^{10}\frac{\theta}{2}\,d\theta$$

13.6.2.

Using the results of **10.4.12**,

$$I_z = 32\pi a^5 \int_0^{\pi} \sin^{10} u\, du$$

$$= 32\pi a^5 \left(-\frac{1}{10}\sin^9 u\cos u - \frac{9}{10\cdot 8}\sin^7 u\cos u\right.$$

$$-\frac{9\cdot 7}{10\cdot 8\cdot 6}\sin^5 u\cos u - \frac{9\cdot 7\cdot 5}{10\cdot 8\cdot 6\cdot 4}\sin^3 u\cos u$$

$$\left.-\frac{9\cdot 7\cdot 5\cdot 3}{10\cdot 8\cdot 6\cdot 4\cdot 2}\sin u\cos u + \frac{9\cdot 7\cdot 5\cdot 3\cdot 1}{10\cdot 8\cdot 6\cdot 4\cdot 2}u\right)\Big|_0^{\pi}$$

$$= \frac{63}{8}\pi^2 a^5 = \frac{63}{40}a^2 V$$

13.6.3. Find the moment of inertia about the axis of revolution of the volume generated when $\dfrac{x^2}{a^2}+\dfrac{y^2}{b^2}=1$ is revolved about the x axis.

Solution. Using disks,

$$I_z = \frac{\pi}{2}\int_{-a}^{a} y^4\, dx = \frac{\pi}{2}\int_{-a}^{a} b^4\left(1-\frac{x^2}{a^2}\right)^2 dx$$

$$= \frac{\pi}{2}b^4\left(x - \frac{2x^3}{3a^2} + \frac{x^5}{5a^4}\right)\Big|_{-a}^{a} = \frac{\pi}{2}ab^4\left(\frac{16}{15}\right) = \frac{8\pi ab^4}{15}$$

We could also have computed this using $x = a\cos\theta$, $y = b\sin\theta$, and a ring element with axis on the x axis. Here the altitude of the ring is $2x$.

$$I_z = 4\pi \int y^2 x\, dy = 4\pi b^4 \int_0^{\pi/2} a \sin^3\theta \cos^2\theta\, d\theta$$

$$= 4\pi a b^4 \int_0^{\pi/2} (\cos^2\theta - \cos^4\theta) \sin\theta\, d\theta$$

$$= 4\pi a b^4 \left(-\frac{1}{3}\cos^3\theta + \frac{1}{5}\cos^5\theta\right)\Big|_0^{\pi/2} = \frac{8\pi a b^4}{15}$$

13.6.4. Find the moment of inertia with respect to the x axis of the surface generated by rotating $x^{2/3} + y^{2/3} = a^{2/3}$ about the x axis.

Solution

$$I_z = 2\pi \int y^3\, ds = 2\pi \int_0^a y^3 \sqrt{\frac{x^{2/3}}{y^{2/3}} + 1}\, dy$$

$$= 2\pi a^{1/3} \int_0^a y^{8/3}\, dy = \frac{6}{11}\pi a^{1/3} y^{11/3}\Big|_0^a = \frac{6}{11}\pi a^4$$

13.6.5. Find the moment of inertia with respect to the x axis of the surface generated by rotating one arch of the cycloid $x = a(\theta - \sin\theta)$, $y = a(1 - \cos\theta)$ about the x axis.

Solution. Here $ds = a\sqrt{2 - 2\cos\theta}\, d\theta$.

$$I_z = 2\pi \int y^3\, ds = 2\sqrt{2}\pi a^4 \int_0^{2\pi} (1 - \cos\theta)^{7/2}\, d\theta$$

$$= 32\pi a^4 \int_0^{2\pi} \sin^7\frac{\theta}{2}\, d\theta$$

$$= 32\pi a^4 \int_0^{2\pi} \left(1 - 3\cos^2\frac{\theta}{2} + 3\cos^4\frac{\theta}{2} - \cos^6\frac{\theta}{2}\right) \sin\frac{\theta}{2}\, d\theta$$

$$= 32\pi a^4 \left(-2\cos\frac{\theta}{2} + 2\cos^3\frac{\theta}{2} - \frac{6}{5}\cos^5\frac{\theta}{2} + \frac{2}{7}\cos^7\frac{\theta}{2}\right)\Big|_0^{2\pi}$$

$$= 32\pi a^4 \times \frac{64}{35} = \frac{2048\pi a^4}{35} = \frac{96}{35} a^2 A \quad \text{See 12. 4. 1.}$$

13.6.6. Find the moment of inertia with respect to the axis of revolution when $y = \sqrt{x}$ is revolved about the x axis from $x = 0$ to 2.

Solution

$$I_z = 2\pi \int_0^2 y^3\, ds = 2\pi \int_0^2 x^{3/2}\sqrt{1+\frac{1}{4x}}\, dx$$

$$= 2\pi \int_0^2 x\sqrt{x+\frac{1}{4}}\, dx$$

$$= 2\pi \left(x\cdot\frac{2}{3}\left(x+\frac{1}{4}\right)^{3/2} - \frac{4}{15}\left(x+\frac{1}{4}\right)^{5/2}\right)\Big|_0^2 = \frac{149\pi}{30}$$

$$A = 2\pi \int_0^2 y\, ds = 2\pi \int_0^2 \sqrt{x}\sqrt{1+\frac{1}{4x}}\, dx$$

$$= 2\pi \int_0^2 \sqrt{x+\frac{1}{4}}\, dx = \frac{4}{3}\pi \left(x+\frac{1}{4}\right)^{3/2}\Big|_0^2 = \frac{13\pi}{3}$$

$$I_z = \frac{149}{130} A$$

3.7. The theorems of Pappus. The two theorems of Pappus an be stated:

1. If the arc of a plane curve is revolved about an axis in its lane and not crossing the arc, the area generated is equal to he product of the length of arc and the length of the path escribed by its center of gravity.

2. If a plane area is revolved about an axis in its plane and ot crossing the area, the volume generated is equal to the roduct of the area and the length of path described by its enter of gravity.

13.7.1. The cardioid $r = a(1 + \cos\theta)$ is revolved about the nitial line. Find the volume generated.

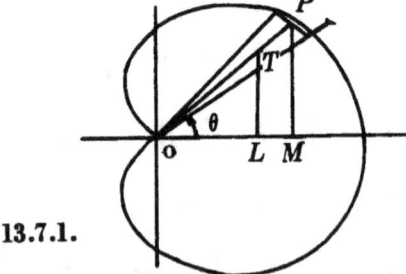

13.7.1.

Solution. We choose for our element of volume the conical shell cut out by the element of area as shown.

The volume of the conical shell may be found by means of the theorem of Pappus. Then $PM = r \sin \theta$, and TL, distance of centroid from the initial line, by similar triangles is $\tfrac{2}{3} r \sin \theta$.

$$\Delta V = 2\pi \bar{y}\, \Delta A = \Delta A \tfrac{2}{3}\pi r \sin \theta = \tfrac{2}{3}\pi r^2 \sin \theta\, \Delta \theta$$

$$V = \frac{2}{3}\pi a^3 \int_0^\pi (1+\cos\theta)^3 \sin\theta\, d\theta$$

$$= -\frac{2}{3}\pi a^3 \left. \frac{(1+\cos\theta)^4}{4}\right|_0^\pi = \frac{8}{3}\pi a^3$$

13.7.2. Find the volume generated when the leaf of the curve $r = \cos 2\theta$ from $\theta = -\tfrac{\pi}{4}$ to $\theta = \tfrac{\pi}{4}$ is revolved about the initial line.

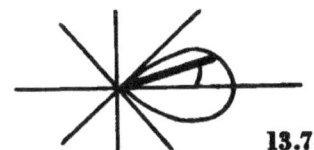

13.7.2.

Solution. Again the radius for the centroid is $\tfrac{2}{3} r \sin \theta$.

$$V = \frac{2\pi}{3}\int_0^{\pi/4} r^3 \sin\theta\, d\theta = \frac{2\pi}{3}\int_0^{\pi/4} (2\cos^2\theta - 1)^3 \sin\theta\, d\theta$$

$$= \frac{2\pi}{3}\int_0^{\pi/4} (8\cos^6\theta - 12\cos^4\theta + 6\cos^2\theta - 1) \sin\theta\, d\theta$$

$$= \frac{2\pi}{3}\left(-\frac{8}{7}\cos^7\theta + \frac{12}{5}\cos^5\theta - 2\cos^3\theta + \cos\theta\right)\bigg|_0^{\pi/4}$$

$$= \frac{2\pi}{3}\left(-\frac{1}{7\sqrt{2}} + \frac{3}{5\sqrt{2}} - \frac{1}{\sqrt{2}} + \frac{1}{\sqrt{2}} + \frac{8}{7} - \frac{12}{5} + 2 - 1\right)$$

$$= \left(\frac{16\sqrt{2}}{105} - \frac{6}{35}\right)\pi$$

13.7.3. Use the theorems of Pappus to find the centroid of the semicircular arc of $x^2 + y^2 = a^2$ above the x axis.

Solution. By symmetry, we know $\bar{x} = 0$. The area generated y revolving the arc about the x axis is $4\pi a^2$; the length of arc πa.

Using the first theorem of Pappus,

$$A = 2\pi \bar{y} s; \qquad 4\pi a^2 = 2\pi \bar{y} \pi a; \qquad \bar{y} = \frac{2a}{\pi}$$

Chapter 14

INFINITE SERIES

14.1. Convergence and divergence of infinite series. We denote by S_n the sum of the first n terms of the infinite series

$$a_0 + a_1 + a_2 + \ldots + a_n + \ldots$$

The series is said to converge to S if $\lim_{n \to \infty} S_n = S$. If the series does not converge, it is said to diverge. Consider first the geometric series

$$a + ar + ar^2 + \ldots + ar^n + \ldots = \sum_{n=0}^{\infty} ar^n$$

whose partial sums are given as

$$S_n = a \frac{1 - r^n}{1 - r} = \frac{a}{1 - r} - a \frac{r^n}{1 - r}$$

If $|r| < 1$, $\lim_{n \to \infty} r^n = 0$, and $\lim_{n \to \infty} S_n = \frac{a}{1 - r}$. If $|r| > 1$, $\lim_{n \to \infty} r^n$ does not exist, and the series diverges. If $r = 1$, our formula for s_n given above is invalid.

$$S_n = a + a + \ldots + a = na$$

and so the series diverges. If $r = -1$, then $S_n = a - a + a - \ldots + (-1)^{n-1} a$. Now $S_n = 0$ if n is even, $S_n = a$ if n is odd. This series diverges in the sense that $\lim_{n \to \infty} S_n$ does not exist. Notice, however, that the partial sums S_n oscillate between 0 and a and do not become infinitely large as when $r \geqslant 1$. Consider now the harmonic series,

$$1 + \frac{1}{2} + \frac{1}{3} + \ldots + \frac{1}{n} + \ldots = \sum_{1}^{\infty} \frac{1}{n}$$

Group the terms as follows:

$$1 + (\tfrac{1}{2}) + (\tfrac{1}{3} + \tfrac{1}{4}) + (\tfrac{1}{5} + \tfrac{1}{6} + \tfrac{1}{7} + \tfrac{1}{8})$$
$$+ (\tfrac{1}{9} + \tfrac{1}{10} + \tfrac{1}{11} + \tfrac{1}{12} + \tfrac{1}{13} + \tfrac{1}{14} + \tfrac{1}{15} + \tfrac{1}{16}) + \ldots$$

the sum in each parenthesis is greater than $\frac{1}{2}$, since

$$\tfrac{1}{3}+\tfrac{1}{4} > \tfrac{1}{4}+\tfrac{1}{4} = \tfrac{1}{2}$$

$$\tfrac{1}{5}+\tfrac{1}{6}+\tfrac{1}{7}+\tfrac{1}{8} > \tfrac{1}{8}+\tfrac{1}{8}+\tfrac{1}{8}+\tfrac{1}{8} = \tfrac{1}{2}$$

$$+\tfrac{1}{10}+\tfrac{1}{11}+\tfrac{1}{12}+\tfrac{1}{13}+\tfrac{1}{14}+\tfrac{1}{15}+\tfrac{1}{16}$$

$$> \tfrac{1}{16}+\tfrac{1}{16}+\tfrac{1}{16}+\tfrac{1}{16}+\tfrac{1}{16}+\tfrac{1}{16}+\tfrac{1}{16}+\tfrac{1}{16} = \tfrac{1}{2}$$

$$\vdots$$

$$\frac{1}{2^{k-1}+1} + \frac{1}{2^{k-1}+2} + \ldots + \frac{1}{2^k} > \frac{1}{2^k} + \frac{1}{2^k} + \ldots + \frac{1}{2^k} = \frac{1}{2}$$

$$S_{2^k} > 1 + \frac{1}{2} + \frac{1}{2} + \ldots + \frac{1}{2} = \frac{k+2}{2}$$

Hence S_n has no limit as n increases, and the series is divergent.

4.2. Tests for convergence, the integral test. Since $\lim_{n\to\infty} a_n = \lim_{n\to\infty} (S_n - S_{n-1}) = S - S$, we see that a necessary condition for convergence is that $\lim_{n\to\infty} a_n = 0$. This rule immediately identifies the series

$$\frac{1}{2} - \frac{2}{3} + \frac{3}{4} - \frac{4}{5} + \ldots + (-1)^n \frac{n-1}{n} + \ldots$$

as divergent as $\lim_{n\to\infty} \frac{n-1}{n} = 1$. On the other hand, the condition $\lim_{n\to\infty} a_n = 0$ is not sufficient, as shown by the example above of the harmonic series.

The integral test can be stated as follows: If the terms a_n of a positive series can be stated as a function of n, $f(n)$, and if $f(x)$ is continuous for all $x \geq a$, where a is fixed, and if $f(x)$ never increases for $x \geq a$; then the series $\sum_{1}^{\infty} a_n$ converges if $\int_a^\infty f(x)\, dx$ exists and diverges if the integral does not exist.

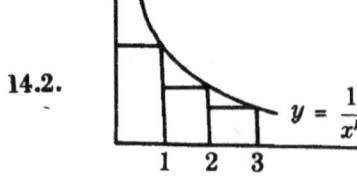

14.2.

Example: Discuss the convergence of $\sum \frac{1}{n^k}$. From the figure it can be seen that $y = \frac{1}{x^k}$ does not approximate the terms of the series very closely for $0 \leq x \leq 1$, and hence we integrate from a lower limit of 1 instead of 0.

If $k \neq 1$, $\int_1^\infty \frac{dx}{x^k} = \frac{-1}{k-1} \cdot \frac{1}{x^{k-1}} \Big|_1^\infty = \frac{+1}{k-1}$ if $k > 1$

The integral does not exist if $0 \leq k < 1$. If $k = 1$, we have the harmonic series, or $\int_1^\infty \frac{dx}{x} = \ln x \Big|_1^\infty$ which does not exist. If $k < 0$ the conditions of our test are not satisfied, since the terms are now increasing. However, if $k < 0$, $\lim_{n \to \infty} a_n = \lim_{n \to \infty} n^{-k} \neq 0$, and the series diverges. To sum up,

$$1 + \frac{1}{2^k} + \frac{1}{3^k} + \ldots + \frac{1}{n^k} + \ldots$$

converges when $k > 1$ and diverges for $k \leq 1$.

14.2.1. Consider the series

$$\tfrac{1}{2} + \tfrac{1}{5} + \tfrac{1}{10} + \tfrac{1}{17} + \ldots = \sum_{n=1}^\infty \frac{1}{n^2+1}$$

$$\int_0^\infty \frac{dx}{x^2+1} = \text{Arc tan } x \Big|_0^\infty = \frac{\pi}{2}$$

and the series converges. Notice the series $100 + 1000 - 15 + \tfrac{1}{16} + \tfrac{1}{17} + \ldots$ would be treated by $\int_3^\infty \frac{dx}{x^2+1}$ and would also converge.

14.2.2. $\quad \frac{1}{e} + \frac{4}{e^2} + \ldots + \frac{n^2}{e^n} + \ldots = \sum \frac{n^2}{e^n}$

$$\int_1^\infty z^2 e^{-z} \, dz = -z^2 e^{-z} \Big|_1^\infty + 2 \int_1^\infty z e^{-z} \, dz$$

$$= (-z^2 e^{-z} - 2z e^{-z} - 2e^{-z}) \Big|_1^\infty = \frac{5}{e}$$

and the series converges.

14.2.3. $\quad \dfrac{1}{1\cdot 2}+\dfrac{1}{2\cdot 3}+\ldots = \sum \dfrac{1}{n(n+1)}$

$$\int_1^\infty \dfrac{dx}{x(x+1)} = \int_1^\infty \dfrac{dx}{x} - \int_1^\infty \dfrac{dx}{x+1}$$

$$= \left[\ln x - \ln(x+1)\right]_1^\infty = \ln\dfrac{x}{x+1}\bigg|_1^\infty = \ln 2$$

and the series converges.

14.2.4. $\quad \dfrac{1}{1+9}+\dfrac{2}{16+9}+\ldots+\dfrac{n}{n^4+9}+\ldots$

$$\int_1^\infty \dfrac{x\,dx}{x^4+9} = \dfrac{1}{6}\operatorname{Arc\,tan}\dfrac{x^2}{3}\bigg|_1^\infty = \dfrac{\pi}{12}-\dfrac{1}{6}\operatorname{Arc\,tan}\dfrac{1}{3}$$

and the series converges.

14.2.5. $\quad \sum_1^\infty \dfrac{n}{n^2+4} = \dfrac{1}{5}+\dfrac{2}{8}+\dfrac{3}{13}+\ldots$

$$\int_1^\infty \dfrac{x\,dx}{x^2+1} = \dfrac{1}{2}\ln(x^2+1)\bigg|_1^\infty$$

This integral does not exist and the series diverges.

14.2.6. $1+\dfrac{1}{3}+\dfrac{1}{5}+\ldots+\dfrac{1}{2n-1}+\ldots = \sum_1^\infty \dfrac{1}{2n-1}$

$$\int_1^\infty \dfrac{dx}{2x-1} = \dfrac{1}{2}\ln|2x-1|\bigg|_1^\infty$$

The integral does not exist and the series diverges.

14.2.7. $\quad \sum_2^\infty \dfrac{1}{n\ln n} = \dfrac{1}{2\ln 2}+\dfrac{1}{3\ln 3}+\ldots$

$$\int_2^\infty \dfrac{dx}{x\ln x} = \ln\ln x\bigg|_2^\infty$$

The integral does not exist and the series diverges.

14.2.8. $\sum_{2}^{\infty} \frac{1}{n \ln^2 n} = \frac{1}{2 \ln^2 2} + \frac{1}{3 \ln^2 3} + \cdots$

$$\int_2^\infty \frac{dx}{x \ln^2 x} = -\frac{1}{\ln x}\Big|_2^\infty = \frac{1}{\ln 2}$$

and the series converges.

14.3. Comparison test. If the terms a_n of a positive series are all less than those c_n of a positive series known to be convergent then the series $\sum_{1}^{\infty} a_n$ converges. If the terms are greater than those of a known divergent series, $\sum_{1}^{\infty} a_n$ diverges. An often more useful statement of this theorem is that, if $\lim_{n \to \infty} \frac{a_n}{c_n}$ exists then $\sum_{1}^{\infty} a_n$ and $\sum_{1}^{\infty} c_n$ converge or (providing the limit does not equal zero) diverge together. Note that c_n, the nth term of the known series, is the denominator in the limit. In the following comparisons, we shall regard the convergence of $\sum_{n=1}^{\infty} \frac{1}{n^k}, k > 1$; $\sum_{n=1}^{\infty} ar^n, |r| < 1$; and the divergence of $\sum_{n=1}^{\infty} \frac{1}{n^k}, 0 < k \leq 1$, as known.

14.3.1. $\frac{1}{1 \cdot 2} + \frac{1}{2 \cdot 3} + \cdots = \sum_{n=1}^{\infty} \frac{1}{n(n+1)}$ is convergent, for we have $\frac{1}{n(n+1)} < \frac{1}{n^2}$ for every n, and $\sum_{n=1}^{n} \frac{1}{n^2}$ converges. $\sum \frac{1}{n(n-1)}$ cannot be so treated as $\frac{1}{n(n-1)} > \frac{1}{n^2}$. However,

$$\lim_{n \to \infty} \frac{\frac{1}{n^2 - n}}{\frac{1}{n^2}} = \lim_{n \to \infty} \frac{n^2}{n^2 - n} = 1,$$

and both series converge. It is also interesting to note that, since $\frac{1}{n(n+1)} = \frac{1}{n} - \frac{1}{n+1}$, the partial sums of $\sum_{1}^{\infty} \frac{1}{n(n+1)}$ can be expressed as

$$S_n = 1 - \frac{1}{2} + \frac{1}{2} - \frac{1}{3} + \frac{1}{3} - \frac{1}{4} + \cdots + \frac{1}{n} - \frac{1}{n+1}$$

$$= 1 - \frac{1}{n+1}$$

and $\lim_{n \to \infty} S_n = 1$. Hence we know directly from the definition that the series converges and that the sum it converges to is $S = 1$.

14.3.2. $1 + \frac{1}{2^2} + \frac{1}{3^3} + \cdots = \sum_{1}^{\infty} \frac{1}{n^n}$ converges, since $\frac{1}{n^n} < \frac{1}{2^n}$ if $n > 2$, and $\sum_{1}^{\infty} \frac{1}{2^n}$ is the geometric series when $r = \frac{1}{2}$. Notice here we could have compared $\sum \frac{1}{n^n}$ with $\sum \frac{1}{10^n}$ if we added the explanation $\frac{1}{n^n} < \frac{1}{10^n}$ if $n > 10$. Here again the behavior of the first few terms does not influence the convergence. Also note that the form of the test used is simpler than the evaluation of

$$\lim_{n \to \infty} \frac{\frac{1}{n^n}}{\frac{1}{2^n}} = \lim_{n \to \infty} \left(\frac{2}{n}\right)^n.$$

14.3.3. $\sum \frac{1}{\sqrt{n(n+2)}} = \frac{1}{\sqrt{1 \cdot 3}} + \frac{1}{\sqrt{2 \cdot 4}} + \cdots$

Solution. Here $\lim_{n \to \infty} \frac{\frac{1}{\sqrt{n(n+2)}}}{\frac{1}{n}} = \lim_{n \to \infty} \frac{n}{\sqrt{n(n+2)}} =$

$\lim_{n \to \infty} \frac{1}{\sqrt{1 + \frac{2}{n}}} = 1$; <u>the series diverges.</u>

14.3.4. $\sum_{n=1}^{\infty} \frac{n+7}{n(n+1)(n+2)}$ $\sum \frac{1}{n^2}$ converges.

Here $\lim\limits_{n\to\infty} \dfrac{\dfrac{n+7}{n(n+1)(n+2)}}{\dfrac{1}{n^2}} = \lim\limits_{n\to\infty} \dfrac{n^2+7n}{(n+1)(n+2)} = 1$; the series converges.

14.3.5. $1 + \dfrac{3}{5} + \dfrac{3^4}{5^4} + \ldots = \sum\limits_{n=0}^{\infty} \left(\dfrac{3}{5}\right)^{n^2}$

Since $\left(\frac{3}{5}\right)^{n^2} < \left(\frac{3}{5}\right)^n$ if $n > 1$, and $\sum \left(\frac{3}{5}\right)^n$ is known to converge, the series converges. Also

$$\lim\limits_{n\to\infty} \dfrac{\left(\dfrac{3}{5}\right)^{n^2}}{\left(\dfrac{3}{5}\right)^n} = \lim\limits_{n\to\infty} \left(\dfrac{3}{5}\right)^{n^2-n} = 0.$$

14.3.6. $\dfrac{(1001)^2}{2} + \dfrac{(1002)^2}{2^2} + \dfrac{(1003)^2}{2^3} + \ldots = \sum \dfrac{(n+1000)^2}{2^n}$

But $\sum\limits_{1}^{\infty} \dfrac{1}{n^2}$ converges and

$$\lim\limits_{n\to\infty} \dfrac{\dfrac{(n+1000)^2}{2^n}}{\dfrac{1}{n^2}} = \lim\limits_{n\to\infty} \dfrac{n^4 + 2000n^3 + 1000000n^2}{2^n} = 0$$

Hence the series converges.

14.3.7. $1 + \dfrac{1}{2!} + \dfrac{1}{3!} + \ldots = \sum \dfrac{1}{n!}, \dfrac{1}{n!} < \dfrac{1}{2^n}$ for $n > 3$, and the series converges.

14.3.8. $\sum\limits_{n=1}^{\infty} \dfrac{1}{n^{1+k/n}}$

Comparing this with

$$\sum\limits_{n=1}^{\infty} \dfrac{1}{n}, \quad \lim\limits_{n\to\infty} \dfrac{\dfrac{1}{n^{1+k/n}}}{\dfrac{1}{n}} = \lim\limits_{n\to\infty} n^{-k/n}$$

and using L'Hospital's rule

$$\ln \lim_{n\to\infty} n^{-k/n} = \lim_{n\to\infty} -\frac{k}{n}\ln n = \lim_{n\to\infty} \frac{-\frac{k}{n}}{1} = 0,$$

so $\lim_{n\to\infty} n^{-k/n} = 1$ and our series diverge together.

14.4. The ratio test. If $\lim_{n\to\infty}\left|\frac{a_{n+1}}{a_n}\right| < 1$, the series converges. If $\lim_{n\to\infty}\left|\frac{a_{n+1}}{a_n}\right| > 1$ or ∞, the series diverges; if $\lim_{n\to\infty}\left|\frac{a_{n+1}}{a_n}\right| = 1$, the test gives no information. This test should really be applied first, and then, if $\lim_{n\to\infty}\left|\frac{a_{n+1}}{a_n}\right| = 1$, other tests can be tried.

Notice for the series $\sum_1^\infty \frac{1}{n^k} = 1 + \frac{1}{2^k} + \frac{1}{3^k} + \cdots$,

$$\lim_{n\to\infty}\left|\frac{a_{n+1}}{a_n}\right| = \lim_{n\to\infty}\frac{\frac{1}{(n+1)^k}}{\frac{1}{n^k}} = \lim_{n\to\infty}\left(\frac{n}{n+1}\right)^k = 1$$

so that the test gives no information for any k.

14.4.1. $\quad \sum \frac{2^n}{n!} = 2 + \frac{2^2}{2!} + \frac{2^3}{3!} + \cdots$

$$\lim_{n\to\infty}\left|\frac{a_{n+1}}{a_n}\right| = \lim_{n\to\infty}\frac{\frac{2^{n+1}}{(n+1)!}}{\frac{2^n}{n!}} = \lim_{n\to\infty}\frac{2^{n+1}}{2^n}\cdot\frac{n!}{(n+1)!} = \lim_{n\to\infty}\frac{2}{n+1} = 0$$

The series converges.

14.4.2. $\quad \sum \frac{n^3}{2^n} = \frac{1}{2} + \frac{8}{4} + \frac{27}{8} + \frac{64}{16} + \cdots$

$$\lim_{n\to\infty}\left|\frac{a_{n+1}}{a_n}\right| = \lim_{n\to\infty}\frac{\frac{(n+1)^3}{2^{n+1}}}{\frac{n^3}{2^n}} = \lim_{n\to\infty}\frac{(n+1)^3}{2^{n+1}}\cdot\frac{2^n}{n^3}$$

$$= \lim_{n\to\infty}\left[\frac{1}{2}\left(1+\frac{1}{n}\right)^3\right] = \frac{1}{2}$$

The series converges.

14.4.3. $\sum \dfrac{n^n}{n!} = 1 + \dfrac{4}{2!} + \dfrac{3^3}{3!} + \ldots$

$$\lim_{n\to\infty} \dfrac{\dfrac{(n+1)^{n+1}}{(n+1)!}}{\dfrac{n^n}{n!}} = \lim_{n\to\infty} \dfrac{(n+1)^n}{n^n} = \lim_{n\to\infty}\left(1+\dfrac{1}{n}\right)^n = e$$

The series diverges.

14.4.4. $1 + \dfrac{1}{100} + \dfrac{2!}{100^2} + \ldots = \sum \dfrac{n!}{100^n}$

$$\lim_{n\to\infty} \dfrac{\dfrac{(n+1)!}{100^{n+1}}}{\dfrac{n!}{100^n}} = \lim_{n\to\infty}\dfrac{n+1}{100} = \infty$$

The series diverges.

14.4.5. $1 + \dfrac{1\cdot 2}{1\cdot 3} + \dfrac{1\cdot 2\cdot 3}{1\cdot 3\cdot 5} + \ldots + \dfrac{n!}{1\cdot 3\cdot 5 \cdots (2n-1)} + \ldots$

$$\lim_{n\to\infty} \dfrac{\dfrac{(n+1)!}{1\cdot 3\cdot 5\cdots(2n-1)(2n+1)}}{\dfrac{n!}{1\cdot 3\cdot 5\cdots(2n-1)}} = \lim_{n\to\infty}\dfrac{n+1}{2n+1} = \dfrac{1}{2}$$

The series converges.

14.4.6. $\dfrac{4}{3} + \dfrac{4\cdot 7}{3\cdot 8} + \dfrac{4\cdot 7\cdot 10}{3\cdot 8\cdot 13}$

$+ \ldots + \dfrac{4\cdot 7\cdot 10 \cdots (3n+1)}{3\cdot 8\cdot 13 \cdots (5n-2)} + \ldots$

$$\lim_{n\to\infty} \dfrac{\dfrac{4\cdot 7\cdot 10 \cdots (3n+4)}{3\cdot 8\cdot 13 \cdots (5n+3)}}{\dfrac{4\cdot 7\cdot 10 \cdots (3n+1)}{3\cdot 8\cdot 13 \cdots (5n-2)}} = \lim_{n\to\infty}\dfrac{3n+4}{5n+3} = \dfrac{3}{5}$$

The series converges.

14.4.7. $\dfrac{3}{2} + \dfrac{9}{4\cdot 4} + \dfrac{27}{9\cdot 8} + \ldots = \sum_{1}^{\infty}\dfrac{3^n}{n^2 2^n}$

$$\lim_{n\to\infty} \dfrac{\dfrac{3^{n+1}}{(n+1)^2 2^{n+1}}}{\dfrac{3^n}{n^2 2^n}} = \lim_{n\to\infty}\dfrac{3}{2}\left(\dfrac{n}{n+1}\right)^2 = \dfrac{3}{2}$$

The series diverges.

4.5. Alternating series. An alternating series is one whose terms are alternately positive and negative.

Example: $1 - \frac{1}{2^2} + \frac{1}{3^2} - \frac{1}{4^2} + \ldots = \sum_{n=1}^{\infty} (-1)^{n-1} \frac{1}{n^2}$

An alternating series converges if each term is less than the preceding in absolute value and $\lim_{n \to \infty} a_n = 0$. Again notice that the first few terms are unimportant to convergence; $1 + 2 + 1000 - \frac{1}{2^2} + \frac{1}{3^2} - \frac{1}{4^2} + \ldots$ is convergent. The error incurred in summing only the first n terms of an alternating series is less than the $(n+1)$st term.

14.5.1. $1 - \frac{1}{2} + \frac{1}{3} - \frac{1}{4} + \ldots + (-1)^{n-1} \frac{1}{n} + \ldots$ <u>converges</u>.

14.5.2. $1 - \frac{1}{2!} + \frac{1}{3!} - \frac{1}{4!} + \ldots + (-1)^{n-1} \frac{1}{n!} + \ldots$ <u>converges</u>.

14.5.3. $1 - \frac{1}{\sqrt{2}} + \frac{1}{\sqrt{3}} - \frac{1}{\sqrt{4}} + \ldots + (-1)^{n-1} \frac{1}{\sqrt{n}} + \ldots$ <u>converges</u>.

14.5.4. $\frac{1}{\ln 2} - \frac{1}{\ln 3} + \frac{1}{\ln 4} - \ldots + (-1)^{n-1} \frac{1}{\ln(n+1)} + \ldots$ <u>converges</u>.

14.5.5. The series $\frac{1}{3} - \frac{2}{5} + \frac{3}{7} - \ldots = \sum_{1}^{\infty} (-1)^{n+1} \frac{n}{2n+1}$ diverges as $\lim_{n \to \infty} a_n \neq 0$; also $a_{n+1} > a_n$.

Notice that in the series of problem **14.5.1**, we must sum 99 terms to be sure of 2 decimal places, whereas the first 4 terms suffice in the series of **14.5.2**.

14.5.6. $-\frac{3}{8} + \frac{5}{16} - \frac{7}{24} + \ldots = \sum_{1}^{\infty} (-1)^n \frac{2n+1}{8n}$

Here $a_{n+1} < a_n$ but $\lim_{n \to \infty} a_n \neq 0$ <u>and the series diverges</u>.

14.6. Absolute convergence and conditional convergence. If the sum of the absolute values of the terms of a series converge the series is absolutely convergent. If the series converges, but not absolutely, it is conditionally convergent. If a series converges absolutely, it is also convergent.

14.6.1. $1 - \frac{1}{2} + \frac{1}{3} - \frac{1}{4} + \ldots = \sum (-1)^{n-1} \frac{1}{n}$, we have seen, converges. However, $1 + \frac{1}{2} + \frac{1}{3} + \frac{1}{4} + \ldots = 1 + |-\frac{1}{2}| + |\frac{1}{3}| + |-\frac{1}{4}| + \ldots$ is divergent. Hence $\sum_{1}^{\infty} (-1)^{n-1} \frac{1}{n}$ is conditionally convergent.

14.6.2. $1 - \frac{1}{2!} + \frac{1}{3!} - \frac{1}{4!} + \ldots = \sum (-1)^{n-1} \frac{1}{n!}$ is absolutely convergent as $1 + \frac{1}{2!} + \frac{1}{3!} + \ldots + \frac{1}{n!} + \ldots$ converges. It is trivially true that any positive series that converges is also absolutely convergent.

14.6.3. $1 - \frac{1}{2^k} + \frac{1}{3^k} - \frac{1}{4^k} + \ldots = \sum (-1)^{n-1} \frac{1}{n^k}$ is absolutely convergent when $k > 1$, since $\sum \frac{1}{n^k}$ then converges. The series is conditionally convergent when $0 < k \leq 1$.

14.6.4. $\frac{1}{\ln 2} - \frac{1}{\ln 3} + \frac{1}{\ln 4} - \ldots = \sum_{1}^{\infty} \frac{(-1)^{n+1}}{\ln(n+1)}$ is conditionally convergent, since $\sum_{1}^{\infty} \frac{1}{\ln(n+1)}$ is divergent.

CHAPTER 15

POWER SERIES

5.1. Interval of convergence of a power series

15.1.1. $1 + x + \frac{1}{2}x^2 + \ldots + \frac{1}{n}x^n + \ldots$

$$\lim_{n \to \infty} \left| \frac{u_{n+1}}{u_n} \right| = \lim_{n \to \infty} |x| \frac{n}{n+1} = |x| \lim_{n \to \infty} \frac{n}{n+1} = |x|$$

so that, by the ratio test, the series converges, $-1 < x < +1$. Since $1 + 1 + \frac{1}{2} + \frac{1}{3} + \ldots + \frac{1}{n} + \ldots$ diverges by the integral test, and $1 - 1 + \frac{1}{2} - \frac{1}{3} + \ldots + (-1)^n \frac{1}{n} + \ldots$ converges by the alternating series test, the interval of convergence is $-1 \leq x < 1$.

15.1.2. $\sum_{n=1}^{\infty} \frac{1}{n^3} x^n$

$$\lim_{n \to \infty} \left| \frac{u_{n+1}}{u_n} \right| = \lim_{n \to \infty} |x| \frac{n^3}{(n+1)^3} = |x| \lim_{n \to \infty} \left(\frac{n}{n+1} \right)^3 = |x|$$

Since $\sum_{n=1}^{\infty} \frac{1}{n^3}$ and $\sum_{n=1}^{\infty} (-1)^n \frac{1}{n^3}$ both converge, the interval of convergence is $-1 \leq x \leq 1$.

15.1.3. $\sum_{n=1}^{\infty} \frac{1}{n^2} (x + 5)^n$

$$\lim_{n \to \infty} \left| \frac{u_{n+1}}{u_n} \right| = \lim_{n \to \infty} |x + 5| \frac{n^2}{(n+1)^2}$$
$$= |x + 5| \lim_{n = \infty} \left(\frac{n}{n+1} \right)^2 = |x + 5|$$

$|x + 5| < 1$ if $-6 < x < -4$. At $x = -6$ and $x = -4$, we have the series $\sum_{n=1}^{\infty} (-1)^n \frac{1}{n^2}$ and $\sum_{n=1}^{\infty} \frac{1}{n^2}$, which converge, and so the interval of convergence is $-6 \leq x \leq -4$.

15.1.4. $\sum_{n=0}^{\infty} n! x^n$

$$\lim_{n \to \infty} \left| \frac{u_{n+1}}{u_n} \right| = \lim_{n \to \infty} |x| \frac{(n+1)!}{n!} = |x| \lim_{n \to \infty} (n + 1)$$

and this limit does not exist unless $x = 0$; i.e., the series is $1 + 0 + 0 + 0 + \ldots$, and the interval of convergence is the point $x = 0$.

15.1.5. $\sum_{n=1}^{\infty} \dfrac{x^n}{n!}$

$$\lim_{n \to \infty} \left| \dfrac{u_{n+1}}{u_n} \right| = \lim_{n \to \infty} |x| \dfrac{n!}{(n+1)!} = |x| \lim_{n \to \infty} \dfrac{1}{n+1}$$

and since $\lim_{n \to \infty} \dfrac{1}{n+1} = 0$, this series will converge for any finite value of x.

15.1.6. $\sum_{n=1}^{\infty} \dfrac{1}{\sqrt{n}\, x^n}$

This is a power series in $\dfrac{1}{x}$.

$$\lim_{n \to \infty} \left| \dfrac{u_{n+1}}{u_n} \right| = \lim_{n \to \infty} \left| \dfrac{1}{x} \right| \dfrac{\sqrt{n}}{\sqrt{n+1}} = \left| \dfrac{1}{x} \right| \lim_{n \to \infty} \sqrt{\dfrac{n}{n+1}} = \left| \dfrac{1}{x} \right|$$

and $\left| \dfrac{1}{x} \right| < 1$ if $x < -1$ or $x > 1$. The series $\sum_{1}^{\infty} \dfrac{1}{\sqrt{n}}$ diverges and $\sum_{1}^{\infty} \dfrac{(-1)^n}{\sqrt{n}}$ converges, so that the series converges whenever $x \leqslant -1$ and $x > 1$.

15.1.7. $\sum_{n=1}^{\infty} \dfrac{1}{2^n} x^n$

$$\lim_{n \to \infty} \left| \dfrac{u_{n+1}}{u_n} \right| = \lim_{n \to \infty} |x| \dfrac{2^n}{2^{n+1}} = \dfrac{|x|}{2}$$

and the interval of convergence is $-2 < x < 2$.

15.2. Taylor's series. If we assume that a function can be written as a power series,

$$f(x) = a_0 + a_1(x - b) + a_2(x - b)^2 + \ldots + a_n(x - b)^n + \ldots$$

the nth coefficient may be found from the formula $a_n = \dfrac{f^{(n)}(b)}{n!}$.

This series is known as Taylor's expansion, and b is called the point around which the function is expanded.

15.2.1. Find the Taylor's expansion of e^{-x} around the point $+2$.

Solution.

$$f'(x) = -e^{-x}; \quad f''(x) = e^{-x}; \quad f'''(x) = -e^{-x}; \quad \ldots;$$
$$f^{(n)}(x) = (-1)^n e^{-x}$$

so that $a_n = \dfrac{(-1)^n e^{-2}}{n!}$, and

$$e^{-x} = e^{-2} - \frac{e^{-2}}{1!}(x-2) + \frac{e^{-2}}{2!}(x-2)^2$$
$$+ \ldots + \frac{(-1)^n}{n!} e^{-2}(x-2)^n + \ldots$$

$$= e^{-2}\left[1 - \frac{(x-2)}{1!} + \frac{(x-2)^2}{2!}\right.$$
$$\left. + \ldots + \frac{(-1)^n (x-2)^n}{n!} + \ldots \right]$$

$$= e^{-2} \sum_{n=0}^{\infty} (-1)^n \frac{(x-2)^n}{n!}$$

Notice that the series converges everywhere.

15.2.2. Find the Taylor's expansion of $\ln x$ about the point $x = +3$.

Solution

$$f(3) = \ln(3); \quad f'(x) = \frac{1}{x}; \quad f''(x) = -\frac{1}{x^2}; \quad \ldots;$$
$$f^{(n)}(x) = (-1)^{n-1} \frac{(n-1)!}{x^n} \ldots$$

so that

$$a_n = \frac{(-1)^{n-1}(n-1)!}{n!} \cdot \frac{1}{3^n} = \frac{(-1)^{n-1}}{3^n n}; \quad n \neq 0$$

and

$$\ln x = \ln 3 + \frac{1}{3}(x-3) - \frac{1}{18}(x-3)^2$$
$$+ \ldots + \frac{(-1)^{n-1}}{3^n n}(x-3)^n + \ldots$$

Notice the series converges for $0 < x \leqslant 6$. The Taylor's expansion about the point 0 is called the Maclaurin's series.

15.2.3. Find the Maclaurin's series for $\sin x$.
Solution

$$f(x) = \sin x; \quad f'(x) = \cos x; \quad f''(x) = -\sin x;$$
$$f'''(x) = -\cos x \quad f^{(4)}(x) = \sin x; \quad \ldots$$

It can easily be seen that for even integers $f^{(2n)}(0) = 0$, and for odd integers $f^{(2n+1)}(0) = (-1)^n$; $\quad a_{2n+1} = \dfrac{(-1)^n}{(2n+1)!}$

$$\sin x = x - \frac{x^3}{3!} + \frac{x^5}{5!} + \ldots + (-1)^n \frac{x^{2n+1}}{(2n+1)!} + \ldots$$

The series converges everywhere.

15.2.4. Find the Maclaurin's series for e^x.

$$y = e^x, \quad f^{(n)}(0) = 1 \quad a_n = \frac{1}{n!}$$

$$y' = e^x, \quad e^x = 1 + x + \frac{x^2}{2!} + \frac{x^3}{3!} + \ldots + \frac{x^n}{n!} + \ldots$$

$$f^{(n)}(x) = e^x \quad = \sum_{n=0}^{\infty} \frac{x^n}{n!}$$

15.2.5. Find the Maclaurin's series for Arc tan x.

$$f(x) = \text{Arc tan } x$$

$$f'(x) = \frac{1}{1+x^2} = 1 - x^2 + x^4 - x^6 + \ldots$$

$$f''(x) = -2x + 4x^3 + \ldots$$

$$f'''(x) = -2 + 4 \times 3x^2 + \ldots$$

It can be seen that

$$f^{(2n)}(0) = 0; \quad f^{(2n+1)}(0) = (-1)^n(2n)!; \quad a_{2n+1} = \frac{(-1)^n(2n)!}{(2n+1)!}$$

Our series becomes

$$\text{Arc tan } x = x - \frac{x^3}{3} + \frac{x^5}{5} + \ldots + (-1)^n \frac{x^{2n+1}}{2n+1} + \ldots$$

Here the interval of convergence is $-1 < x \leqslant 1$. Notice $\dfrac{\pi}{4} =$

$$1 - \frac{1}{3} + \frac{1}{5} - \ldots + \frac{(-1)^n}{2n+1} + \ldots .$$

15.2.6. Find the Maclaurin's series for $\ln(x+1)$.

Solution

$$f(x) = \ln(x+1),$$

$$f'(x) = \frac{1}{x+1},$$

$$f''(x) = \frac{-1}{(x+1)^2},$$

$$\vdots$$

$$f^{(n)}(x) = \frac{(-1)^{n-1}(n-1)!}{(x+1)^n},$$

$$\vdots$$

$$f^{(n)}(0) = (-1)^{n-1}(n-1)! \quad \text{and} \quad a_n = \frac{(-1)^{n-1}}{n}, \quad a_0 = 0.$$

$$\ln(x+1) = x - \frac{x^2}{2} + \frac{x^3}{3} - \frac{x^4}{4} + \ldots + (-1)^{n-1}\frac{x^n}{n} + \ldots$$

The interval of convergence is $-1 < x \leq 1$. Notice that we can not find a Maclaurin's series for $\ln x$, since $f(0)$ does not exist in this case.

15.2.7. Find the Maclaurin's series for $\frac{1}{1-x}$.

Solution

$$f(x) = \frac{1}{1-x},$$

$$f'(x) = \frac{1}{(1-x)^2},$$

$$f''(x) = \frac{2}{(1-x)^3},$$

$$\vdots$$

$$f^{(n)}(x) = \frac{n!}{(1-x)^{n+1}},$$

$$f^{(n)}(0) = n! \quad \text{and} \quad a_n = 1 \quad \text{for all } n.$$

$$f(x) = 1 + x + x^2 + \ldots + x^n + \ldots$$

15.2.8. Find the Maclaurin's series for $\cos x$.
Solution
$$f(x) = \cos x,$$
$$f'(x) = -\sin x,$$
$$f''(x) = -\cos x,$$
$$f'''(x) = \sin x,$$
$$f^{(4)}(x) = \cos x,$$
$$\vdots$$

It can be seen that $f^{(2n+1)}(0) = 0$,
$$f^{(2n)}(0) = (-1)^n, \qquad a_{2n} = \frac{(-1)^n}{(2n)!}.$$
$$\cos x = 1 - \frac{x^2}{2!} + \frac{x^4}{4!} + \ldots + (-1)^n \frac{x^{2n}}{(2n)!} + \ldots .$$

15.3. Taylor's series and remainder. Though we have found the interval of convergence for power series in preceding paragraphs, it is also useful to know the error incurred by adding the first, say, 12 terms of the series and neglecting the rest. If the error is small we can approximate the sum of an infinite series by adding the first few terms. To this end, we have the following:

Theorem: If a function $f(x)$ and its first $n - 1$ derivatives are continuous in the closed interval $[a, x]$, and it has the nth derivative at every interior point; then

$$f(x) = f(a) + (x - a)f'(a) + \ldots + \frac{f^{(n-1)}(a)}{(n - 1)!}(x - a)^{n-1} + R_n$$

where $R_n = \frac{1}{n!} f^{(n)}(\xi)(x - a)^n$, and ξ is an appropriate value between a and x.

The proof of this theorem, which we shall not give here, follows from the mean value theorem of **8.1**. Indeed the theorem itself might be said to be an extended mean value theorem.

15.3.1. How many terms of the series $1 + x + \frac{x^2}{2!} + \ldots + \frac{x^n}{n!} + \ldots$ must be used to compute $e^{1/10}$ to ten decimal places?

Solution. Here we want to find n, so $R_n < 10^{-10}$. Hence $\frac{1}{n!} e^\xi \left(\frac{1}{10}\right)^n < 10^{-10}$, since $f^{(n)}(x) = e^x$, and a, the point of expansion, is 0. As e^x is increasing for $0 < \xi < \frac{1}{10}$,

$$\frac{1}{n!} e^\xi \left(\frac{1}{10}\right)^n < \frac{1}{n!} e^{1/10} \left(\frac{1}{10}\right)^n < \frac{2}{n!} \left(\frac{1}{10}\right)^n$$

since $e^{1/10} < 2$. This last inequality follows readily from the obvious inequality $e < 2^{10}$. By trial and error, we find $R_7 < 10^{-10}$, and we need add only the first seven terms of the series. In the same way, if three-place accuracy is desired $\frac{2}{n!}\left(\frac{1}{10}\right)^n < 10^{-3}$, $n = 3$.

$$e^{1/10} = 1 + \tfrac{1}{10} + \tfrac{1}{2}(\tfrac{1}{10})^2 = 1.105 \text{ to three places.}$$

15.3.2. Find $\sin 1°$ to six decimal places. Remember $1° = \frac{\pi}{180}$ radians.

Solution

$$|R_{2n+1}| = \frac{|(-1)^n|}{(2n+1)!} \cos\xi \cdot \left(\frac{\pi}{180}\right)^{2n+1} < \frac{|(-1)^n|}{(2n+1)!}\left(\frac{2}{100}\right)^{2n+1}$$

where $0 < \xi < \frac{\pi}{180}$

For $n = 2$, $\frac{1}{5!}\left(\frac{2}{100}\right)^5 < 10^{-6}$. The term

$$|R_{2n}| < \frac{(-1)^n}{(2n)!} \sin\frac{\pi}{180}\left(\frac{\pi}{180}\right)^{2n} < 10^{-6} \text{ for } n = 2.$$

$$\sin 1° = \sin 0.017453 = 0.017453 - \frac{(0.017453)^3}{3!}$$

$$= 0.017453 - 0.000001$$

$$= 0.017452$$

15.3.3. Calculate $\sin 46°$ to four decimal places.

Solution. First we expand $\sin x$ about $\dfrac{\pi}{4}$.

$f(x) = \sin x,$

$f'(x) = \cos x,$

$f''(x) = -\sin x,$

$f'''(x) = -\cos x,$

$f\left(\dfrac{\pi}{4}\right) = \dfrac{\sqrt{2}}{2}, \quad f'\left(\dfrac{\pi}{4}\right) = \dfrac{\sqrt{2}}{2}, \ldots$

$$a_{2n} = \dfrac{(-1)^n}{(2n)!}\dfrac{\sqrt{2}}{2}, \quad a_{2n+1} = \dfrac{(-1)^n}{(2n+1)!}\dfrac{\sqrt{2}}{2}.$$

$$\sin x = \dfrac{\sqrt{2}}{2} + \dfrac{\sqrt{2}}{2}\left(x - \dfrac{\pi}{4}\right) - \dfrac{\sqrt{2}}{2} \times \dfrac{1}{2!}\left(x - \dfrac{\pi}{4}\right)^2$$

$$- \dfrac{\sqrt{2}}{2} \times \dfrac{1}{3!}\left(x - \dfrac{\pi}{4}\right)^3 + \cdots$$

$$= \dfrac{\sqrt{2}}{2}\left[1 + \left(x - \dfrac{\pi}{4}\right) - \dfrac{\left(x - \dfrac{\pi}{4}\right)^2}{2!} - \dfrac{\left(x - \dfrac{\pi}{3}\right)^3}{3!} + \cdots\right]$$

Here $R_n = \dfrac{f^{(n)}(\xi)}{n!}\left(x - \dfrac{\pi}{4}\right)^n$ where $f^{(n)}(\xi)$ may be $\sin \xi$ or $\cos$

ξ and $\dfrac{\pi}{4} < \xi < \dfrac{46\pi}{180}$. In any case $f^{(n)}(\xi) < 1$; $|R_n| < \dfrac{1}{n!}\left(\dfrac{\pi}{180}\right)^n$.
For four decimals $n = 3$.

$$\sin 46° = \sin \dfrac{46\pi}{180} = \dfrac{\sqrt{2}}{2} + \dfrac{\sqrt{2}}{2} \times \dfrac{\pi}{180} - \dfrac{\sqrt{2}}{2}\left(\dfrac{\pi}{180}\right)^2 \times \dfrac{1}{2!}$$

$$= 0.7071 + 0.0124 - 0.0001 = 0.7194$$

15.3.4. Using $e^2 = 7.3891$, find $e^{2.1}$ to four decimal places.
$|R_n| = \dfrac{f^{(n)}(\xi)}{n!}(0.1)^n$, and, by choosing $n = 4$, we have an error $< 10^{-4}$.

$$e^x = e^2\left[1 + (x-2) + \dfrac{(x-2)^2}{2!} + \dfrac{(x-2)^3}{3!} + \cdots\right]$$

$$e^{2.1} = e^2(1 + .1 + .005 + .0002)$$
$$= 7.3891 \times 1.1052 = 8.1664$$

15.3.5. For what values of x can $\sin x$ be approximated by with an error < 0.001?

Solution. Since $\sin x = x - \dfrac{x^3}{3!} + \dfrac{x^5}{5!} - \ldots$, this means $\left|\dfrac{x^3}{3!} \sin \xi\right| < 0.001$ or $x^3 < 0.006$,

$$x < 0.18 \text{ radian or } x < 10°.$$

15.3.6. Suppose we expand $f(x)$ about point x $(a = x)$ and find the value at $x + \Delta x$, $(x - a) = \Delta x$.

Solution

(1)
$$f(x + \Delta x) = f(x) + \frac{f'(x)}{1!}\Delta x + \frac{f''(x)}{2!}\overline{\Delta x^2}$$
$$+ \ldots + \frac{f^{(n)}(x)}{n!}\overline{\Delta x^n} + \ldots$$

In previous chapters, we have approximated $f(x + \Delta x) - f(x)$ by $f'(x)\Delta x$. That is, since $dy = f'(x)\,dx$, we assumed $\Delta y = f'(x)\Delta x$. From the theory of Taylor's series, we now have a means of finding how accurate an approximation we have made. The remainder formula tells us

$$R_n = \frac{f''(\xi)}{2!}\Delta\bar{x}^2 \quad \text{where } x < \xi < x + \Delta x$$

As an example, find $\sqrt{101}$.

$$y = \sqrt{x}; \quad y' = \tfrac{1}{2}x^{-1/2}; \quad y'' = -\tfrac{1}{4}x^{-3/2}$$

If $x = 100$, $\Delta x = 1$.

$$\Delta y = \tfrac{1}{20} \times 1 = 0.05; \quad \sqrt{101} = 10 + 0.05 = 10.05$$
$$|R_n| = \tfrac{1}{8}\xi^{-3/2}(1)^2 \quad \text{where } 100 < \xi < 101$$
$$|R_n| < \tfrac{1}{8} \times \tfrac{1}{1000}$$

Our result is accurate to three decimal places, $\sqrt{101} = 10.050$. If we had asked for $\sqrt{99}$, our error would still be $< -\tfrac{1}{8} \times$

$\frac{1}{1000}$, but in this case not from the remainder, but since $\Delta x = -1$ now, and (1) is thereby an alternating series. Otherwise our accuracy would be

$$|R_n| = +\frac{1}{8}\xi^{-3/2}(-1)^2 < \frac{1}{8} \times \frac{1}{99^{3/2}}$$

15.3.7. Find ln 1.01
Solution

$$y = \ln x; \qquad y' = \frac{1}{x}; \qquad y'' = -\frac{1}{x^2}$$

If $\Delta x = 0.01;\qquad \Delta y = \frac{0.01}{1} = 0.01$

$y + \Delta y = 0 + 0.01 = 0.01;\qquad \ln 1.01 = 0.01$

$$R_n = -\frac{1}{2!} \times \frac{1}{\xi^2}(0.01)^2$$

and since $1 < \xi < 1.01$,

$|R_n| < \frac{1}{2} \times (.01)^2 = 0.00005; \qquad \ln 1.01 = 0.0100$

15.4. Computations with series. The intervals of convergence of a power series and those series obtained from it by integrating term by term or differentiating term by term are the same, with the possible exception of the end points. It is also true that a power series representing a function is the Taylor series of that function, regardless of how that series is obtained. In other words, a power series expansion about a point is unique for any particular function.

15.4.1. Use $\int_0^x \frac{dx}{1+x^2}$ to find a series for Arc tan x.
Solution

$$\text{Arc tan } x = \int_0^x \frac{dx}{1+x^2} = \int_0^x (1 - x^2 + x^4 - \ldots)\, dx$$

$$= x - \frac{x^3}{3} + \frac{x^5}{5} + \ldots + (-1)^n \frac{x^{2n+1}}{2n+1} + \ldots$$

The interval of convergence for the series for $\frac{1}{1+x^2}$ is $-1 < x < 1$, and for the series for Arc tan x, $-1 < x \leq 1$.

15.4.2. Use series for $(1 + x)^{-1}$ to find an expansion of $(1 + x)^{-2}$ and of $\ln(1 + x)$.

Solution

$$(1+x)^{-1} = 1 - x + x^2 - x^3 + \ldots + (-1)^n x^n + \ldots$$

Differentiating term by term

$$(1+x)^{-2} = 1 - 2x + 3x^2 + \ldots + (-1)^{n-1} n x^{n-1} + \ldots$$

$$\ln(1+x) = \int_0^x \frac{dx}{1+x} = \int_0^x (1 - x + x^2 - x^3 + \ldots$$

$$+ (-1)^n x^n + \ldots) \, dx$$

$$= x - \frac{x^2}{2} + \frac{x^3}{3} - \ldots + (-1)^{n-1} \frac{x^n}{n} + \ldots$$

15.4.3. From the series for $\sin x$ find that of $\cos x$.

Solution

$$\sin x = x - \frac{x^3}{3!} + \frac{x^5}{5!} - \ldots + (-1)^n \frac{x^{2n+1}}{(2n+1)!} + \ldots$$

Differentiating term by term

$$\cos x = 1 - \frac{x^2}{2!} + \frac{x^4}{4!} - \ldots + (-1)^n \frac{x^{2n}}{(2n)!} + \ldots$$

Both series converge everywhere. Notice the same result could have been obtained by integrating term by term.

15.4.4. Find series for $\ln(1-x)$ and $\ln \frac{1-x}{1+x}$

Solution

$$\int_0^x \frac{dx}{1-x} = \int_0^x (1 + x + x^2 + \ldots + x^n + \ldots) \, dx$$

$$-\ln(1-x) = x + \frac{x^2}{2} + \ldots + \frac{x^n}{n} + \ldots$$

$$\ln \frac{1-x}{1+x} = \ln(1-x) - \ln(1+x)$$

$$= -\left(x + \frac{x^2}{2} + \ldots + \frac{x^n}{n} + \ldots\right)$$

$$- \left(x - \frac{x^2}{2} + \ldots + (-1)^{n-1} \frac{x^n}{n} + \ldots\right)$$

$$= -2\left(x + \frac{x^3}{3} + \ldots + \frac{x^{2n-1}}{2n+1} + \ldots\right)$$

For the second part of this problem, notice that, since the interval of convergence for the series for $\ln(1-x)$ is $-1 \leqslant x < 1$ and of $\ln(1+x)$, $-1 < x \leqslant 1$, the interval for $\ln \dfrac{1-x}{1+x}$ is the common interval, or $-1 < x < 1$.

15.4.5. Find the power series for Arc $\sin x$.

Solution

$$\text{Arc } \sin x = \int_0^x \frac{dx}{\sqrt{1-x^2}} = \int_0^x \left(1 + \frac{1}{2}x^2 + \frac{3}{8}x^4 + \frac{5}{16}x^6 \right.$$

$$+ \ldots + \frac{1\cdot 3 \cdot 5 \ldots (2n-1)}{2^n n!} x^{2n} + \ldots \bigg) dx$$

$$= x + \frac{x^3}{6} + \frac{3}{40}x^5 + \ldots + \frac{1\cdot 3 \cdot 5 \ldots (2n-1)}{2\cdot 4 \cdot 6 \ldots 2n} \frac{x^{2n+1}}{2n+1} + \ldots$$

Here we used Newton's expansion for $(1-x^2)^{-1/2}$. For the integrated series, the interval of convergence is $-1 \leqslant x < 1$.

15.4.6. Calculate $\sin\left(x + \dfrac{\pi}{4}\right)$ from the identity

$$\sin\left(x + \frac{\pi}{4}\right) = \frac{\sqrt{2}}{2}(\sin x + \cos x)$$

Solution

$$\sin\left(x + \frac{\pi}{4}\right) = \frac{\sqrt{2}}{2}\left(1 + x - \frac{x^2}{2!} - \frac{x^3}{3!} + \frac{x^4}{4!} + \frac{x^5}{5!} - \ldots\right)$$

All series involved converge everywhere.

15.4.7. Power series can be multiplied by multiplying each term in the first by each term in the second and collecting terms of like powers.

Solution

$$(a_0 + a_1 x + \ldots + a_n x^n + \ldots)(b_0 + b_1 x + \ldots + b_n x^n + \ldots)$$

$$= a_0 b_0 + (a_0 b_1 + a_1 b_0) x$$

$$+ (a_0 b_2 + a_1 b_1 + a_2 b_0) x^2 + \ldots + (a_0 b_n + \ldots + a_n b_0) x^n + \ldots$$

For example, find $e^x e^{-x}$.

$$e^x e^{-x} = \left(1 + x + \frac{x^2}{2!} + \ldots + \frac{x^n}{n!} + \ldots\right)$$

$$\left(1 - x + \frac{x^2}{2!} - \frac{x^3}{3!} + \ldots - (-1)^n \frac{x^n}{n!} + \ldots\right)$$

$$= 1 + (1 - 1)x + \left(\frac{1}{2!} - 1 \cdot 1 + \frac{1}{2!}\right)x^2$$

$$+ \left(\frac{1}{3!} - \frac{1}{2!} \cdot 1 + \frac{1}{2!} \cdot 1 - \frac{1}{3!}\right)x^3 + \ldots$$

$$+ \left(\frac{1}{n!} - \frac{1}{(n-1)!} + \frac{1}{(n-2)!2!} - \ldots\right.$$

$$\left. - \frac{1}{(n-2)!2!} + \frac{1}{(n-1)!} - \frac{1}{n!}\right)x^n + \ldots = 1$$

Of course, we can obtain this result much more readily from the law of exponents.

15.4.8. Find a series for $(\sin x)(\cos x)$.

Solution

$$(\sin x)(\cos x) = \left(x - \frac{x^3}{3!} + \ldots + (-1)^n \frac{x^{2n+1}}{(2n+1)!} + \ldots\right)$$

$$\left(1 - \frac{x^2}{2!} + \frac{x^4}{4!} - \ldots + (-1)^n \frac{x^{2n}}{(2n)!} + \ldots\right)$$

$$= x + \left(-\frac{1}{3!} - \frac{1}{2!}\right)x^3 + \left(\frac{1}{4!} + \frac{1}{3!2!} + \frac{1}{5!}\right)x^5$$

$$+ \ldots + \left(\frac{(-1)^n}{(2n+1)!} + \frac{(-1)^{n-1}(-1)}{(2n-1)!2!} + \ldots\right.$$

$$\left. + \frac{(-1)^n}{1!(2n)!}\right)x^{2n+1} + \ldots = x - \frac{4}{3!}x^3 + \frac{16}{5!}x^5$$

$$- \ldots + (-1)^n \frac{2^{2n}}{(2n+1)!}x^{2n+1} + \ldots$$

This follows since

$$\left(\frac{1}{(2n+1)!} + \frac{1}{(2n-1)!2!} + \cdots + \frac{1}{1!(2n)!}\right)$$
$$= \frac{1}{(2n+1)!} \cdot \left[1 + \frac{(2n+1)2n}{2!} + \cdots + \frac{(2n+1)!}{(2n)!}\right]$$

and

$$1 + \frac{(2n+1)2n}{2!} z^2 + \cdots + \frac{(2n+1)!}{(2n)!} z^{2n}$$
$$= \tfrac{1}{2}[(1+z)^{2n+1} + (1-z)^{2n+1}]$$

Substituting

$$z = 1, \text{ then } 1 + \frac{(2n+1)2n}{2!} + \cdots + \frac{(2n+1)!}{(2n)!} = 2^{2n}.$$

The student can compare the final result with the series for $\tfrac{1}{2} \sin 2x$ obtained by substituting $w = 2x$ in the series for $\tfrac{1}{2} \sin w$.

CHAPTER 16

HYPERBOLIC FUNCTIONS

16.1. Definitions of the hyperbolic functions and their inverses. The following equations are used to define the hyperbolic functions.

$$\sinh x = \frac{e^x - e^{-x}}{2}; \qquad \cosh x = \frac{e^x + e^{-x}}{2}$$

$$\tanh x = \frac{\sinh x}{\cosh x}; \qquad \coth x = \frac{\cosh x}{\sinh x}$$

$$\operatorname{sech} x = \frac{1}{\cosh x}; \qquad \operatorname{csch} x = \frac{1}{\sinh x}$$

The inverse hyperbolic functions are defined by the following equations.

$$\sinh^{-1} x = \ln(x + \sqrt{x^2 + 1}); \qquad \cosh^{-1} x = \ln(x + \sqrt{x^2 - 1});$$
$$x \geq 1$$

$$\tanh^{-1} x = \frac{1}{2} \ln \frac{1+x}{1-x}, \, x^2 < 1; \qquad \coth^{-1} x = \frac{1}{2} \ln \frac{x+1}{x-1},$$
$$x^2 > 1$$

$$\operatorname{sech}^{-1} x = \ln \left(\frac{1}{x} + \sqrt{\frac{1}{x^2} - 1} \right); \qquad 0 < x \leq 1;$$

$$\operatorname{csch}^{-1} x = \ln \left(\frac{1}{x} + \sqrt{\frac{1}{x^2} + 1} \right); \qquad x \neq 0$$

16.2. Differentiation formulas. We have the following formulas:

$$\frac{d}{dx} \sinh x = \cosh x; \qquad \frac{d}{dx} \cosh x = \sinh x$$

$$\frac{d}{dx} \tanh x = \operatorname{sech}^2 x; \qquad \frac{d}{dx} \operatorname{sech} x = -\operatorname{sech} x \tanh x$$

$$\frac{d}{dx} \coth x = -\operatorname{csch}^2 x; \qquad \frac{d}{dx} \operatorname{csch} x = -\operatorname{csch} x \coth x$$

$$\frac{d}{dx}(\sinh^{-1} x) = \frac{1}{\sqrt{1+x^2}} \; ; \quad \frac{d}{dx}(\coth^{-1} x) = \frac{1}{1-x^2} \; ; \quad x^2 > 1$$

$$\frac{d}{dx}(\cosh^{-1} x) = \frac{1}{\sqrt{x^2-1}} \; ; \; x^2 > 1; \quad \frac{d}{dx}(\text{sech}^{-1} x) = \frac{-1}{\sqrt{x^2-x^4}} \; ;$$
$$0 < x < 1$$

$$\frac{d}{dx}(\tanh^{-1} x) = \frac{1}{1-x^2} \; ; \quad x^2 < 1 \; ;$$

$$\frac{d}{dx}(\text{csch}^{-1} x) = \frac{-1}{\sqrt{x^2+x^4}} \; ; \; x \neq 0$$

16.2.1. Differentiate $y = \sqrt{1 + \tanh 2x}$.
Solution

$$y' = \tfrac{1}{2}(1 + \tanh 2x)^{-1/2} \cdot 2 \, \text{sech}^2 \, 2x = \frac{\text{sech}^2 \, 2x}{\sqrt{1 + \tanh 2x}}$$

16.2.2. Differentiate $y = \sinh(1 - x^2)$
Solution

$$y' = -2x \cosh(1 - x^2).$$

16.2.3. Evaluate $\int x \cosh x^2 \, dx$.
Solution

$$\int x \cosh x^2 \, dx = \tfrac{1}{2} \int \cosh x^2 \cdot 2x \, dx = \tfrac{1}{2} \sinh x^2 + c$$

16.2.4. Evaluate $\int x \cosh x \, dx$.
Solution. By parts

$$\int x \cosh x \, dx = x \sinh x - \int \sinh x \, dx$$
$$= x \sinh x - \cosh x + c$$

16.2.5. Evaluate $\int \text{sech} \, x \, dx$.
Solution

$$\int \text{sech} \, x \, dx = \int \frac{2 \, dx}{e^x + e^{-x}} = 2 \int \frac{e^x \, dx}{e^{2x} + 1} = 2 \, \text{Arc} \tan e^x + c$$

16.2.6. Evaluate $\int e^x \cosh x \, dx$.

Solution

$$\int e^x \cosh x \, dx = \int \frac{e^{2x} + 1}{2} dx = \frac{1}{4} e^{2x} + \frac{1}{2} x + c$$

16.2.7. $I = \int \tanh^3 2x \, \text{sech}^2 2x \, dx$

Solution
Let $\tanh 2x = u$.

$$I = \frac{1}{2} \int u^3 \, du = \frac{1}{8} u^4 + c = \frac{1}{8} \tanh^4 2x + c$$

16.2.8. Differentiate $y = \sinh^{-1} 2x$.
Solution

$$\frac{dy}{dx} = \frac{2}{\sqrt{1 + 4x^2}}$$

16.2.9. Differentiate $y = \tanh^{-1} x^2$.
Solution

$$\frac{dy}{dx} = \frac{2x}{1 - x^4}$$

16.2.10. Differentiate
Solution

$$y = e^{\tanh^{-1} x};$$

$$y' = e^{\tanh^{-1} x} \frac{1}{1 - x^2}$$

16.2.11. Differentiate $y = \text{csch}^{-1} e^x$.
Solution

$$y' = \frac{-e^x}{\sqrt{e^{2x} + e^{4x}}} = -\frac{1}{\sqrt{1 + e^{2x}}}$$

16.2.12. Evaluate $\int \frac{dx}{\sqrt{9x^2 + 4}}$

Solution. Let $\tfrac{3}{2}x = u$.

$$\int \frac{dx}{\sqrt{9x^2+4}} = \frac{1}{2}\int \frac{dx}{\sqrt{\tfrac{9}{4}x^2+1}}$$

$$= \frac{1}{3}\int \frac{du}{\sqrt{u^2+1}} = \frac{1}{3}\sinh^{-1}\tfrac{3}{2}x + c$$

16.2.13. Evaluate $\int \dfrac{x\,dx}{9-x^4}$.

Solution. Let $\dfrac{x^2}{3} = u$.

$$\int \frac{x\,dx}{9-x^4} = \frac{1}{9}\int \frac{x\,dx}{1-\dfrac{x^4}{9}} = \frac{1}{6}\int \frac{du}{1-u^2} = \frac{1}{6}\tanh^{-1} u + c$$

$$= \frac{1}{6}\tanh^{-1}\frac{x^2}{3} + c; \qquad x^4 < 9$$

$$= \frac{1}{6}\coth^{-1} u + c$$

$$= \frac{1}{6}\coth^{-1}\frac{x^2}{3} + c; \qquad x^4 > 9$$

16.2.14. Evaluate $\int \dfrac{dx}{5+4x-x^2}$.

Solution. Let $\dfrac{x-2}{3} = u$.

$$\int \frac{dx}{5+4x-x^2} = \int \frac{dx}{9-(x-2)^2} = \frac{1}{9}\int \frac{dx}{1-\left(\dfrac{x-2}{3}\right)^2}$$

$$= \frac{1}{3}\int \frac{du}{1-u^2} = \frac{1}{3}\tanh^{-1}\frac{x-2}{3} + c; \quad \left(\frac{x-2}{3}\right)^2 < 1$$

$$= \frac{1}{3}\coth^{-1}\frac{x-2}{3} + c; \quad \left(\frac{x-2}{3}\right)^2 > 1$$

16.2.15. Evaluate $\int \dfrac{dx}{\sqrt{10+4x+4x^2}}$.

Solution. Let $\dfrac{1+2x}{3} = u$.

$$\int \frac{dx}{\sqrt{10 + 4x + 4x^2}} = \int \frac{dx}{\sqrt{9 + (1 + 2x)^2}}$$
$$= \frac{1}{3} \int \frac{dx}{\sqrt{1 + \left(\frac{1 + 2x}{3}\right)^2}} = \frac{1}{2} \int \frac{du}{\sqrt{1 + u^2}}$$
$$= \frac{1}{2} \sinh^{-1} u + c = \frac{1}{2} \sinh^{-1}\left(\frac{1 + 2x}{3}\right) + c$$

CHAPTER 17

PARTIAL DIFFERENTIATION

17.1. Functions of several variables. If z is a function of x and y, we write $z = f(x, y)$. We can consider y a constant, and then z will be a function of x alone, and the derivative of z with respect to x will be the partial derivative. Some of the symbols for the partial derivative with respect to x are: $\dfrac{\partial z}{\partial x}$, z_x, $f_x(x, y)$, $\dfrac{\partial f}{\partial x}$.

Find the partial derivatives of each of the following functions.

17.1.1. $z = \dfrac{x}{y} + \dfrac{y}{x}$

Solution

$$\frac{\partial z}{\partial x} = \frac{1}{y} - \frac{y}{x^2}, \quad \text{considering } y \text{ a constant}$$

$$\frac{\partial z}{\partial y} = -\frac{x}{y^2} + \frac{1}{x}, \quad \text{considering } x \text{ a constant}$$

17.1.2. $z = e^{x^2+y^2}$

Solution

$$\frac{\partial z}{\partial x} = 2xe^{x^2+y^2}; \quad \frac{\partial z}{\partial y} = 2ye^{x^2+y^2}$$

17.1.3. $z = \ln(x^2 + y)$

Solution

$$\frac{\partial z}{\partial x} = \frac{2x}{x^2+y}; \quad \frac{\partial z}{\partial y} = \frac{1}{x^2+y}$$

17.1.4. $x^4y^2 + yz^2 + 3xz - 2y^2 = 0$

Solution. Keeping y a constant and differentiating with respect to x,

$$4x^3y^2 + 2yz\frac{\partial z}{\partial x} + 3z + 3x\frac{\partial z}{\partial x} = 0$$

or

$$\frac{\partial z}{\partial x} = -\frac{3z + 4x^3 y^2}{2yz + 3x}$$

Similarly,

$$2x^4 y + z^2 + 2yz\frac{\partial z}{\partial y} + 3x\frac{\partial z}{\partial y} - 4y = 0$$

$$\frac{\partial z}{\partial y} = \frac{4y - z^2 - 2x^4 y}{2yz + 3x}$$

17.1.5. $z = \text{Arc tan}\,\dfrac{y}{x}$

Solution

$$\frac{\partial z}{\partial x} = \frac{-\dfrac{y}{x^2}}{1 + \dfrac{y^2}{x^2}} = \frac{-y}{x^2 + y^2}; \qquad \frac{\partial z}{\partial y} = \frac{\dfrac{1}{x}}{1 + \dfrac{y^2}{x^2}} = \frac{x}{x^2 + y^2}$$

17.1.6. Partial derivatives of functions of several variables can be defined by keeping all the variables but one constant and differentiating.

$$u = x^2 + \frac{yz}{x}; \qquad \frac{\partial u}{\partial x} = 2x - \frac{yz}{x^2};$$

$$\frac{\partial u}{\partial y} = \frac{z}{x}; \qquad \frac{\partial u}{\partial z} = \frac{y}{x}$$

17.2. Partial derivatives of higher order. If we now differentiate $\dfrac{\partial z}{\partial x}$ with respect to x, we shall obtain the second partial with respect to x, written $\dfrac{\partial^2 z}{\partial x^2}$. Likewise we can differentiate $\dfrac{\partial z}{\partial x}$ with respect to y and obtain the second partial with respect to x and y, written $\dfrac{\partial^2 z}{\partial x\, \partial y}$. It can be shown in more advanced work that $\dfrac{\partial^2 z}{\partial x\, \partial y} = \dfrac{\partial^2 z}{\partial y\, \partial x}$ if certain conditions are fulfilled.

17.2.1. Show $\dfrac{\partial^2 z}{\partial x\, \partial y} = \dfrac{\partial^2 z}{\partial y\, \partial x}$ for $z = \text{Arc tan}\,\dfrac{y}{x}$.

Solution. From 17.1.5,

$$\frac{\partial z}{\partial x} = \frac{-y}{x^2 + y^2}; \qquad \frac{\partial z}{\partial y} = \frac{x}{x^2 + y^2}$$

$$\frac{\partial^2 z}{\partial x\, \partial y} = \frac{-(x^2 + y^2) + y \cdot 2y}{(x^2 + y^2)^2} = \frac{y^2 - x^2}{(x^2 + y^2)^2};$$

$$\frac{\partial^2 z}{\partial y\, \partial x} = \frac{(x^2 + y^2) - 2x \cdot x}{(x^2 + y^2)^2} = \frac{y^2 - x^2}{(x^2 + x^2)^2}$$

17.2.2. Show $\dfrac{\partial^3 z}{\partial^2 x\, \partial y} = \dfrac{\partial^3 z}{\partial x\, \partial y\, \partial x} = \dfrac{\partial^3 z}{\partial y\, \partial x^2}$ for $z = e^{x^2 + y^2}$.

Solution

$$\frac{\partial z}{\partial x} = 2xe^{x^2+y^2}; \qquad \frac{\partial z}{\partial y} = 2ye^{x^2+y^2}$$

$$\frac{\partial^2 z}{\partial x^2} = 2e^{x^2+y^2} + 4x^2 e^{x^2+y^2}; \qquad \frac{\partial^2 z}{\partial y\, \partial x} = 4xye^{x^2+y^2}$$

$$\frac{\partial^3 z}{\partial x^2\, \partial y} = 4ye^{x^2+y^2} + 8x^2 y e^{x^2+y^2}; \qquad \frac{\partial^3 z}{\partial y\, \partial x^2} = 4ye^{x^2+y^2} + 8x^2 y e^{x^2+y^2}$$

$$\frac{\partial z}{\partial x} = 2xe^{x^2+y^2}; \qquad \frac{\partial^2 z}{\partial x\, \partial y} = 4xye^{x^2+y^2}$$

$$\frac{\partial^3 z}{\partial x\, \partial y\, \partial x} = 4ye^{x^2+y^2} + 8x^2 y e^{x^2+y^2}$$

17.2.3. Show $z = \dfrac{x}{y} + \dfrac{y}{x}$ satisfies $x \dfrac{\partial z}{\partial x} + y \dfrac{\partial z}{\partial y} = 0$.

Solution

$$\frac{\partial z}{\partial x} = \frac{1}{y} - \frac{y}{x^2}; \qquad \frac{\partial z}{\partial y} = -\frac{x}{y^2} + \frac{1}{x}$$

$$x \frac{\partial z}{\partial x} + y \frac{\partial z}{\partial y} = \frac{x}{y} - \frac{y}{x} + \left(-\frac{x}{y} + \frac{y}{x}\right) = 0$$

17.2.4. $\dfrac{\partial}{\partial x}\left(z \dfrac{\partial y}{\partial x} - y \dfrac{\partial z}{\partial x}\right)$

$$= \frac{\partial z}{\partial x} \cdot \frac{\partial y}{\partial x} + z \frac{\partial^2 y}{\partial x^2} - \frac{\partial y}{\partial x} \cdot \frac{\partial z}{\partial x} - y \frac{\partial^2 z}{\partial x^2} = z \frac{\partial^2 y}{\partial x^2} - y \frac{\partial^2 z}{\partial x^2}$$

17.2.5. Show $\dfrac{\partial^2 z}{\partial x^2} + \dfrac{\partial^2 z}{\partial y^2} = 0$ if $z = \ln \sqrt{(x^2 + y^2)}$.

Solution

$$\frac{\partial z}{\partial x} = \frac{x}{x^2 + y^2}; \quad \frac{\partial z}{\partial y} = \frac{y}{x^2 + y^2}$$

$$\frac{\partial^2 z}{\partial x^2} = \frac{x^2 + y^2 - 2x^2}{(x^2 + y^2)^2} = \frac{y^2 - x^2}{(x^2 + y^2)^2}$$

$$\frac{\partial^2 z}{\partial y^2} = \frac{x^2 + y^2 - 2y^2}{(x^2 + y^2)^2} = \frac{x^2 - y^2}{(x^2 + y^2)^2}$$

Therefore $\quad \dfrac{\partial^2 z}{\partial x^2} + \dfrac{\partial^2 z}{\partial y^2} = 0$

17.2.6. If $z = \cos(x + y) + \cos(x - y)$, then $\dfrac{\partial^2 z}{\partial x^2} - \dfrac{\partial^2 z}{\partial y^2} = 0$.

$$\frac{\partial z}{\partial x} = -\sin(x + y) - \sin(x - y);$$

$$\frac{\partial z}{\partial y} = -\sin(x + y) + \sin(x - y)$$

$$\frac{\partial^2 z}{\partial x^2} = -\cos(x + y) - \cos(x - y);$$

$$\frac{\partial^2 z}{\partial y^2} = -\cos(x + y) - \cos(x - y)$$

$$\frac{\partial^2 z}{\partial x^2} - \frac{\partial^2 z}{\partial y^2} = 0$$

17.2.7. If $u = z \operatorname{Arc} \tan \dfrac{x}{y}$; then $\dfrac{\partial^2 u}{\partial x^2} + \dfrac{\partial^2 u}{\partial y^2} + \dfrac{\partial^2 u}{\partial z^2} = 0$.

$$\frac{\partial u}{\partial x} = \frac{yz}{x^2 + y^2}; \quad \frac{\partial u}{\partial y} = \frac{-xz}{x^2 + y^2}; \quad \frac{\partial u}{\partial z} = \operatorname{Arc} \tan \frac{x}{y}$$

$$\frac{\partial^2 u}{\partial x^2} = \frac{-2xyz}{(x^2 + y^2)^2}; \quad \frac{\partial^2 u}{\partial y^2} = \frac{2xyz}{(x^2 + y^2)^2}; \quad \frac{\partial^2 u}{\partial z^2} = 0;$$

$$\frac{\partial^2 u}{\partial x^2} + \frac{\partial^2 u}{\partial y^2} + \frac{\partial^2 u}{\partial z^2} = 0$$

17.2.8. If $u = 13x^2 + 10xy + 2y^2$, then $\dfrac{\partial^2 u}{\partial x^2} - 5 \dfrac{\partial^2 u}{\partial x \, \partial y} + 6 \dfrac{\partial^2 u}{\partial y^2} = 0$.

$$\frac{\partial u}{\partial x} = 26 + 10y; \quad \frac{\partial^2 u}{\partial x\, \partial y} = 10; \quad \frac{\partial u}{\partial y} = 10x + 4y; \quad \frac{\partial^2 u}{\partial y^2} = 4$$

$$26 - 5\cdot 10 + 6\cdot 4 = 0$$

17.3. The total differential. The total differential of $z = f(x, y)$ is given by $dz = \dfrac{\partial z}{\partial x}\, dx + \dfrac{\partial z}{\partial y}\, dy$. This equation is true even if x and y are not independent variables. The total differential can be used to approximate errors in the same way that the differential can be used for functions of a single variable.

17.3.1. If $z = \ln(x^2 + y^2)$, find the total differential.

Solution

$$\frac{\partial z}{\partial x} = \frac{2x}{x^2 + y^2}; \quad \frac{\partial z}{\partial y} = \frac{2y}{x^2 + y^2}; \quad dz = \frac{2x}{x^2 + y^2}\, dx + \frac{2y}{x^2 + y^2}\, dy$$

17.3.2. Find the total differential $z = 3x^2 + 4xy^2$.

Solution

$$\frac{\partial z}{\partial x} = 6x + 4y^2; \quad \frac{\partial z}{\partial y} = 8xy; \quad dz = (6x + 4y^2)\, dx + 8xy\, dy$$

17.3.3. Find the approximate error in computing the area of a rectangle of length 100 feet and width 25 feet if the length is accurate to $\frac{1}{2}$ foot and width to $\frac{1}{4}$ foot.

Solution

$$A = xy; \quad dA = y\, dx + x\, dy = 25 \times \tfrac{1}{2} + 100 \times \tfrac{1}{4} = 37\tfrac{1}{2}\text{ ft}^2$$

17.3.4. Find the approximate error in computing the area of a triangle by the formula $A = \tfrac{1}{2} bc \sin \theta$ if $b = 10$ ft, $c = 24$ ft, $\theta = \dfrac{\pi}{6}$, and b is accurate to $\tfrac{1}{10}$ foot, c to $\tfrac{1}{4}$ foot, and θ to 1 minute. 1 minute $= 0.000291$ radians.

Solution

$$dA = \tfrac{1}{2} c \sin \theta\, db + \tfrac{1}{2} b \sin \theta\, dc + \tfrac{1}{2} bc \cos \theta\, d\theta$$

$$= \tfrac{1}{2} \times 24 \times \tfrac{1}{2} \times 0.1 + \tfrac{1}{2} \times 10 \times \tfrac{1}{2} \times \tfrac{1}{4}$$

$$+ \tfrac{1}{2} \times 10 \times 24 \times \frac{\sqrt{3}}{2} \times 0.000291$$

$$= 0.600 + 0.417 + 0.030 = 1.047 \text{ ft}^2$$

17.3.5. Find the approximate volume of a rectangular box whose dimensions are 10.98 ft, 7.01 ft., and 14.03 ft.
Solution

$v = xyz$. We consider $x = 11$, $y = 7$, $z = 14$; $dx = -0.02$,
$$dy = 0.01, dz = 0.03.$$

$dv = yz\,dx + xz\,dy + xy\,dz$

$\quad = 7 \times 14 \times -0.02 + 11 \times 14 \times 0.01 + 11 \times 7 \times 0.03$

$\quad = -1.96 + 1.54 + 2.31 = 1.89$ ft³

$v = 11 \times 7 \times 14 + 1.89 = 1078 + 1.89 = 1079.89$ ft³

17.3.6. If g is computed from the formula $s = \frac{1}{2}gt^2$, find the maximum percentage error in g due to errors of 1% in s and t.
Solution

$$g = \frac{2s}{t^2}$$

$$dg = \frac{2}{t^2}ds - \frac{4s}{t^3}dt = \frac{2s}{t^2}\frac{ds}{s} + \frac{2s}{t^2}\left(-2\frac{dt}{t}\right)$$

$$\frac{dg}{g} = \frac{ds}{s} - 2\frac{dt}{t}$$

Then $\dfrac{dg}{g}$ has a maximum value of 0.03 by taking $\dfrac{ds}{s} = 0.01$, $\dfrac{dt}{t} = -0.01$.

17.4. Total derivative. If x and y are functions of t, and z is a function of x and y,

$$\frac{\partial z}{\partial t} = \frac{\partial z}{\partial x} \cdot \frac{\partial x}{\partial t} + \frac{\partial z}{\partial y} \cdot \frac{\partial y}{\partial t} \qquad (17.4.1)$$

If x and y are functions of t *alone*, we may write

$$\frac{dz}{dt} = \frac{\partial z}{\partial x} \cdot \frac{dx}{dt} + \frac{\partial z}{\partial y} \cdot \frac{dy}{dt}$$

We have a special case of this formula when y is a function o x so that z is, in effect, a function of x alone:

$$\frac{dz}{dx} = \frac{\partial z}{\partial x} + \frac{\partial z}{\partial y} \cdot \frac{dy}{dx}$$

Suppose $z = f(x, y)$ where x and y are functions of r and s. $x = g_1(r, s)$, $y = g_2(r, s)$ where r and s are independent.

$$dz = \frac{\partial z}{\partial x} dx + \frac{\partial z}{\partial y} dy$$

$$dx = \frac{\partial x}{\partial r} dr + \frac{\partial x}{\partial s} ds; \quad dy = \frac{\partial y}{\partial r} dr + \frac{\partial y}{\partial s} ds$$

Hence

$$dz = \frac{\partial z}{\partial x}\left(\frac{\partial x}{\partial r} dr + \frac{\partial x}{\partial s} ds\right) + \frac{\partial z}{\partial y}\left(\frac{\partial y}{\partial r} dr + \frac{\partial y}{\partial s} ds\right)$$

$$= \left(\frac{\partial z}{\partial x} \cdot \frac{\partial x}{\partial r} + \frac{\partial z}{\partial y} \cdot \frac{\partial y}{\partial r}\right) dr + \left(\frac{\partial z}{\partial x} \cdot \frac{\partial x}{\partial s} + \frac{\partial z}{\partial y} \cdot \frac{\partial y}{\partial s}\right) ds$$

Using formula (17.4.1),

$$dz = \frac{\partial z}{\partial r} dr + \frac{\partial z}{\partial s} ds.$$

17.4.1. Find $\frac{dz}{dt}$, where $z = \text{Arc tan } \frac{y}{x}$, $x = \ln t$, $y = e^t$.

Solution

$$\frac{dz}{dt} = \frac{-y}{x^2 + y^2} \cdot \frac{dx}{dt} + \frac{x}{x^2 + y^2} \cdot \frac{dy}{dt} = \frac{-e^t}{\ln^2 t + e^{2t}} \cdot \frac{1}{t}$$

$$+ \frac{\ln t}{\ln^2 t + e^{2t}} \cdot e^t = \frac{e^t}{\ln^2 t + e^{2t}} \left(\ln t - \frac{1}{t}\right)$$

17.4.2. Find $\frac{dz}{dx}$ when $z = (x^2 + y^2 + xy)^{3/2}$, and $y = x^2$.

Solution

$$\frac{dz}{dx} = \tfrac{3}{2}(2x + y)(x^2 + y^2 + xy)^{1/2} + \tfrac{3}{2}(2y + x)$$

$$\cdot (x^2 + y^2 + xy)^{1/2} \cdot 2x$$

$$= \tfrac{3}{2}(x^4 + x^3 + x^2)^{1/2}(4x^3 + 3x^2 + 2x)$$

Second solution

$$z = (x^4 + x^3 + x^2)^{3/2}$$

$$\frac{dz}{dx} = \tfrac{3}{2}(x^4 + x^3 + x^2)^{1/2}(4x^3 + 3x^2 + 2x)$$

17.4.3. Find $\dfrac{dz}{dx}$ when $z = \tan \dfrac{y}{x}$, and $y = e^x$.

Solution

$$\frac{dz}{dx} = -\frac{y}{x^2}\sec^2\frac{y}{x} + \frac{1}{x}\sec^2\frac{y}{x} \cdot e^x = e^x\left(\frac{1}{x} - \frac{1}{x^2}\right)\sec^2\frac{e^x}{x}$$

17.4.4. If $z = 4x^2 - 9y^2$ and $x = se^t$, $y = se^{-t}$, find $\dfrac{\partial z}{\partial s}, \dfrac{\partial z}{\partial t}$.

Solution

$$= \frac{\partial z}{\partial x}\cdot\frac{\partial x}{\partial s} + \frac{\partial z}{\partial y}\cdot\frac{\partial y}{\partial s} = 8xe^t - 18ye^{-t} = 8se^{2t} - 18se^{-2t}$$

$$= \frac{\partial z}{\partial x}\cdot\frac{\partial x}{\partial t} + \frac{\partial z}{\partial y}\cdot\frac{\partial y}{\partial t} = 8xse^t - 18y(-se^{-t}) = 8s^2 e^{2t} + 18s^2 e^{-2t}$$

17.4.5. If $z = \ln\sqrt{x^2 + y^2}$ where $x = \dfrac{s}{t}$, $y = s + t$, find $\dfrac{\partial z}{\partial s}, \dfrac{\partial z}{\partial t}$.

Solution

$$\frac{\partial z}{\partial s} = \frac{x}{x^2 + y^2}\cdot\frac{1}{t} + \frac{y}{x^2 + y^2}\cdot 1 = \frac{\dfrac{s}{t^2}}{\dfrac{s^2}{t^2} + (s+t)^2}$$

$$+ \frac{s+t}{\dfrac{s^2}{t^2} + (s+t)^2} = \frac{\dfrac{s}{t^2} + s + t}{\dfrac{s^2}{t^2} + (s+t)^2}$$

$$\frac{\partial z}{\partial t} = \frac{x}{x^2 + y^2}\left(-\frac{s}{t^2}\right) + \frac{y}{x^2 + y^2}\cdot 1 = \frac{-\dfrac{s^2}{t^3} + s + t}{\dfrac{s^2}{t^2} + (s+t)^2}$$

17.4.6. If $u = f(x - y, y - x)$, show $\dfrac{\partial u}{\partial x} + \dfrac{\partial u}{\partial y} = 0$.

Solution. Let $x - y = r$, $y - x = s$, and $u = f(r, s)$. Thus, if we use our formula (17.4.1),

$$\frac{\partial u}{\partial x} = \frac{\partial f}{\partial r} \cdot 1 + \frac{\partial f}{\partial s}(-1); \qquad \frac{\partial u}{\partial y} = \frac{\partial f}{\partial r}(-1) + \frac{\partial f}{\partial s} \cdot 1;$$

$$\frac{\partial u}{\partial x} + \frac{\partial u}{\partial y} = 0$$

17.4.7. If we have a function $z = f(x, y)$ such that $f(tx, ty) = t^n f(x, y)$, then $f(x, y)$ is called a homogeneous function. Prove for such a homogeneous function that

$$x f_x(x, y) + y f_y(x, y) = n f(x, y)$$

Solution. First, let $u = f(tx, ty)$, and by substituting, $tx = r$, $ty = s$.

$$\frac{du}{dt} = f_r(r, s) \frac{\partial r}{\partial t} + f_s(r, s) \frac{\partial s}{\partial t} = x f_r(r, s) + y f_s(r, s)$$

In this differentiation, we have considered r and s as functions of t alone (x and y constants). On the other hand, from

$$u = t^n f(x, y)$$

$$\frac{du}{dt} = n t^{n-1} f(x, y)$$

So

$$x f_r(r, s) + y f_s(r, s) = n t^{n-1} f(x, y)$$

Setting $t = 1$, $r = x$, $s = y$,

$$x f_x(x, y) + y f_y(x, y) = n f(x, y)$$

This is known as Euler's formula for homogeneous functions. Notice in particular that if $f(tx, ty) = f(x, y)$, then

$$x f_x(x, y) + y f_y(x, y) = 0$$

17.4.8. If $z = f(x, y)$, and $x = r\cos\theta$, $y = r\sin\theta$, show that

$$\frac{\partial z}{\partial x} = \frac{\partial z}{\partial r}\cos\theta - \frac{\partial z}{\partial \theta}\cdot\frac{\sin\theta}{r}$$

$$\frac{\partial z}{\partial y} = \frac{\partial z}{\partial r}\sin\theta + \frac{\partial z}{\partial \theta}\cdot\frac{\cos\theta}{r}$$

Solution. Let $z = \varphi(r, \theta)$ where $r = \sqrt{x^2 + y^2}$ and $\theta = \text{rc tan}\frac{y}{x}$.

$$\frac{\partial z}{\partial x} = \frac{\partial z}{\partial r}\cdot\frac{\partial r}{\partial x} + \frac{\partial z}{\partial \theta}\cdot\frac{\partial \theta}{\partial x}$$

$$= \frac{\partial z}{\partial r}\cdot\frac{x}{\sqrt{x^2+y^2}} + \frac{\partial z}{\partial \theta}\cdot\frac{-y}{x^2+y^2}$$

ince $x = r\cos\theta$, $y = r\sin\theta$, and $r = \sqrt{x^2+y^2}$,

$$\frac{\partial z}{\partial x} = \frac{\partial z}{\partial r}\cos\theta - \frac{\partial z}{\partial \theta}\cdot\frac{\sin\theta}{r}$$

$$\frac{\partial z}{\partial y} = \frac{\partial z}{\partial r}\cdot\frac{y}{\sqrt{x^2+y^2}} + \frac{\partial z}{\partial \theta}\cdot\frac{x}{x^2+y^2} = \frac{\partial z}{\partial r}\sin\theta + \frac{\partial z}{\partial \theta}\cdot\frac{\cos\theta}{r}$$

Also notice

$$\frac{\partial z}{\partial r} = \frac{\partial z}{\partial x}\cos\theta + \frac{\partial z}{\partial y}\sin\theta$$

$$\frac{\partial z}{\partial \theta} = \frac{\partial z}{\partial x}(-r\sin\theta) + \frac{\partial z}{\partial y}(r\cos\theta)$$

17.5. Implicit functions. For a function of the form $f(x, y) = 0$, $\frac{dy}{dx} = -\frac{f_x}{f_y}$; if we have

$$F(x, y, z) = 0, \text{ then } \frac{\partial z}{\partial x} = -\frac{F_x}{F_z}, \frac{\partial z}{\partial y} = -\frac{F_y}{F_z}.$$

17.5.1. Find $\frac{dy}{dx}$ if $x^3 + y^3 - 3axy = 0$.

Solution

$$f_x = 3x^2 - 3ay; \quad f_y = 3y^2 - 3ax; \quad \frac{dy}{dx} = -\frac{x^2 - ay}{y^2 - ax}$$

17.5.2. Find $\dfrac{dy}{dx}$ if $\sin \dfrac{x}{y} + \sin \dfrac{y}{x} = 1$.

Solution

$$f_x = \frac{1}{y}\cos\frac{x}{y} - \frac{y}{x^2}\cos\frac{y}{x}, \quad f_y = -\frac{x}{y^2}\cos\frac{x}{y} + \frac{1}{x}\cos\frac{y}{x};$$

$$\frac{dy}{dx} = -\frac{x^2 y \cos\dfrac{x}{y} - y^3 \cos\dfrac{y}{x}}{-x^3 \cos\dfrac{x}{y} + xy^2 \cos\dfrac{y}{x}}$$

17.5.3. Find $\dfrac{\partial z}{\partial x}, \dfrac{\partial z}{\partial y}$ if $x^2 + y^2 - 2z^2 = a^2$.

Solution

$$F_x = 2x; \quad F_y = 2y; \quad F_z = -4z \quad \frac{\partial z}{\partial x} = \frac{x}{2z}; \quad \frac{\partial z}{\partial y} = \frac{y}{2z}$$

17.5.4. Find $\dfrac{\partial z}{\partial x}, \dfrac{\partial z}{\partial y}$ if $\sin(x+y) + \sin(y+z) + \sin(z+x) = 1$.

Solution

$$F_x = \cos(x+y) + \cos(x+z);$$
$$F_y = \cos(x+y) + \cos(y+z)$$
$$F_z = \cos(y+z) + \cos(x+z);$$

$$\frac{\partial z}{\partial x} = -\frac{\cos(x+y) + \cos(z+x)}{\cos(y+z) + \cos(z+x)}$$

$$\frac{\partial z}{\partial y} = -\frac{\cos(x+y) + \cos(y+z)}{\cos(y+z) + \cos(z+x)}$$

17.5.5. Find $\dfrac{\partial z}{\partial x}, \dfrac{\partial z}{\partial y}$ if $xe^{yz} + ye^{zx} + ze^{xy} = 3$.

Solution. Here

$$F_x = e^{yz} + yze^{zx} + yze^{xy}$$
$$F_y = e^{zx} + xze^{yz} + xze^{xy}$$
$$F_z = e^{xy} + xye^{yz} + xye^{zx}$$

$$\frac{\partial z}{\partial x} = -\frac{e^{yz} + yze^{zx} + yze^{xy}}{e^{xy} + xye^{yz} + xye^{zx}}$$

$$\frac{\partial z}{\partial y} = -\frac{e^{zx} + xze^{yz} + xze^{xy}}{e^{xy} + xye^{yz} + xye^{zx}}$$

17.6. Exact differentials. $P\,dx + Q\,dy$ is the total differential of some function or an exact differential if, and only if, $\frac{\partial P}{\partial y} = \frac{\partial Q}{\partial x}$.

17.6.1. Determine if the following are exact differentials.

(a) $(x^2 + y^2)\,dx + 2xy\,dy$
Solution

$$\frac{\partial P}{\partial y} = 2y, \qquad \frac{\partial Q}{\partial x} = 2y, \qquad \text{an exact differential}$$

(b) $(3x + 5y)\,dx + (4x - 3y)\,dy$
Solution

$$\frac{\partial P}{\partial y} = 5, \qquad \frac{\partial Q}{\partial x} = 4, \qquad 5 \neq 4 \qquad \text{not an exact differential}$$

(c) $\frac{y}{x^2 + y^2}\,dx - \frac{x}{x^2 + y^2}\,dy$
Solution

$$\frac{\partial P}{\partial y} = \frac{x^2 - y^2}{(x^2 + y^2)^2}, \qquad \frac{\partial Q}{\partial x} = \frac{x^2 - y^2}{(x^2 + y^2)^2}, \qquad \text{an exact differential}$$

(d) $e^x \cos y\,dx + e^x \sin y\,dy$
Solution

$$\frac{\partial P}{\partial y} = -e^x \sin y, \qquad \frac{\partial Q}{\partial x} = e^x \sin y, \qquad \text{not an exact differential}$$

(e) $P(x)\,dx + Q(y)\,dy$
Solution

$$\frac{\partial P(x)}{\partial y} = 0, \qquad \frac{\partial Q(y)}{\partial x} = 0, \qquad \text{an exact differential}$$

17.7. The directional derivative. We find the limit of the ratio $\frac{EQ}{EP}$ as EP has limit zero. Line EP makes an angle α with the x axis, and the resulting derivative is called the directional derivative.

$$\frac{dz}{dl} = \frac{\partial z}{\partial x}\cos \alpha + \frac{\partial z}{\partial y}\sin \alpha$$

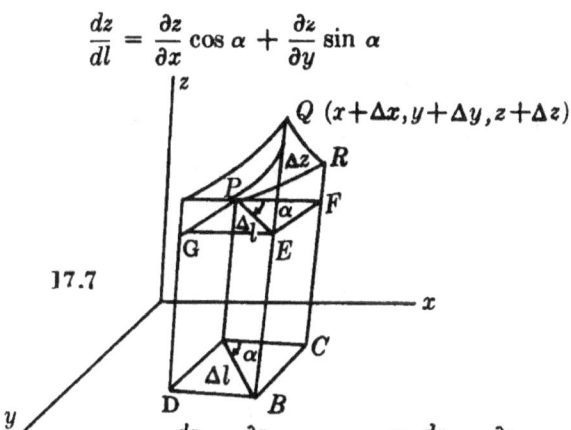

17.7

In the special case $\alpha = 0$, $\frac{dz}{dl} = \frac{\partial z}{\partial x}$, and $\alpha = \frac{\pi}{2}$, $\frac{dz}{dl} = \frac{\partial z}{\partial y}$.

17.7.1. Find the derivative of $z = \ln \sqrt{x^2 + y^2}$ at $(\frac{3}{5}, \frac{4}{5}, 0)$ in a direction of 60° from the x axis.

Solution

$$\frac{\partial z}{\partial x} = \frac{x}{x^2 + y^2}; \quad \frac{\partial z}{\partial y} = \frac{y}{x^2 + y^2}$$

$$\frac{\partial z}{\partial l} = \frac{\frac{3}{5}}{\frac{9}{25} + \frac{16}{25}} \cos 60° + \frac{\frac{4}{5}}{\frac{9}{25} + \frac{16}{25}} \sin 60°$$

$$= \frac{3}{10} + \frac{4\sqrt{3}}{10} = \frac{3 + 4\sqrt{3}}{10}$$

For the direction of zero change,

$$\tfrac{3}{5} \cos \alpha + \tfrac{4}{5} \sin \alpha = 0; \quad \tan \alpha = -\tfrac{3}{4}$$

The curve formed by the intersection of the surface and the plane $y = -\tfrac{3}{4} x + \tfrac{3}{4}$ will have a horizontal tangent at this point.

17.7.2. Find the directional derivative of $x^3 + y^3 - 3xy = z$ at $(2, 1, 3)$ in the direction of Arc tan $\tfrac{3}{4}$, first quadrant.

Solution

$$\frac{\partial z}{\partial x} = 3x^2 - 3y; \quad \frac{\partial z}{\partial y} = 3y^2 - 3x$$

$$\frac{dz}{dl} = (3\cdot 4 - 3\cdot 1)\frac{4}{5} + (3\cdot 1 - 3\cdot 2)\frac{3}{5} = \frac{27}{5}$$

17.7.3. Show that the maximum value of the directional derivative at (x_1, y_1) is $\sqrt{\left(\frac{\partial z}{\partial x}\right)_1^2 + \left(\frac{\partial z}{\partial y}\right)_1^2}$ where the subscript 1 indicates the partials are evaluated at (x_1, y_1).

Solution. We have

$$\left(\frac{dz}{dl}\right)_1 = f(\alpha) = \left(\frac{\partial z}{\partial x}\right)_1 \cos \alpha + \left(\frac{\partial z}{\partial y}\right)_1 \sin \alpha$$

and

$$f'(\alpha) = -\left(\frac{dz}{dx}\right)_1 \sin \alpha + \left(\frac{\partial z}{\partial y}\right)_1 \cos \alpha$$

or

$$\tan \alpha_{max} = \frac{\left(\frac{\partial z}{\partial y}\right)_1}{\left(\frac{\partial z}{\partial x}\right)_1}$$

We can easily check that this value gives a maximum.

$$\cos \alpha_{max} = \frac{\left(\frac{\partial z}{\partial x}\right)_1}{\sqrt{\left(\frac{\partial z}{\partial x}\right)_1^2 + \left(\frac{\partial z}{\partial y}\right)_1^2}}, \sin \alpha_{max} = \frac{\left(\frac{\partial z}{\partial y}\right)_1}{\sqrt{\left(\frac{\partial z}{\partial x}\right)_1^2 + \left(\frac{\partial z}{\partial y}\right)_1^2}}$$

$$\left(\frac{dz}{dl}\right)_{max} = \frac{\left(\frac{\partial z}{\partial x}\right)_1^2}{\sqrt{\left(\frac{\partial z}{\partial x}\right)_1^2 + \left(\frac{\partial z}{\partial y}\right)_1^2}} + \frac{\left(\frac{\partial z}{\partial y}\right)_1^2}{\sqrt{\left(\frac{\partial z}{\partial x}\right)_1^2 + \left(\frac{\partial z}{\partial y}\right)_1^2}}$$

$$= \sqrt{\left(\frac{\partial z}{\partial x}\right)_1^2 + \left(\frac{\partial z}{\partial y}\right)_1^2}$$

Then $\alpha_{max} + 180°$ will give us $\left(\dfrac{dz}{dl}\right)_{min} = -\left(\dfrac{dz}{dl}\right)_{max}$

Notice, if $f(\alpha) = 0$, $\tan \alpha_0 = -\dfrac{\left(\dfrac{\partial z}{\partial x}\right)_1}{\left(\dfrac{\partial z}{\partial y}\right)_1}$, and $\tan \alpha_0 \tan \alpha_{max} = -1$, so that the directions of zero increase and maximum increase are perpendicular.

17.8. Tangent plane and normal line. If our function is $z = f(x, y)$, the tangent plane has the form

$$\left(\dfrac{\partial z}{\partial x}\right)_1 (x - x_1) + \left(\dfrac{\partial z}{\partial y}\right)_1 (y - y_1) - (z - z_1) = 0.$$

If our function is implicit, $F(x, y, z) = 0$, we can write the equation of the tangent plane as

$$F_{x_1}(x - x_1) + F_{y_1}(y - y_1) + F_{z_1}(z - z_1) = 0$$

The line perpendicular to the tangent plane at (x_1, y_1, z_1), or normal line, can be written

$$\dfrac{x - x_1}{\left(\dfrac{\partial z}{\partial x}\right)_1} = \dfrac{y - y_1}{\left(\dfrac{\partial z}{\partial y}\right)_1} = \dfrac{z - z_1}{-1}, \quad \text{or} \quad \dfrac{x - x_1}{F_{x_1}} = \dfrac{y - y_1}{F_{y_1}} = \dfrac{z - z_1}{F_{z_1}}$$

17.8.1. Find the equation of the tangent plane and normal line to the ellipsoid $x^2 + 9y^2 + z^2 = 61$ at $(4, 2, -3)$.

Solution

$$\dfrac{\partial F}{\partial x} = 2x; \quad \dfrac{\partial F}{\partial y} = 18y; \quad \dfrac{\partial F}{\partial z} = 2z$$

The equations are

$$8(x - 4) + 36(y - 2) - 6(z + 3) = 0$$

$$\dfrac{x - 4}{8} = \dfrac{y - 2}{36} = \dfrac{z + 3}{-6}$$

Upon simplifying, the equation of the plane is

$$4x + 18y - 3z = 61$$

17.8.2. Find the equation of the tangent plane and normal ne to $z = xy$ at a point (x_1, y_1, z_1).

Solution

$$\frac{\partial z}{\partial x} = y; \qquad \frac{\partial z}{\partial y} = x$$

$$y_1(x - x_1) + x_1(y - y_1) - (z - z_1) = 0$$

$$\frac{x - x_1}{y_1} = \frac{y - y_1}{x_1} = \frac{z - z_1}{-1}$$

'he x intercept of the tangent plane, setting $y = z = 0$, is

$$y_1(x - x_1) = x_1 y_1 - z_1 = 0; \qquad x = x_1.$$

Jow $(x_1, 0, 0)$ is the x intercept; $(0, y_1, 0)$ is the y intercept; $0, 0, -z_1)$ is the z intercept. The plane may be written

$$\frac{x}{x_1} + \frac{y}{y_1} - \frac{z}{z_1} = 1$$

17.8.3. Find the equation of the tangent plane and normal ine to $xy + xz + yz = 1$ at $(3, 2, -1)$.

Solution

$$F_x = y + z; \qquad F_y = x + z; \qquad F_z = x + y$$

$$(x - 3) + 2(y - 2) + 5(z + 1) = 0;$$

$$\frac{x - 3}{1} = \frac{y - 2}{2} = \frac{z + 1}{5}$$

17.8.4. Find the tangent plane to $x^{2/3} + y^{2/3} + z^{2/3} = 21$ it $(-8, 1, 64)$.

Solution.

$$F_x = \tfrac{2}{3} x^{-1/3}; \qquad F_y = \tfrac{2}{3} y^{-1/3}; \qquad F_z = \tfrac{2}{3} z^{-1/3}$$

$$-\tfrac{1}{2}(x + 8) + \tfrac{2}{3}(y - 1) + \tfrac{1}{6}(z - 64) = 0$$

17.8.5. Two surfaces are said to be tangent to each other at a point if they have a common tangent plane at the point. Show $2x^2 + 2y^2 - z^2 = 25$ and $x^2 + y^2 = 5z$ are tangent to each other at $(4, 3, 5)$.

Solution. For $2x^2 + 2y^2 - z^2 = 25$; $F_x = 4x$, $F_y = 4y$, $F_z = -2z$

For $x^2 + y^2 = 5z$; $F_x = 2x, F_y = 2y, F_z = -5$

At (4, 3, 5) the tangent planes are

$16(x - 4) + 12(y - 3) - 10(z - 5) = 0$ and $8(x - 4) + 6(y - 3) - 5(z - 5) =$

Dividing through the first by 2, we see they are identical.

17.9. Maxima and minima. For $z = f(x, y)$ a necessary condition for a maximum or minimum is $\left(\dfrac{\partial z}{\partial x}\right)_1 = 0, \left(\dfrac{\partial z}{\partial y}\right)_1 = 0$
This condition is not sufficient, as can be seen from the example of the saddle surface $x^2 - y^2 = z$. For this surface $\dfrac{\partial z}{\partial x} = 2x$, $\dfrac{\partial z}{\partial y} = 2y$, and at (0, 0, 0), $\dfrac{\partial z}{\partial x} = \dfrac{\partial z}{\partial y} = 0$, but the point (0, 0, 0) is neither a maximum nor a minimum, as can be seen from the figure. In more advanced books on mathematics the following rule is derived.

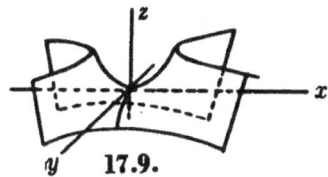

17.9.

Suppose $z = f(x, y)$. Find values of x and y for which $\dfrac{\partial z}{\partial x} = 0$, $\dfrac{\partial z}{\partial y} = 0$. Calculate $\Delta = \left(\dfrac{\partial^2 z}{\partial x\, \partial y}\right)^2 - \dfrac{\partial^2 z}{\partial x^2} \cdot \dfrac{\partial^2 z}{\partial y^2}$ for these values. The corresponding value of z will be

1. A maximum if $\Delta < 0$, and $\dfrac{\partial^2 z}{\partial x^2}$ or $\dfrac{\partial^2 z}{\partial y^2}$ is negative.

2. A minimum if $\Delta < 0$, and $\dfrac{\partial^2 z}{\partial x^2}$ or $\dfrac{\partial^2 z}{\partial y^2}$ is positive.

3. Neither a maximum nor a minimum if $\Delta > 0$.

If $\Delta = 0$, the rule is not sufficient, and the value of z may be a maximum, minimum, or neither.

17.9.1. Find any maxima or minima for $z = xy$.

Solution

$$\frac{\partial z}{\partial x} = y, \quad \frac{\partial z}{\partial y} = x$$

For $x = 0, y = 0$, we calculate Δ:

$$\frac{\partial^2 z}{\partial x^2} = 0; \quad \frac{\partial^2 z}{\partial y^2} = 0; \quad \frac{\partial^2 z}{\partial x\, \partial y} = 1; \quad \Delta = 1;$$

and there is no maximum or minimum.

17.9.2. Find any maxima or minima for $z = x^3 + y^2 - 3x$.

Solution

$$\frac{\partial z}{\partial x} = 3x^2 - 3; \quad \frac{\partial z}{\partial y} = 2y$$

Now $(1, 0, -2)$, and $(-1, 0, 2)$ are to be examined.

$$\frac{\partial^2 z}{\partial x\, \partial y} = 0; \quad \frac{\partial^2 z}{\partial x^2} = 6x; \quad \frac{\partial^2 z}{\partial y^2} = 2; \quad \Delta = -12x$$

For $x = 1, \Delta < 0$, and $\frac{\partial^2 z}{\partial x^2} > 0$, which means $(1, 0, -2)$ is a minimum. For $x = -1, \Delta > 0$, neither maximum nor minimum.

17.9.3. Find any maxima or minima for $z = x^2 + xy + y^2 - 2x - 6y$.

Solution

$$\frac{\partial z}{\partial x} = 2x + y - 2; \quad \frac{\partial z}{\partial y} = x + 2y - 6$$

Since

$$2x + y - 2 = 0; \quad x + 2y - 6 = 0; \quad y = \frac{10}{3}; \quad x = -\frac{2}{3}$$

$$\frac{\partial^2 z}{\partial x\, \partial y} = 1; \quad \frac{\partial^2 z}{\partial x^2} = 2; \quad \frac{\partial^2 z}{\partial y^2} = 2; \quad \Delta = (1)^2 - 4 = -3$$

$$\Delta < 0, \quad \frac{\partial^2 z}{\partial x^2} > 0, \quad \frac{\partial^2 z}{\partial y^2} > 0$$

Then $(-\frac{2}{3}, \frac{10}{3}, -\frac{84}{9})$ is the minimum.

17.9.4. Find the dimensions of the rectangular parallelepiped of maximum volume with faces parallel to the coordinate axes and inscribed in

$$\frac{x^2}{a^2} + \frac{y^2}{b^2} + \frac{z^2}{c^2} = 1$$

One-eighth of the volume and the ellipsoid is shown in the figure.
Solution

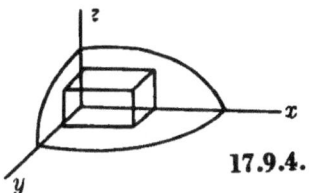

17.9.4.

$V = 8xyz$

$$\frac{\partial V}{\partial x} = 8yz + 8xy\frac{\partial z}{\partial x}$$

$$\frac{\partial V}{\partial y} = 8xz + 8xy\frac{\partial z}{\partial y}$$
(17.9.1)

Since

$$\frac{x^2}{a^2} + \frac{y^2}{b^2} + \frac{z^2}{c^2} = 1 \qquad (17.9.2)$$

$$\frac{2x}{a^2} + \frac{2z}{c^2}\cdot\frac{\partial z}{\partial x} = 0; \qquad \frac{\partial z}{\partial x} = -\frac{c^2 x}{a^2 z}$$

Likewise $\frac{\partial z}{\partial y} = -\frac{c^2 y}{b^2 z}$, and substituting in (17.9.1),

$$\frac{\partial V}{\partial x} = 8y\left(z - x\frac{c^2 x}{a^2 z}\right); \qquad \frac{\partial V}{\partial y} = 8x\left(z - y\frac{c^2 y}{b^2 z}\right)$$

Setting $\frac{\partial V}{\partial x} = 0$, $\frac{\partial V}{\partial y} = 0$, we have

$$z = \frac{c^2 x^2}{a^2 z}, \quad \text{or} \quad x = \frac{a}{c}z, \quad \text{and} \quad z = \frac{c^2 y^2}{b^2 z}, \quad \text{or} \quad y = \frac{b}{c}z$$

Substituting in (17.9.2),

$$\frac{\frac{a^2}{c^2}z^2}{a^2} + \frac{\frac{b^2}{c^2}z^2}{b^2} + \frac{z^2}{c^2} = 1, \quad \text{or} \quad z = \frac{c\sqrt{3}}{3};$$

hence $x = \dfrac{a\sqrt{3}}{3}$, $y = \dfrac{b\sqrt{3}}{3}$

To check that this is a maximum value,

$$\frac{\partial^2 V}{\partial x \, \partial y} = 8\left(z - \frac{c^2 x^2}{a^2 z}\right) + 8y\left(1 + \frac{c^2 x^2}{a^2 z^2}\right)\frac{\partial z}{\partial y}$$

$$= 8\left(z - \frac{c^2 x^2}{a^2 z}\right) - 8\frac{c^2 y^2}{b^2 z}\left(1 + \frac{c^2 x^2}{a^2 z^2}\right)$$

For $z = \dfrac{c\sqrt{3}}{3}$, $x = \dfrac{a\sqrt{3}}{3}$, $y = \dfrac{b\sqrt{3}}{3}$,

$$\frac{\partial^2 V}{\partial x \, \partial y} = 8\left(\frac{c}{\sqrt{3}} - \frac{c}{\sqrt{3}}\right) - \frac{8c}{\sqrt{3}}(1 + 1) = -\frac{16c}{\sqrt{3}}$$

$$\frac{\partial^2 V}{\partial x^2} = 8y\left(-3\frac{c^2 x}{a^2 z} + \frac{c^2 x^2}{a^2 z^2}\frac{\partial z}{\partial x}\right) = 8y\left(-3\frac{c^2 x}{a^2 z} - \frac{c^4 x^3}{a^4 z^3}\right)$$

For $z = \dfrac{c\sqrt{3}}{3}$, $x = \dfrac{a\sqrt{3}}{3}$, $y = \dfrac{b\sqrt{3}}{3}$,

$$\frac{\partial^2 V}{\partial x^2} = \frac{8b\sqrt{3}}{3}\left(-3\frac{c}{a} - \frac{c}{a}\right) = \frac{-32bc\sqrt{3}}{3a};$$

$$\frac{\partial^2 V}{\partial y^2} = \frac{-32ac\sqrt{3}}{3b}; \quad \frac{\partial^2 V}{\partial x^2} \cdot \frac{\partial^2 V}{\partial y^2} = \frac{1024}{3}c^2;$$

$$\Delta = \frac{256}{3}c^2 - \frac{1024}{3}c^2 < 0$$

Since $\dfrac{\partial^2 V}{\partial x^2} < 0$, we have a maximum.

$$V = \frac{8abc}{3\sqrt{3}}$$

Chapter 18

MULTIPLE INTEGRALS

18.1. Double integrals and plane areas. The area between $y = f(x)$ and $y = F(x)$, $x = a$, $x = b$, is given by $A = \int_a^b \int_{f(x)}^{F(x)} dy\, dx$, where the integration is first carried out with respect to y.

18.1.1. Find $\int_1^2 \int_0^y xy\, dx\, dy$.

Solution

$$\int_1^2 \int_0^y xy\, dx\, dy = \int_1^2 \frac{x^2}{2} y \Big|_0^y dy = \int_1^2 \frac{y^3}{2} dy = \frac{y^4}{8}\Big|_1^2 = 2 - \frac{1}{8} = \frac{15}{8}$$

18.1.1.

Notice that the area over which the function is integrated is the trapezoid shown. If we wished to integrate first with respect to y, the value of the integral over $ABCE$ would be $\int_0^1 \int_1^2 xy\, dy\, dx$, and over ADE, $\int_1^2 \int_x^2 xy\, dy\, dx$.

$$\int_0^1 \int_1^2 xy\, dy\, dx = \int_0^1 \frac{xy^2}{2}\Big|_1^2 dx = \frac{3}{2}\int_0^1 x\, dx = \frac{3}{4}x^2\Big|_0^1 = \frac{3}{4}$$

$$\int_1^2 \int_x^2 xy\, dy\, dx = \int_1^2 \frac{xy^2}{2}\Big|_x^2 dx = \int_1^2 \left(2x - \frac{x^3}{2}\right) dx$$

$$= \left(x^2 - \frac{x^4}{8}\right)\Big|_1^2 = \frac{9}{8}$$

So $\frac{3}{4} + \frac{9}{8} = \frac{15}{8}$, giving the same total value as before.

18.1.2. Find $\int_a^{2a} \int_0^{\sqrt{x^2-a^2}} \frac{y}{\sqrt{x^2+y^2+a^2}} dy\, dx$

$$= \int_a^{2a} \sqrt{x^2+y^2+a^2}\,\Big|_0^{\sqrt{x^2-a^2}} dx$$

Solution

$$\int_a^{2a} (\sqrt{2}x - \sqrt{x^2 + a^2})\, dx$$

$$\left(\frac{\sqrt{2}}{2} x^2 - \frac{a^2}{2} \ln |x + \sqrt{x^2 + a^2}| - \frac{x}{2}\sqrt{x^2+a^2} \right)_a^{2a}$$

$$\frac{a^2}{2} \ln[(1+\sqrt{2})a] + 2\sqrt{2}a^2 - \frac{a^2}{2}\ln[(2+\sqrt{5})a] - \sqrt{5}a^2$$

$$\frac{a^2}{2} \ln \frac{1+\sqrt{2}}{2+\sqrt{5}} + (2\sqrt{2} - \sqrt{5})a^2$$

18.1.2.

If we examine our area of integration bounded by the x axis, the hyperbola $y = \sqrt{x^2 - a^2}$, and $x = a$, $x = 2a$, we see the integral could also be written

$$\int_0^{\sqrt{3}a} \int_{\sqrt{y^2+a^2}}^{2a} \frac{y}{\sqrt{x^2+y^2+a^2}}\, dx\, dy$$

18.1.3. $\int_0^\pi \int_0^{a(1+\cos\theta)} r^2 \sin\theta\, dr\, d\theta = \int_0^\pi \frac{r^3}{3} \sin\theta \Big|_0^{a(1+\cos\theta)} d\theta$

$= \frac{a^3}{3} \int_0^\pi (1+\cos\theta)^3 \sin\theta\, d\theta = \frac{-a^3}{12}(1+\cos\theta)^4 \Big|_0^\pi = \frac{4a^3}{3}$

18.1.4. Find the area bounded by the curve $y^2 = x^3$ and the line $y = x$.

18.1.4.

Solution. If we integrate first with respect to x, the limits will be y and $y^{2/3}$.

$$A = \int_0^1 \int_y^{y^{2/3}} dx\, dy = \int_0^1 x \Big|_y^{y^{2/3}} dy = \int_0^1 (y^{2/3} - y)\, dy$$

$$= \left(\frac{3}{5} y^{5/3} - \frac{y^2}{2}\right)\Big|_0^1 = \frac{1}{10}$$

This area could also be written $\int_0^1 \int_{x^{3/2}}^x dy\, dx.$

18.1.5. Find the area bounded by $y = x^2$ and $y = \dfrac{2}{x^2+1}$.

Solution

$$A = \int_{-1}^1 \int_{x^2}^{2/(x^2+1)} dy\, dx = \int_{-1}^1 y \Big|_{x^2}^{2/(x^2+1)} dx$$

$$= \int_{-1}^1 \left(\frac{2}{x^2+1} - x^2\right) dx = \left(2 \arctan x - \frac{x^3}{3}\right)\Big|_{-1}^1 = \pi -$$

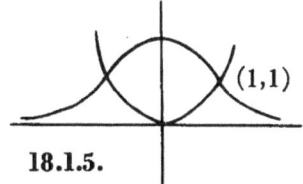

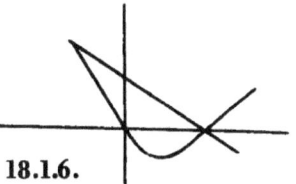

18.1.5. 18.1.6.

18.1.6. Find the area bounded by $y + x = 1$ and $y = x^2 - $
Solution. Here y varies from $x^2 - x$ to $1 - x$, and x varies from -1 to 1.

$$A = \int_{-1}^1 \int_{x^2-x}^{1-x} dy\, dx = \int_{-1}^1 y \Big|_{x^2-x}^{1-x} dx = \int_{-1}^1 (1 - x^2)\, dx$$

$$= \left(x - \frac{x^3}{3}\right)\Big|_{-1}^1 = \frac{4}{3}$$

18.1.7. Find the first quadrant area bounded by the curve $y = \sin x$, $y = \cos x$, and the y axis.

$$A = \int_0^{\pi/4} \int_{\sin x}^{\cos x} dy\, dx = \int_0^{\pi/4} y \Big|_{\sin x}^{\cos x} dx$$

$$= \int_0^{\pi/4} (\cos x - \sin x)\, dx = (\sin x + \cos x)\Big|_0^{\pi/4} = \sqrt{2} - 1$$

The formula for the area in polar coordinates is $A = \iint r\, dr\, d\theta$.

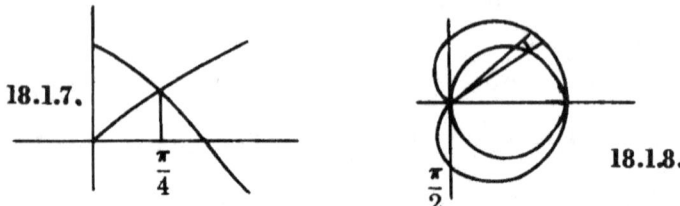

18.1.7.

18.1.8.

18.1.8. Find the area outside the circle $r = 2\cos\theta$ and inside the cardiod $r = 1 + \cos\theta$.

Solution. Here r varies from $2\cos\theta$ to $1 + \cos\theta$, and by integrating with respect to θ from 0 to $\frac{\pi}{2}$, we shall have the area between the two curves in the first quadrant.

$$A_1 = \int_0^{\pi/2}\int_{2\cos\theta}^{1+\cos\theta} r\, dr\, d\theta = \int_0^{\pi/2} \frac{1}{2}r^2\Big|_{2\cos\theta}^{1+\cos\theta} d\theta$$

$$= \frac{1}{2}\int_0^{\pi/2} [1 + 2\cos\theta - 3\cos^2\theta]\, d\theta$$

$$= \frac{1}{2}\left(\theta + 2\sin\theta - \frac{3}{2}\theta - \frac{3}{4}\sin 2\theta\right)_0^{\pi/2} = 1 - \frac{\pi}{8}$$

The area in the second quadrant is

$$A_2 = \int_{\pi/2}^{\pi}\int_0^{1+\cos\theta} r\, dr\, d\theta = \int_{\pi/2}^{\pi} \frac{1}{2}r^2\Big|_0^{1+\cos\theta} d\theta$$

$$= \frac{1}{2}\int_{\pi/2}^{\pi} (1 + 2\cos\theta + \cos^2\theta)\, d\theta$$

$$= \frac{1}{2}\left(\theta + 2\sin\theta + \frac{\theta}{2} + \frac{1}{4}\sin 2\theta\right)\Big|_{\pi/2}^{\pi} = \frac{3\pi}{8} - 1$$

By symmetry, $A = 2\left(1 - \frac{\pi}{8} + \frac{3\pi}{8} - 1\right) = \frac{\pi}{2}$.

We had to break our integral up since $r = 2 \cos \theta$ traces out the entire circle as θ varies from 0 to π, while $r = 1 + \cos \theta$ traces out only the first two quadrants.

18.1.9. Find the area outside the circle $r = \dfrac{a}{2}$ and inside the rose $r = a \cos 2\theta$.

18.1.9.

Solution. Inside one leaf, r varies from $\dfrac{a}{2}$ to $a \cos 2\theta$, and when $\dfrac{a}{2} = a \cos 2\theta$, $\theta = -\dfrac{\pi}{6}$ or $\dfrac{\pi}{6}$.

$$A = 4 \int_{-\pi/6}^{\pi/6} \int_{a/2}^{a \cos 2\theta} r \, dr \, d\theta = 2 \int_{-\pi/6}^{\pi/6} r^2 \Big|_{a/2}^{a \cos 2\theta} d\theta$$

$$= 2a^2 \int_{-\pi/6}^{\pi/6} \left(\cos^2 2\theta - \frac{1}{4} \right) d\theta = 2a^2 \left(\frac{\theta}{2} + \frac{\sin 4\theta}{8} - \frac{\theta}{4} \right)_{-\pi/6}^{\pi/6}$$

$$= 2a^2 \left(\frac{\pi}{12} + \frac{\sqrt{3}}{8} \right)$$

18.2. Double integrals and volume under a surface. Suppose we wish to find the volume under a surface $z = f(x, y)$ and over a specified region in the xy plane; thus we might ask for the volume under the sphere $z = \sqrt{1 - x^2 - y^2}$ over the region in the xy plane bounded by $y = x^2$ and the line $y = \frac{1}{4}$. Observe that the volume to be found is bounded by the given surface, one of the coordinate planes, and a cylindrical surface.

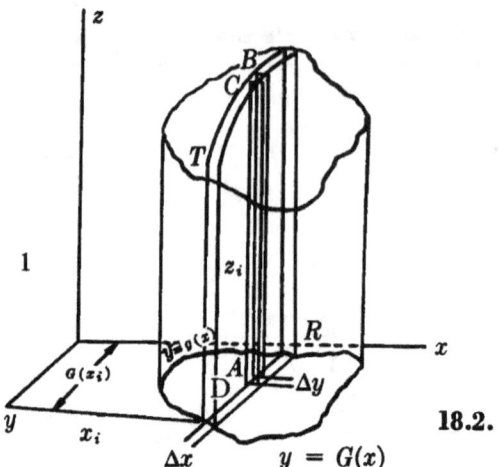

18.2.

To do this we divide our volume into elementary slabs. In Fig. 18.2 one of these, RT, of thickness Δx is shown. We shall then add up the volumes of these slabs and so obtain the total volume. First, however, we must find the volume of a slab for any position x_i; this volume is approximately $A_i \Delta x$, where A_i is the area of the face of the ith slab. To do this we divide the face up into approximately rectangular elementary areas $ABCD$ of base Δy and height z_i; the area of the face is then given by $\int_{g(x_i)}^{G(x_i)} z_i \, dy$. Here by z_i we denote $f(x_i, y)$, that is, x is held constant during the integration, and z is considered a function of y alone. The limits on y are determined by the value of the bounding curves $g(x)$ and $G(x)$ at x_i. As we vary x_i, A_i is a function of x, $A(x)$, and we may find the volume by summing the volumes of the slabs, that is, by

$$\int_a^b A(x) \, dx \quad \text{or} \quad \int_a^b dx \left[\int_{g(x)}^{G(x)} f(x, y) \, dy \right]$$

Our second expression is often written $\int_a^b \int_{g(x)}^{G(x)} f(x, y) \, dy \, dx$, where it is understood that x is treated as a constant during the first integration with respect to y. The student will see that the process is only a modification of the one by which we found the volume of a solid of known cross section. Here we have had to find the area of the cross section by integration first.

18.2.1. Find the volume in the first octant under the saddle surface $z = xy$ bounded by the right circular cylinder $(x-a)^2 + y^2 = a^2 (a > 0)$.

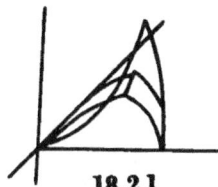

18.2.1

$$V = \int_0^{2a} \int_0^{\sqrt{2ax-x^2}} xy\, dy\, dx = \frac{1}{2}\int_0^{2a} xy^2 \Big|_0^{\sqrt{2ax-x^2}}$$

$$= \int_0^{2a} \frac{1}{2}x(2ax - x^2)\, dx = \left(\frac{ax^3}{3} - \frac{1}{8}x^4\right)\Big|_0^{2a} = \frac{2}{3}a^4$$

Notice that x varies from $a - \sqrt{a^2 - y^2}$ to $a + \sqrt{a^2 - y^2}$, and our problem could be solved as follows:

$$V = \int_0^a \int_{a-\sqrt{a^2-y^2}}^{a+\sqrt{a^2-y^2}} xy\, dx\, dy = \frac{1}{2}\int_0^a x^2 y \Big|_{a-\sqrt{a^2-y^2}}^{a+\sqrt{a^2-y^2}}\, dy$$

$$= 2a\int_0^a \sqrt{a^2-y^2}\, y\, dy = -\frac{2}{3}a(a^2-y^2)^{3/2}\Big|_0^a = \frac{2}{3}a^4$$

In the first case, the limits 0 to $2a$ were chosen because, as x varies from 0 to $2a$, we add up all possible slabs parallel to the yz plane and having corners on the x axis and on the circle.

18.2.2. Find the volume of the wedge cut from the cylinder $x^2 + y^2 = a^2$ by the planes $z = 0$ and $-x + z = a$.

Solution. Here y varies along the strip from $-\sqrt{a^2-x^2}$ to $\sqrt{a^2-x^2}$.

$$V = \int_{-a}^a \int_{-\sqrt{a^2-x^2}}^{\sqrt{a^2-x^2}} (a+x)\, dy\, dx = \int_{-a}^a (a+x)y\Big|_{-\sqrt{a^2-x^2}}^{\sqrt{a^2-x^2}}\, dx$$

$$= 2\int_{-a}^a (a+x)\sqrt{a^2-x^2}\, dx$$

$$= \left[2a\left(\frac{x}{2}\sqrt{a^2-x^2} + \frac{a^2}{2}\arcsin\frac{x}{a}\right) - \frac{2}{3}(a^2-x^2)^{3/2}\right]\Big|_{-a}^a = \pi a^3$$

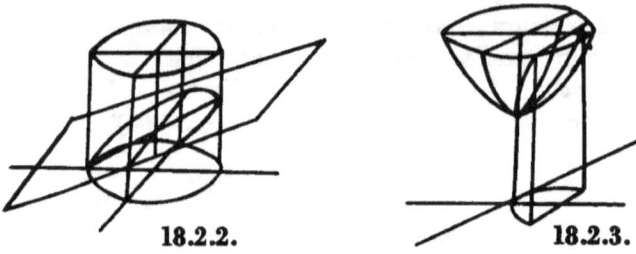

18.2.2.   18.2.3.

18.2.3. Find the volume bounded by the paraboloid $x^2 + y^2 + 2 = z$, the cylinder $y^2 = x$, and the planes $x = 1$, $y = 0$ and $z = 0$.

Solution. Here, choosing slabs parallel to the xz plane, x varies from y^2 to 1, y varies from 0 to 1.

$$V = \int_0^1 \int_{y^2}^1 (x^2 + y^2 + 2)\, dx\, dy$$

$$= \int_0^1 \left[\frac{1}{3}x^3 + y^2 x + 2x\right]_{y^2}^1 dy$$

$$= \int_0^1 \left(\frac{7}{3} - y^2 - y^4 - \frac{1}{3}y^6\right) dy$$

$$= \left(\frac{7}{3}y - \frac{y^3}{3} - \frac{1}{5}y^5 - \frac{1}{21}y^7\right)\Big|_0^1 = \frac{184}{105}$$

18.2.4. Find the volume made by the intersection of the two cylinders $x^2 + z^2 = a^2$ and $y^2 + z^2 = a^2$.

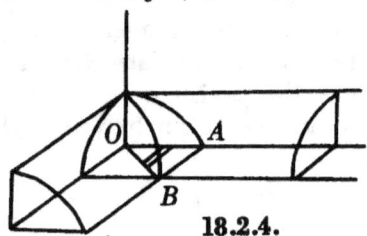

18.2.4.

Solution. In order to integrate under one surface, $y^2 + z^2 = a^2$, we integrate over the triangle OAB. Now x varies from 0 to y, and y varies from 0 to a.

$$V = 16 \int_0^a \int_0^y \sqrt{a^2 - y^2}\, dx\, dy = 16 \int_0^a x\sqrt{a^2 - y^2}\Big|_0^y dy$$

$$= 16 \int_0^a y\sqrt{a^2 - y^2}\, dy = -\frac{16}{3}(a^2 - y^2)^{3/2}\Big|_0^a = \frac{16}{3}a^3$$

18.2.5. Find the volume cut off above the xy plane by the sphere $x^2 + y^2 + z^2 = 1$.

Solution

$$z = \sqrt{1 - x^2 - y^2}; \quad V = \int_{-1}^{1} \int_{-\sqrt{1-x^2}}^{\sqrt{1-x^2}} \sqrt{1 - x^2 - y^2}\, dy\, dx$$

In the first integration, we make the substitution $y = \sqrt{1 - x^2} \sin \theta$. Then $\sqrt{1 - x^2 - y^2}$ becomes

$$\sqrt{1 - x^2 - (1 - x^2) \sin^2 \theta} = \sqrt{1 - x^2}\sqrt{1 - \sin^2 \theta}$$
$$= \sqrt{1 - x^2} \cos \theta$$

When $y = \sqrt{1 - x^2};\ \sin \theta = 1, \theta = \dfrac{\pi}{2}, y = -\sqrt{1 - x^2},$

$$\theta = -\frac{\pi}{2}, dy = \sqrt{1 - x^2} \cos \theta\, d\theta.$$

$$V = \int_{-1}^{1} \int_{-\pi/2}^{\pi/2} \sqrt{1 - x^2} \cos \theta \sqrt{1 - x^2} \cos \theta\, d\theta\, dx$$

$$= \int_{-1}^{1} \int_{-\pi/2}^{\pi/2} (1 - x^2) \cos^2 \theta\, d\theta\, dx$$

$$= \int_{-1}^{1} (1 - x^2) \left(\frac{\theta}{2} + \frac{1}{4} \sin 2\theta\right) \bigg|_{-\pi/2}^{\pi/2} dx$$

$$= \frac{\pi}{2} \int_{-1}^{1} (1 - x^2)\, dx = \frac{\pi}{2} \left(x - \frac{x^3}{3}\right) \bigg|_{-1}^{1} = \frac{2}{3} \pi$$

18.2.6. Find the volume in the first octant bounded by the surface $x^2 + y = z$ and the planes $y = x,\ x = 1$.

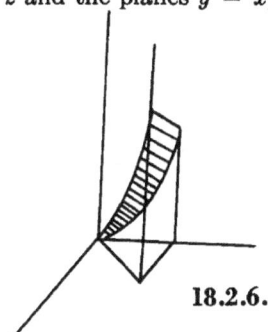

18.2.6.

Solution. Here, if we integrate first with respect to y, 0 and x will be the limits.

$$V = \int_0^1 \int_0^x (x^2 + y)\, dy\, dx = \int_0^1 \left[x^2 y + \frac{y^2}{2}\right]_0^x dx$$

$$= \int_0^1 \left(x^3 + \frac{x^2}{2}\right) dx = \left(\frac{x^4}{4} + \frac{x^3}{6}\right)\Big|_0^1 = \frac{5}{12}$$

18.3. Cylindrical coordinates. To convert to cylindrical coordinates from rectangular, set $x = r\cos\theta$, $y = r\sin\theta$, $z = z$. Care must be taken to replace $dx\, dy$ by $r\, dr\, d\theta$ as $r\, dr\, d\theta$ is now the element of area in the xy-plane.

18.3.1. Find the volume bounded by the xy plane, the paraboloid $az = x^2 + y^2$, and the cylinder $x^2 + y^2 = a^2$.

Solution. In cylindrical coordinates, the equation of the surface is $az = r^2$, and

$$V = \int_0^{2\pi} \int_0^a \frac{r^2}{a} r\, dr\, d\theta = \int_0^{2\pi} \frac{r^4}{4a}\Big|_0^a d\theta = \frac{a^3}{4} \int_0^{2\pi} d\theta = \frac{\pi}{2} a^3$$

18.3.1.   18.3.2.

18.3.2. Find the volume under the cone $z = r$ inside the cylinder $r = a(1 + \cos\theta)$ and above the xy plane. Then r varies from 0 to $a(1 + \cos\theta)$.

$$V = \int_0^{2\pi} \int_0^{a(1+\cos\theta)} r^2\, dr\, d\theta = \int_0^{2\pi} \frac{r^3}{3}\Big|_0^{a(1+\cos\theta)} d\theta$$

$$= \frac{a^3}{3} \int_0^{2\pi} (1 + \cos\theta)^3\, d\theta$$

$$= \frac{a^3}{3} \int_0^{2\pi} \left(1 + 3\cos\theta + \frac{3(\cos 2\theta + 1)}{2}\right.$$

$$\left. + (1 - \sin^2\theta)\cos\theta\right) d\theta$$

$$= \frac{a^3}{3}\left(\frac{5}{2}\theta + 3\sin\theta + \frac{3}{4}\sin 2\theta + \sin\theta - \frac{1}{3}\sin^3\theta\right)\Big|_0^{2\pi} = 5\pi\frac{a^3}{3}$$

18.3.3. Find the volume above the xy plane inside the cylinder $x^2 + y^2 - ax = 0$ and under the sphere $x^2 + y^2 + z^2 = a^2$.

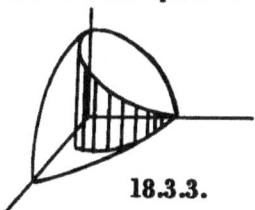

18.3.3.

Solution. In cylindrical coordinates the equation of our surface is $z = \sqrt{a^2 - r^2}$; r varies from 0 to $a \cos \theta$; θ varies from 0 to π.

$$V = \int_0^\pi \int_0^{a\cos\theta} r\sqrt{a^2 - r^2}\, dr\, d\theta = -\int_0^\pi \frac{1}{3}(a^2 - r^2)^{3/2}\Big|_0^{a\cos\theta} d\theta$$

$$= \frac{1}{3}\int_0^\pi [a^3 - (a^2 - a^2\cos^2\theta)^{3/2}]\, d\theta = \frac{a^3}{3}\int_0^\pi (1 - \sin^3\theta)\, d\theta$$

$$= \frac{a^3}{3}\left(\cos\theta - \frac{\cos^3\theta}{3} + \theta\right)_0^\pi = \frac{a^3}{3}\left(\pi - \frac{4}{3}\right)$$

18.3.4. Find the volume between the cone $x^2 + y^2 = z^2$ and the paraboloid $x^2 + y^2 = z$.

Solution. In cylindrical coordinates the equations are $z = r$ and $z = r^2$. The two surfaces intersect in the circle $z = 1, r = 1$. Taking the difference of the two volumes in these limits,

$$V = \int_0^{2\pi}\int_0^1 (z_1 - z_2)r\, dr\, d\theta = \int_0^{2\pi}\int_0^1 (r - r^2)\, r\, dr\, d\theta$$

$$= \int_0^{2\pi}\left(\frac{r^3}{3} - \frac{r^4}{4}\right)\Big|_0^1 d\theta = \frac{1}{12}\int_0^{2\pi} d\theta = \frac{\pi}{6}$$

18.3.5.

18.3.5. Find the volume within the cylinder $x^2 + y^2 - 2ax = 0$ above the xy plane and under the cylinder $z = a^2 - y^2$.

Solution. Here $z = a^2 - r^2 \sin^2 \theta$, and r varies from 0 to $2a \cos \theta$.

$$V = \int_{-\pi/2}^{\pi/2} \int_0^{2a\cos\theta} (a^2 - r^2 \sin^2 \theta) r \, dr \, d\theta$$

$$= \int_{-\pi/2}^{\pi/2} \left(\frac{a^2 r^2}{2} - \frac{r^4}{4} \sin^2 \theta \right) \Big|_0^{2a\cos\theta} d\theta$$

$$= \int_{-\pi/2}^{\pi/2} (2a^4 \cos^2 \theta - 4a^4 \cos^4 \theta \sin^2 \theta) \, d\theta$$

$$= a^4 \int_{-\pi/2}^{\pi/2} \cos^2 \theta (2 - \sin^2 2\theta) \, d\theta$$

$$= a^4 \int_{-\pi/2}^{\pi/2} \frac{1 + \cos 2\theta}{2} (2 - \sin^2 2\theta) \, d\theta$$

$$= \frac{a^4}{2} \int_{-\pi/2}^{\pi/2} \left(2 + 2 \cos 2\theta + \frac{\cos 4\theta - 1}{2} - \sin^2 2\theta \cos 2\theta \right) d\theta$$

$$= \frac{a^4}{2} \left(\frac{3}{2} \theta + \sin 2\theta + \frac{1}{8} \sin 4\theta - \frac{1}{6} \sin^3 2\theta \right) \Big|_{-\pi/2}^{\pi/2} = \frac{3\pi}{4} a^4$$

18.4. Double integration and surface area. For the surface $z = f(x, y)$ the area is given by

$$\int_a^b \int_{y_1(x)}^{y_2(x)} \sqrt{\left(\frac{\partial z}{\partial x}\right)^2 + \left(\frac{\partial z}{\partial y}\right)^2 + 1} \, dy \, dx,$$

or

$$\int_a^b \int_{y_1(x)}^{y_2(x)} \frac{\sqrt{\left(\frac{\partial F}{\partial x}\right)^2 + \left(\frac{\partial F}{\partial y}\right)^2 + \left(\frac{\partial F}{\partial z}\right)^2}}{\left|\frac{\partial F}{\partial z}\right|} \, dy \, dx$$

when the surface is given as $F(x, y, z) = 0$. In cylindrical coordinates

$$A = \int_\alpha^\beta \int_{r_1(\theta)}^{r_2(\theta)} \sqrt{1 + \left(\frac{\partial z}{\partial r}\right)^2 + \frac{1}{r^2}\left(\frac{\partial z}{\partial \theta}\right)^2} \, r \, dr \, d\theta$$

18.4.1. Find the area of the surface cut from the cylinder $y^2 + z^2 = a^2$ by the cylinder $x^2 + y^2 = a^2$.

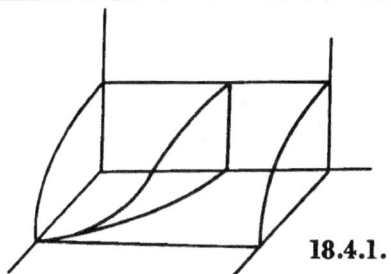

18.4.1.

Solution. Here $F(x, y, z) = y^2 + z^2 - a^2$, and, finding the area in the first octant, $\dfrac{\partial F}{\partial x} = 0, \dfrac{\partial F}{\partial y} = 2y, \dfrac{\partial F}{\partial z} = 2z.$

$$\sqrt{\left(\frac{\partial F}{\partial x}\right)^2 + \left(\frac{\partial F}{\partial y}\right)^2 + \left(\frac{\partial F}{\partial z}\right)^2} = \sqrt{4y^2 + 4z^2} = 2a$$

$$\frac{\sqrt{\left(\dfrac{\partial F}{\partial x}\right)^2 + \left(\dfrac{\partial F}{\partial y}\right)^2 + \left(\dfrac{\partial F}{\partial z}\right)^2}}{\left|\dfrac{\partial F}{\partial z}\right|} = \frac{2a}{2z} = \frac{a}{z} = \frac{a}{\sqrt{a^2 - y^2}}$$

$$A = 8 \int_0^a \int_0^{\sqrt{a^2-y^2}} \frac{a}{\sqrt{a^2 - y^2}} \, dx \, dy$$

$$= 8 \int_0^a \frac{ax}{\sqrt{a^2 - y^2}} \Big|_0^{\sqrt{a^2-y^2}} dy = 8a \int_0^a dy = 8a^2$$

Our limits were chosen since x varies from 0 to $\sqrt{a^2 - y^2}$ and y from 0 to a. Since z is always positive $\left|\dfrac{\partial F}{\partial z}\right| = 2z.$

18.4.2. Find the area of the sphere $x^2 + y^2 + z^2 = a^2$ within the cylinder $x^2 + y^2 = ax.$

Solution. Here we use polar coordinates so that the equation of the sphere is $r^2 + z^2 = a^2, \dfrac{\partial z}{\partial r} = \dfrac{-r}{\sqrt{a^2 - r^2}}$

$$\sqrt{1 + \left(\frac{\partial z}{\partial r}\right)^2 + \frac{1}{r^2}\left(\frac{\partial z}{\partial \theta}\right)^2} = \sqrt{1 + \frac{r^2}{a^2 - r^2}} = \frac{a}{\sqrt{a^2 - r^2}}$$

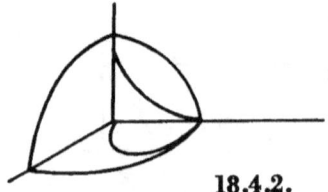

18.4.2.

Since the equation of our bounding cylinder is $r = a\cos\theta$, r varies from 0 to $a\cos\theta$, and θ varies from $-\frac{\pi}{2}$ to $\frac{\pi}{2}$.

$$A = 2\int_{-\pi/2}^{\pi/2}\int_0^{a\cos\theta} \frac{a}{\sqrt{a^2-r^2}}\,r\,dr\,d\theta$$

$$= 2\int_{-\pi/2}^{\pi/2} -a\sqrt{a^2-r^2}\Big|_0^{a\cos\theta} d\theta$$

$$= 2\int_{-\pi/2}^{\pi/2} (-a\sqrt{a^2 - a^2\cos^2\theta} + a^2)\,d\theta$$

$$= 2\int_{-\pi/2}^{\pi/2} (a^2 - a^2\sin\theta)\,d\theta = 2(a^2\cos\theta + a^2\theta)\Big|_{-\pi/2}^{\pi/2} = 2\pi a^2.$$

We multiplied by 2 to account for the two areas above and below the xy plane.

18.4.3. Find the area in the first octant on the surface $x^2 + y = z$ and cut off by the planes $y = x$, $x = 1$.

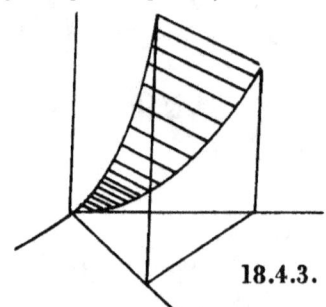

18.4.3.

Solution. Here $\sqrt{1 + \left(\dfrac{\partial z}{\partial x}\right)^2 + \left(\dfrac{\partial z}{\partial y}\right)^2} = \sqrt{2 + 4x^2}$, y varies from 0 to x, and x varies from 0 to 1.

$$A = \int_0^1 \int_0^x \sqrt{2+4x^2}\, dy\, dx = \int_0^1 y\sqrt{2+4x^2}\Big|_0^x dx$$
$$= \int_0^1 x\sqrt{2+4x^2}\, dx = \frac{1}{12}(2+4x^2)^{3/2}\Big|_0^1$$
$$= \tfrac{1}{12}(6^{3/2} - 2^{3/2})$$

If we had integrated first with respect to x, the integrals obtained would have been more difficult to work.

18.4.4. Find the area of the conical surface $x^2 + y^2 = z^2$ above the plane $z = 0$, cut out by the prism whose axis is the z axis, and whose base is a square of side 2.

Solution. Here

$$\sqrt{\left(\frac{\partial F}{\partial x}\right)^2 + \left(\frac{\partial F}{\partial y}\right)^2 + \left(\frac{\partial F}{\partial z}\right)^2} = \sqrt{4x^2 + 4y^2 + 4z^2} = \sqrt{8z^2}$$

$$\frac{\sqrt{\left(\frac{\partial F}{\partial x}\right)^2 + \left(\frac{\partial F}{\partial y}\right)^2 + \left(\frac{\partial F}{\partial z}\right)^2}}{\left|\frac{\partial F}{\partial z}\right|} = \frac{\sqrt{8z^2}}{2z} = \sqrt{2}$$

$$A = \sqrt{2}\int_{-1}^1\int_{-1}^1 dx\, dy = \sqrt{2}\int_{-1}^1 x\Big|_{-1}^1 dy = 2\sqrt{2}\int_{-1}^1 dy = 4\sqrt{2}$$

18.4.5. Find the area on the surface $z = xy$ cut off by the cylinder $x^2 + y^2 = a^2$.

Solution. We change to cylindrical coordinates so that $z = r^2 \sin\theta \cos\theta = \tfrac{1}{2}r^2 \sin 2\theta$, and the bounding cylinder is $r = a$.

$$\sqrt{1 + \left(\frac{\partial z}{\partial r}\right)^2 + \frac{1}{r^2}\left(\frac{\partial z}{\partial \theta}\right)^2} = \sqrt{1 + r^2 \sin^2 2\theta + r^2 \cos^2 2\theta}$$
$$= \sqrt{1 + r^2}$$

$$A = \int_0^{2\pi}\int_0^a \sqrt{1+r^2}\, r\, dr\, d\theta = \frac{1}{3}\int_0^{2\pi}(1+r^2)^{3/2}\Big|_0^a d\theta$$
$$= \frac{2\pi}{3}[(1+a^2)^{3/2} - 1]$$

18.4.6. Find the area of that part of the surface of the paraboloid $x^2 + z^2 = 4ay$ which is inside the parabolic cylinder $x^2 = ay$ and cut off by the plane $y = 8a$.

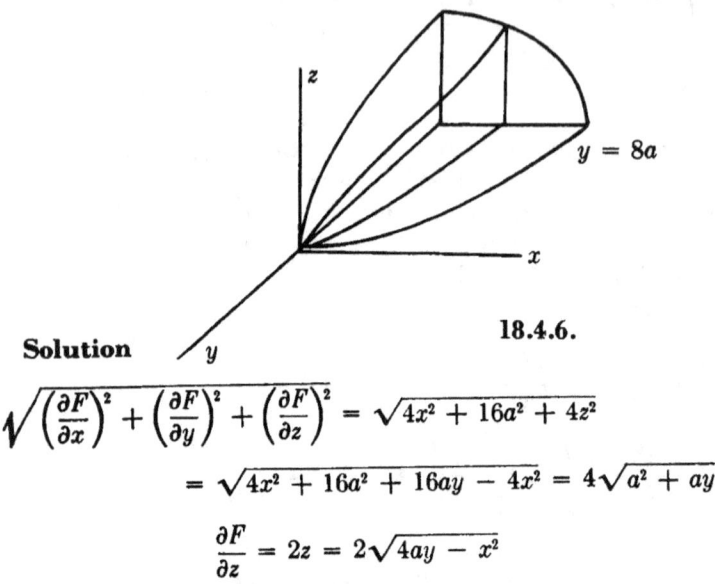

18.4.6.

Solution

$$\sqrt{\left(\frac{\partial F}{\partial x}\right)^2 + \left(\frac{\partial F}{\partial y}\right)^2 + \left(\frac{\partial F}{\partial z}\right)^2} = \sqrt{4x^2 + 16a^2 + 4z^2}$$

$$= \sqrt{4x^2 + 16a^2 + 16ay - 4x^2} = 4\sqrt{a^2 + ay}$$

$$\frac{\partial F}{\partial z} = 2z = 2\sqrt{4ay - x^2}$$

We integrate first with respect to x, so x varies from $-\sqrt{ay}$ to $\sqrt{ay}$.

$$A = 2\int_0^{8a} \int_{-\sqrt{ay}}^{\sqrt{ay}} \frac{2\sqrt{a^2 + ay}}{\sqrt{4ay - x^2}}\, dx\, dy$$

$$= 4\int_0^{8a} \sqrt{a^2 + ay} \, \text{arc sin}\, \frac{x}{2\sqrt{ay}}\bigg|_{-\sqrt{ay}}^{\sqrt{ay}}\, dy$$

$$= 4\int_0^{8a} \sqrt{a^2 + ay}\left(\frac{\pi}{3}\right)\, dy = \frac{4\pi}{3}\cdot\frac{2}{3a}(a^2 + ay)^{3/2}\bigg|_0^{8a}$$

$$= \frac{8\pi}{9a}(27a^3 - a^3) = \frac{208}{9}\pi a^2$$

We took twice the integral in order to include the area both above and below the xy plane.

18.5. Miscellaneous physical applications. In this paragraph, we are going to apply the double integral to certain physical problems. The student must keep in mind that the moment of an elementary area is the product of that area and the moment arm. Moment of inertia is the product of the area and the square of its moment arm.

18.5.1. Find the centroid of the area inside the parabola $y^2 = x$ and inside the circle $x^2 + y^2 = 2x$.

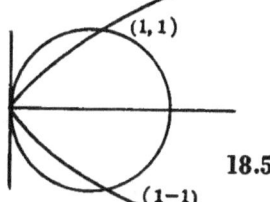

18.5.1.

Solution. Again by symmetry $\bar{y} = 0$. An element of area $\Delta x \, \Delta y$ has a moment $x \, \Delta x \, \Delta y$ about the y axis; $2x - x^2 = x$ at 0, 1.

$$A\bar{x} = \int_{-1}^{1}\int_{y^2}^{1+\sqrt{1-y^2}} x \, dx \, dy = \frac{1}{2}\int_{-1}^{1} x^2 \Big|_{y^2}^{1+\sqrt{1-y^2}} dy$$

$$= \frac{1}{2}\int_{-1}^{1} (2 + 2\sqrt{1-y^2} - y^2 - y^4) \, dy$$

$$= \frac{1}{2}\left(2y - \frac{1}{3}y^3 - \frac{1}{5}y^5 + y\sqrt{1-y^2} + \arcsin y\right)\Big|_{-1}^{1}$$

$$= \frac{22}{15} + \frac{\pi}{2}$$

Notice for the first integration, we choose $1 + \sqrt{1-y^2}$, and not $1 - \sqrt{1-y^2}$, as we are integrating to the right-hand branch of the circle.

$$A = \int_{-1}^{1}\int_{y^2}^{1+\sqrt{1-y^2}} dx \, dy = \int_{-1}^{1} (1 + \sqrt{1-y^2} - y^2) \, dy$$

$$= \left(y - \frac{1}{3}y^3 + \frac{y}{2}\sqrt{1-y^2} + \frac{1}{2}\arcsin y\right)\Big|_{-1}^{1} = 2 + \frac{\pi}{2}$$

$$\bar{x} = \frac{\frac{22}{15} + \frac{\pi}{2}}{2 + \frac{\pi}{2}}$$

18.5.2. Find the centroid of the area bounded by the cardioid $r = 1 + \cos\theta$.

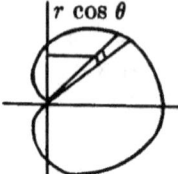

18.5.2.

Solution. By symmetry we see $\bar{y} = 0$. Our element has an area $r\,\Delta r\,\Delta\theta$, and its moment arm with respect to the y axis is $r\cos\theta$.

$$A\bar{x} = \int_0^{2\pi}\int_0^{1+\cos\theta} r\cos\theta\, r\, dr\, d\theta = \frac{1}{3}\int_0^{2\pi}(1+\cos\theta)^3\cos\theta\, d\theta$$

$$= \frac{1}{3}\int_0^{2\pi}(\cos\theta + 3\cos^2\theta + 3\cos^3\theta + \cos^4\theta)\, d\theta$$

$$= \frac{1}{3}\left(\sin\theta + \frac{3}{2}\theta + \frac{3}{4}\sin 2\theta + 3\sin\theta - \sin^3\theta\right.$$

$$\left. + \frac{3}{8}\theta + \frac{1}{4}\sin 2\theta + \frac{1}{32}\sin 4\theta\right)\Big|_0^{2\pi}$$

Then $A\bar{x} = \dfrac{5\pi}{4}$; since $A = \dfrac{3\pi}{2}$, $\bar{x} = \dfrac{5}{6}$.

18.5.3. Find the centroid of the loop in the first quadrant of $r = a\sin 2\theta$.

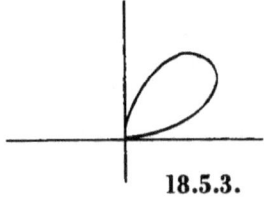

18.5.3.

Solution. It can be seen by symmetry that $\bar{x} = \bar{y}$. Our element of area is $r\,\Delta r\,\Delta\theta$, and its moment arm with respect to the x axis $r\sin\theta$.

$$A\bar{y} = \int_0^{\pi/2} \int_0^{a\sin 2\theta} r\sin\theta\, r\, dr\, d\theta = \frac{1}{3}\int_0^{\pi/2} r^3 \sin\theta \Big|_0^{a\sin 2\theta} d\theta$$

$$= \frac{a^3}{3}\int_0^{\pi/2} 8\sin^4\theta \cos^3\theta\, d\theta = \frac{8a^3}{3}\int_0^{\pi/2}(\sin^4\theta - \sin^6\theta)\cos\theta\, d$$

$$= \frac{8a^3}{3}\left(\frac{1}{5}\sin^5\theta - \frac{1}{7}\sin^2\theta\right)\Big|_0^{\pi/2}$$

$$A\bar{y} = \frac{16a^3}{105}; \qquad A = \frac{\pi a^2}{8}; \qquad \text{and } \bar{y} = \frac{128a}{105}$$

18.5.4. Find the moment of inertia of the area bounded by $y^2 = 4x + 4$ and the y axis with respect to an axis perpendicular to the xy plane at the origin.

18.5.4.

Solution. The length of the moment arm is $\sqrt{x^2 + y^2}$, and the moment of inertia of the element of area will be $(x^2 + y^2)\,\Delta y\, \Delta x$.

$$I = \int_{-2}^{2}\int_{(y^2-4)/4}^{0}(x^2 + y^2)\, dx\, dy = \int_{-2}^{2}\left(\frac{1}{3}x^3 + y^2 x\right)\Big|_{(y^2-4)/4}^{0} dy$$

$$= -\int_{-2}^{2}\left[\frac{1}{192}(y^6 - 12y^4 + 48y^2 - 64) + \frac{1}{4}y^4 - y^2\right] dy$$

$$= \frac{-1}{192}\left(\frac{1}{7}y^7 + \frac{36}{5}y^5 - 48y^3 - 64y\right)\Big|_{-2}^{2}$$

$$= -\frac{1}{3}\left(\frac{4}{7} + \frac{36}{5} - 12 - 4\right) = -\frac{1}{3}\left(-\frac{288}{35}\right) = \frac{96}{35}$$

$$A = \int_{-2}^{2}\int_{(y^2-4)/4}^{0} dx\, dy = -\int_{-2}^{2}\left(\frac{1}{4}y^2 - 1\right) dy$$

$$= -\left(\frac{1}{12}y^3 - y\right)\Big|_{-2}^{2} = \frac{8}{3}; \qquad I = \frac{36}{35}A$$

Suppose now that the area A is endowed with mass of uniform density with d units of mass per unit of area. Then the integral for the moment of inertia I of this mass would have the factor d so that $I = \iint d = \iint \frac{1}{A} A d = \iint M$ where $M = Ad$ is the total mass of the area. Hence, in these cases, we may substitute M for A in the expression for I.

18.5.5. Find the moment of inertia of the plane area bounded by the cardioid $r = a(1 + \cos\theta)$ with respect to an axis perpendicular to the plane and passing through the origin.

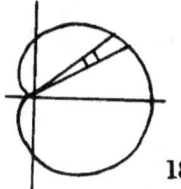

18.5.5.

Solution. The area $r\,\Delta r\,\Delta\theta$ has a moment of inertia of $r^3\,\Delta r\,\Delta\theta$ about this axis since the moment arm is r.

$$I = \int_0^{2\pi} \int_0^{a(1+\cos\theta)} r^3\, dr\, d\theta = \frac{a^4}{4} \int_0^{2\pi} (1+\cos\theta)^4\, d\theta$$

$$= \frac{a^4}{4} \int_0^{2\pi} (1 + 4\cos\theta + 6\cos^2\theta + 4\cos^3\theta + \cos^4\theta)\, d\theta$$

$$= \frac{a^4}{4}\left(\theta + 4\sin\theta + 3\theta + \frac{3}{2}\sin 2\theta + 4\sin\theta - \frac{4}{3}\sin^3\theta + \frac{3}{8}\theta\right.$$

$$\left. + \frac{1}{4}\sin 2\theta + \frac{1}{32}\sin 4\theta\right)\Big|_0^{2\pi} = \frac{35}{16}\pi a^4 = \frac{35}{24}Ma^2.$$

18.5.6. Find the moment of inertia with respect to the z axis of the wedge above the xy plane formed by the cylinder $x^2 + y^2 = a^2$ and the plane $z = x$.

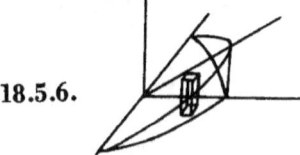

18.5.6.

Solution. Our elementary column has a volume $z\,\Delta x\,\Delta y$, and its moment of inertia about the z axis is $(x^2 + y^2)z\,\Delta x\,\Delta y$; x varies from 0 to $\sqrt{a^2 - y^2}$, and y varies from $-a$ to a.

$$I = \int_{-a}^{a}\int_{0}^{\sqrt{a^2-y^2}} (x^2 + y^2)x\,dx\,dy = \int_{-a}^{a}\left(\frac{x^4}{4} + \frac{x^2 y^2}{2}\right)_{0}^{\sqrt{a^2-y^2}} dy$$

$$= \frac{1}{4}\int_{-a}^{a} (a^4 - y^4)\,dy = \frac{1}{4}\left(a^4 y - \frac{1}{5}y^5\right)_{-a}^{a} = \frac{2}{5}a^5$$

$$V = \int_{-a}^{a}\int_{0}^{\sqrt{a^2-y^2}} x\,dx\,dy = \frac{2}{3}a^3; \qquad I = \frac{3}{5}Ma^2$$

If we had integrated first with respect to y, we would have had odd powers of $\sqrt{a^2 - x^2}$ to deal with in the second integration.

18.6. Triple integration. Just as we speak of the integral of a function of one variable over a range on the axis, or the double integral of a function of two variables over an area, we may also speak of a triple integral of a function of three variables over a volume. Usually the limits of the innermost integral will be a function two variables, the limits of the second functions of one variable, and the outermost limits constants. The differentials are written down in the order of integration.

18.6.1. Evaluate $\int_{0}^{1}\int_{0}^{z}\int_{0}^{y^2+z^2} xz\,dx\,dy\,dz$.

$$\int_{0}^{1}\int_{0}^{z}\int_{0}^{y^2+z^2} xz\,dx\,dy\,dz = \int_{0}^{1}\int_{0}^{z}\frac{1}{2}x^2 z\bigg|_{0}^{y^2+z^2} dy\,dz$$

$$= \int_{0}^{1}\int_{0}^{z}\frac{1}{2}(y^4 + 2y^2 z^2 + z^4)z\,dy\,dz$$

$$= \frac{1}{2}\int_{0}^{1}\left(\frac{1}{5}y^5 z + \frac{2}{3}y^3 z^3 + yz^5\right)\bigg|_{0}^{z} dz = \frac{1}{2}\int_{0}^{1}\frac{28}{15}z^6\,dz$$

$$= \frac{2}{15}z^7\bigg|_{0}^{1} = \frac{2}{15}$$

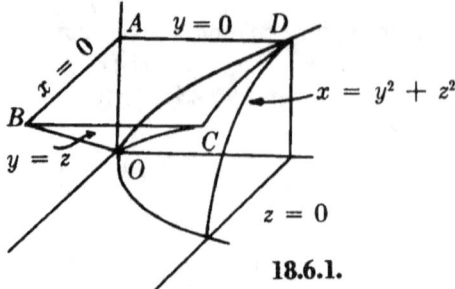

18.6.1.

Notice that in the first integration, x varies from 0 to $y^2 + z^2$ so that the volume is bounded by the plane $x = 0$ and the paraboloid $x = y^2 + z^2$. From the second integration it is bounded by the planes $y = 0$ and $y = z$, and finally by $z = 0$ and $z = 1$. This volume $AOBCD$ is shown in the figure and is the volume over which the function xy^0z, or xz, is integrated.

18.6.2. Evaluate $\displaystyle\int_1^2 \int_0^{\ln y} \int_0^{-y-z} e^{x+y+z}\, dx\, dz\, dy$

$$\int_1^2 \int_0^{\ln y} \int_0^{-y-z} e^{x+y+z}\, dx\, dz\, dy = \int_1^2 \int_0^{\ln y} e^{x+y+z}\Big|_0^{-y-z} dz\, dy$$

$$= \int_1^2 \int_0^{\ln y} (1 - e^{y+z})\, dz\, dy = \int_1^2 (z - e^{y+z})\Big|_0^{\ln y} dy$$

$$= \int_1^2 (\ln y - ye^y + e^y)\, dy = (y \ln y - y - ye^y + e^y + e^y)\Big|_1^2$$

$$= 2 \ln 2 - e - 1$$

The volume of integration is, in this case, bounded by the planes $x = 0$ and $x = -y - z$; the plane $z = 0$ and the cylinder $z = \ln y$; and the planes $y = 1$ and $y = 2$.

18.6.3. Evaluate $\displaystyle\int_1^2 \int_0^1 \int_0^{\sqrt{x^2+z^2}} \frac{x\, dy\, dx\, dz}{x^2 + y^2 + z^2}$

Solution

$$\int_1^2 \int_0^z \int_0^{\sqrt{x^2+z^2}} \frac{x \, dy \, dx \, dz}{x^2 + y^2 + z^2}$$

$$= \int_1^2 \int_0^z \frac{x}{\sqrt{x^2+z^2}} \text{Arc tan} \frac{y}{\sqrt{x^2+z^2}} \Big|_0^{\sqrt{x^2+z^2}} dx \, dz$$

$$= \int_1^2 \int_0^z \frac{\pi}{4} \frac{x}{\sqrt{x^2+z^2}} dx \, dz = \frac{\pi}{4} \int_1^2 \sqrt{x^2+z^2} \Big|_0^z dz$$

$$= \frac{\pi}{4} \int_1^2 (\sqrt{2}-1)z \, dz = \frac{\pi}{8} (\sqrt{2}-1)z^2 \Big|_1^2 = \frac{3\pi}{8} (\sqrt{2}-1)$$

The triple integral can be used in finding volumes. We have seen in a previous section on double integration that a volume could be evaluated by breaking our elementary slabs of thicknesses Δx into columns of base $\Delta x \, \Delta y$. By adding up the columns we obtained the volume. We can further divide the columns into elementary rectangular parallelepipeds of dimensions $\Delta x \, \Delta y \, \Delta z$. By adding these together we obtain the total volume,

$$V = \iiint dx \, dy \, dz.$$

18.6.4. Find the volume of the ellipsoid $\dfrac{x^2}{a^2} + \dfrac{y^2}{b^2} + \dfrac{z^2}{c^2} = 1$ to the right of $x = 0$.

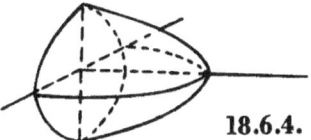

18.6.4.

Solution. Here x varies from 0 to $a\sqrt{1 - \frac{y^2}{b^2} - \frac{z^2}{c^2}}$, y varies from $-b\sqrt{1 - \frac{z^2}{c^2}}$ to $b\sqrt{1 - \frac{z^2}{c^2}}$, and z from $-c$ to c. Other limits could be used, but this order of integration is the most convenient. The limits $-b\sqrt{1 - \frac{z^2}{c^2}}$, $b\sqrt{1 - \frac{z^2}{c^2}}$ are determined by the ellipse in the plane $x = 0$; the elementary columns are contained in slabs perpendicular to this plane.

$$V = \int_{-c}^{c} \int_{-b\sqrt{1-z^2/c^2}}^{b\sqrt{1-z^2/c^2}} \int_0^{a\sqrt{1-y^2/b^2-z^2/c^2}} dx\, dy\, dz$$

$$= a \int_{-c}^{c} \int_{-b\sqrt{1-z^2/c^2}}^{b\sqrt{1-z^2/c^2}} \sqrt{1 - \frac{y^2}{b^2} - \frac{z^2}{c^2}}\, dy\, dz$$

$$= ab \int_{-c}^{c} \left(\frac{y}{2b} \sqrt{1 - \frac{y^2}{b^2} - \frac{z^2}{c^2}}\right.$$

$$\left. + \frac{1 - z^2/c^2}{2} \operatorname{Arc\,sin} \frac{y}{b\sqrt{1-z^2/c^2}}\right)\Big|_{-b\sqrt{1-z^2/c^2}}^{b\sqrt{1-z^2/c^2}}$$

$$= \frac{\pi ab}{2} \int_{-c}^{c} \left(1 - \frac{z^2}{c^2}\right) dz = \frac{\pi ab}{2}\left(z - \frac{z^3}{3c^2}\right)\Big|_{-c}^{c} = \frac{2\pi}{3} abc$$

18.6.5. Find the volume bounded by the paraboloid $x^2 + 4y^2 = z$, the sphere $x^2 + y^2 + z^2 = 4$, and the cylinder $x^2 + y^2 = 4$.

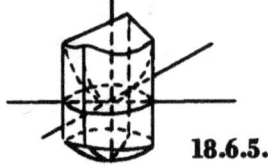

18.6.5.

Solution. Here z varies from $-\sqrt{4 - x^2 - y^2}$ to $x^2 + 4y^2$, y varies from $-\sqrt{4 - x^2}$ to $\sqrt{4 - x^2}$, and x from -2 to 2.

$$V = \int_{-2}^{2} \int_{-\sqrt{4-x^2}}^{\sqrt{4-x^2}} \int_{-\sqrt{4-x^2-y^2}}^{x^2+4y^2} dz\, dy\, dx$$

$$= \int_{-2}^{2} \int_{-\sqrt{4-x^2}}^{\sqrt{4-x^2}} (x^2 + 4y^2 + \sqrt{4 - x^2 - y^2})\, dy\, dx$$

$$= \int_{-2}^{2} \left(x^2 y + \frac{4}{3} y^3 + \frac{y}{2} \sqrt{4 - x^2 - y^2} \right.$$
$$\left. + \frac{4 - x^2}{2} \arcsin \frac{y}{\sqrt{4 - x^2}} \right) \bigg|_{-\sqrt{4-x^2}}^{\sqrt{4-x^2}} dx$$
$$= \int_{-2}^{2} \left[2x^2 \sqrt{4 - x^2} + \frac{8}{3} (4 - x^2)^{3/2} + \frac{\pi}{2} (4 - x^2) \right] dx$$
$$= \left\{ 2 \left(-\frac{x}{4} (4 - x^2)^{3/2} + \frac{x}{2} \sqrt{4 - x^2} + 2 \arcsin \frac{x}{2} \right) \right.$$
$$+ \frac{8}{3} \left[-\frac{x}{8} (2x^2 - 20) \sqrt{4 - x^2} + 6 \arcsin \frac{x}{2} \right]$$
$$\left. + \frac{\pi}{2} \left(4x - \frac{x^3}{3} \right) \right\} \bigg|_{-2}^{2}$$
$$= 20\pi + \frac{16\pi}{3} = \frac{76\pi}{3}$$

18.6.6. Find the volume bounded by the sphere $r^2 + z^2 = 4a^2$ and the cylinder $r = 2a \cos \theta$.

18.6.6.

Solution. In cylindrical coordinates, the element of volume is $r \, \Delta z \, \Delta r \, \Delta \theta$. Now z varies from $-\sqrt{4a^2 - r^2}$ to $\sqrt{4a^2 - r^2}$; r varies from 0 to $2a \cos \theta$; and θ from 0 to π.

$$V = \int_0^{\pi} \int_0^{2a\cos\theta} \int_{-\sqrt{4a^2 - r^2}}^{\sqrt{4a^2 - r^2}} r \, dz \, dr \, d\theta$$
$$= 2 \int_0^{\pi} \int_0^{2a\cos\theta} \sqrt{4a^2 - r^2} \, r \, dr \, d\theta$$
$$= \int_0^{\pi} -\frac{2}{3} (4a^2 - r^2)^{3/2} \bigg|_0^{2a\cos\theta} d\theta$$

$$= \frac{2}{3}\int_0^\pi (8a^3 - 8a^3 \sin^3 \theta)\, d\theta$$

$$= \frac{2}{3}\left(8a^3\theta + 8a^3 \cos\theta - \frac{8}{3}a^3 \cos^3\theta\right)\Big|_0^\pi$$

$$= \frac{16}{3}\pi a^3 - \frac{64}{9}a^3$$

18.7. Physical applications of the triple integral. Centroids can be found by finding the moment of the solid with respect to the coordinate planes and then dividing by the volume. We obtain the moment by multiplying the elementary volume $\Delta x\, \Delta y\, \Delta z$ by the distance from the plane.

$$M_{xy} = \iiint z\, dx\, dy\, dz; \qquad M_{yz} = \iiint x\, dx\, dy\, dz;$$

$$M_{xz} = \iiint y\, dx\, dy\, dz; \qquad V\bar{x} = M_{yz};$$

$$V\bar{y} = M_{xz}; \qquad V\bar{z} = M_{xy}$$

18.7.1. Find the centroid of the volume bounded by the plane $z = 4$ and the paraboloid $x^2 + 4y^2 = z$.

Solution. By symmetry $\bar{x} = \bar{y} = 0$.

$$M_{xy} = \int_{-2}^{2}\int_{-\frac{1}{2}\sqrt{4-x^2}}^{\frac{1}{2}\sqrt{4-x^2}}\int_{x^2+4y^2}^{4} z\, dz\, dy\, dx$$

The limits of the second and first integrations are obtained by considering the trace $z = 4$, $x^2 + 4y^2 = 4$, which is the intersection of the paraboloid and the plane.

$$M_{xy} = \frac{1}{2}\int_{-2}^{2}\int_{-\frac{1}{2}\sqrt{4-x^2}}^{\frac{1}{2}\sqrt{4-x^2}} z^2\Big|_{x^2+4y^2}^{4}\, dy\, dx$$

$$= \frac{1}{2}\int_{-2}^{2}\int_{-\frac{1}{2}\sqrt{4-x^2}}^{\frac{1}{2}\sqrt{4-x^2}} (16 - x^4 - 8x^2y^2 - 16y^4)\, dy\, dx$$

$$= \frac{1}{2}\int_{-2}^{2}\left(16y - x^4 y - \frac{8}{3}x^2 y^3 - \frac{16}{5}y^5\right)\Big|_{-\frac{1}{2}\sqrt{4-x^2}}^{\frac{1}{2}\sqrt{4-x^2}} dx$$

$$= \frac{1}{2}\int_{-2}^{2}\left[16\sqrt{4 - x^2} - x^4\sqrt{4 - x^2}\right.$$

$$-\frac{2}{3}(4x^2 - x^4)\sqrt{4-x^2} - \frac{1}{5}(16 - 8x^2 + x^4)\sqrt{4-x^2}\Big]dx$$

$$= \frac{1}{2}\left\{\frac{64}{5}\left(\frac{x}{2}\sqrt{4-x^2} + 2\arcsin\frac{x}{2}\right)\right.$$

$$-\frac{16}{15}\left[-\frac{x}{4}(4-x^2)^{3/2} + \frac{1}{2}x\sqrt{4-x^2} + \arcsin\frac{x}{2}\right]$$

$$-\frac{8}{15}\left[-\frac{x^3}{6}(4-x^2)^{3/2}\right.$$

$$\left.-\frac{x}{2}(4-x^2)^{3/2} + x\sqrt{4-x^2} + 4\arcsin\frac{x}{2}\right]\Big\}\Big|_{-2}^{2}$$

$$= \frac{1}{2}\left(\frac{128}{5}\pi - \frac{32}{15}\pi - \frac{32}{15}\pi\right) = \frac{160}{15}\pi = \frac{32}{3}\pi$$

$$V = \int_{-2}^{2}\int_{-\frac{1}{2}\sqrt{4-x^2}}^{\frac{1}{2}\sqrt{4-x^2}}\int_{x^2+4y^2}^{4} dz\,dy\,dx$$

$$= \int_{-2}^{2}\int_{-\frac{1}{2}\sqrt{4-x^2}}^{\frac{1}{2}\sqrt{4-x^2}}(4 - x^2 - 4y^2)\,dy\,dx$$

$$= \int_{-2}^{2}\left(4y - x^2 y - \frac{4}{3}y^3\right)\Big|_{-\frac{1}{2}\sqrt{4-x^2}}^{\frac{1}{2}\sqrt{4-x^2}}dx$$

$$= \int_{-2}^{2}\left[4\sqrt{4-x^2} - x^2\sqrt{4-x^2} - \frac{1}{3}(4-x^2)^{3/2}\right]dx$$

$$= \Big[2x\sqrt{4-x^2} + 8\arcsin\frac{x}{2} + \frac{x}{4}(4-x^2)^{3/2} - \frac{x}{2}\sqrt{4-x^2}$$

$$-2\arcsin\frac{x}{2}$$

$$+ \frac{x}{24}(2x^2 - 20)\sqrt{4-x^2} - 2\arcsin\frac{x}{2}\Big]\Big|_{-2}^{2} = 4\pi$$

$$\bar{z} = \frac{8}{3}.$$

18.7.2. Find the centroid of the volume bounded by the surface $r^2 = z$, the cylinder $r = \cos\theta$, and the plane $z = 0$. The moment of the element $r\,\Delta z\,\Delta r\,\Delta\theta$ with respect to the xy plane is $zr\,\Delta z\,\Delta r\,\Delta\theta$, and with respect to the yz plane $xr\,\Delta z\,\Delta r\,\Delta\theta$, or

$r^2 \cos \theta \, \Delta z \, \Delta r \, \Delta \theta$. Now z varies from 0 to r^2; r varies from 0 to $\cos \theta$; and θ from 0 to π.

18.7.2.

$$V = \int_0^\pi \int_0^{\cos\theta} \int_0^{r^2} r \, dz \, dr \, d\theta$$

$$= \int_0^\pi \int_0^{\cos\theta} r^3 \, dr \, d\theta = \frac{1}{4} \int_0^\pi \cos^4 \theta \, d\theta = \frac{3}{32} \pi$$

$$M_{xy} = \int_0^\pi \int_0^{\cos\theta} \int_0^{r^2} zr \, dz \, dr \, d\theta = \frac{1}{2} \int_0^\pi \int_0^{\cos\theta} r^5 \, dr \, d\theta$$

$$= \frac{1}{12} \int_0^\pi \cos^6 \theta \, d\theta = \frac{5\pi}{192}$$

$$M_{yz} = \int_0^\pi \int_0^{\cos\theta} \int_0^{r^2} r^2 \cos \theta \, dz \, dr \, d\theta$$

$$= \int_0^\pi \int_0^{\cos\theta} r^4 \cos \theta \, dr \, d\theta = \frac{1}{5} \int_0^\pi \cos^6 \theta \, d\theta = \frac{\pi}{16}$$

$$\bar{x} = \frac{2}{3}; \quad \bar{y} = 0; \quad \bar{z} = \frac{5}{18}$$

18.7.3. Find the centroid of the volume in the first octant bounded by the surface $z = xy$ and the cylinder $x^2 + y^2 = a^2$.

Solution. z varies from 0 to xy, y from 0 to $\sqrt{a^2 - x^2}$, and x, from 0 to a.

$$V = \int_0^a \int_0^{\sqrt{a^2-x^2}} \int_0^{xy} dz \, dy \, dx = \int_0^a \int_0^{\sqrt{a^2-x^2}} xy \, dy \, dx$$

$$= \frac{1}{2} \int_0^a xy^2 \Big|_0^{\sqrt{a^2-x^2}} dx = \frac{1}{2} \int_0^a x(a^2 - x^2) \, dx$$

$$= -\frac{1}{8} (a^2 - x^2)^2 \Big|_0^a = \frac{1}{8} a^4$$

$$M_{xy} = \int_0^a \int_0^{\sqrt{a^2-x^2}} \int_0^{xy} z\, dz\, dy\, dx = \frac{1}{2} \int_0^a \int_0^{\sqrt{a^2-x^2}} x^2 y^2\, dy\, dx$$

$$= \frac{1}{6} \int_0^a x^2(a^2 - x^2)^{3/2}\, dx = \left\{ \frac{1}{12} a^2 \left[-\frac{x}{4}(a^2 - x^2)^{3/2} \right. \right.$$
$$\left. + \frac{a^2}{8} x\sqrt{a^2 - x^2} + \frac{a^4}{8} \arcsin \frac{x}{a} \right]$$
$$\left. + \frac{x^3}{36}(a^2 - x^2)^{3/2} \right\} \Big|_0^a = \frac{\pi a^6}{192}$$

$$M_{yz} = \int_0^a \int_0^{\sqrt{a^2-x^2}} \int_0^{xy} x\, dz\, dy\, dx = \int_0^a \int_0^{\sqrt{a^2-x^2}} x^2 y\, dy\, dx$$

$$= \frac{1}{2} \int_0^a (a^2 x^2 - x^4)\, dx = \frac{1}{2}\left(\frac{1}{3} a^2 x^3 - \frac{1}{5} x^5 \right)\Big|_0^a = \frac{1}{15} a^5$$

$$M_{xz} = \int_0^a \int_0^{\sqrt{a^2-x^2}} \int_0^{xy} y\, dz\, dy\, dx = \int_0^a \int_0^{\sqrt{a^2-x^2}} xy^2\, dy\, dx$$

$$= \frac{1}{3} \int_0^a x(a^2 - x^2)^{3/2}\, dx = -\frac{1}{15}(a^2 - x^2)^{5/2}\Big|_0^a = \frac{a^5}{15}$$

$$\bar{x} = \bar{y} = \frac{8}{15} a; \qquad \bar{z} = \frac{\pi a^2}{24}$$

We could have foreseen that $z = xy$ is symmetric with respect to the plane $x = y$, and hence that $\bar{x} = \bar{y}$. Since in cylindrical coordinates $x^2 + y^2 = a^2$ is the surface $r = a$, and $z = xy$ is $z = \frac{1}{2} r^2 \sin 2\theta$, this problem could be set up in cylindrical coordinates.

$$M_{xy} = \int_0^{\pi/2} \int_0^a \int_0^{\frac{1}{2} r^2 \sin 2\theta} zr\, dz\, dr\, d\theta = \frac{1}{8} \int_0^{\pi/2} \int_0^a r^5 \sin^2 2\theta\, dr\, d\theta$$

$$= \frac{1}{48} a^6 \int_0^{\pi/2} \sin^2 2\theta\, d\theta = \frac{\pi a^6}{192}$$

In finding moments of inertia, we must remember that the distance of our volume element from the x axis is $\sqrt{y^2 + z^2}$, from the y axis $\sqrt{x^2 + z^2}$, and from the z axis $\sqrt{x^2 + y^2}$. Hence

$$I_x = \iiint (y^2 + z^2)\, dx\, dy\, dz; \qquad I_y = \iiint (x^2 + z^2)\, dx\, dy\, dz;$$

$$I_z = \iiint (x^2 + y^2)\, dx\, dy\, dz$$

18.7.4. Find I_x, I_z for the first octant portion of the solid common to the two cylinders $x^2 + z^2 = a^2$ and $y^2 + z^2 = a^2$.

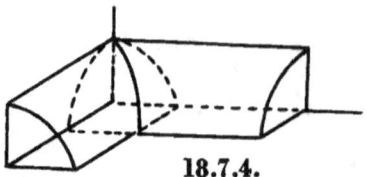

18.7.4.

Solution. We integrate first with respect to y, and y varies from 0 to $\sqrt{a^2 - z^2}$; x also varies from 0 to $\sqrt{a^2 - z^2}$, and z from 0 to a.

$$I_x = \int_0^a \int_0^{\sqrt{a^2-z^2}} \int_0^{\sqrt{a^2-z^2}} (y^2 + z^2)\, dy\, dx\, dz$$

$$= \int_0^a \int_0^{\sqrt{a^2-z^2}} (\tfrac{1}{3}y^3 + z^2 y)\Big|_0^{\sqrt{a^2-z^2}} dx\, dz$$

$$= \int_0^a \int_0^{\sqrt{a^2-z^2}} [\tfrac{1}{3}(a^2 - z^2)^{3/2} + z^2(a^2 - z^2)^{1/2}]\, dx\, dz$$

$$= \int_0^a [\tfrac{1}{3}(a^2 - z^2)^2 + z^2(a^2 - z^2)]\, dz$$

$$= \int_0^a (\tfrac{1}{3}a^4 + \tfrac{1}{3}a^2 z^2 - \tfrac{2}{3}z^4)\, dz = (\tfrac{1}{3}a^4 z + \tfrac{1}{9}a^2 z^3 - \tfrac{2}{15}z^5)\Big|_0^a$$

$$= \tfrac{16}{45}a^5$$

By symmetry, it is obvious $I_y = I_x$.

$$I_z = \int_0^a \int_0^{\sqrt{a^2-z^2}} \int_0^{\sqrt{a^2-z^2}} (x^2 + y^2)\, dy\, dx\, dz$$

$$= \int_0^a \int_0^{\sqrt{a^2-z^2}} (x^2 y + \tfrac{1}{3}y^3)\Big|_0^{\sqrt{a^2-z^2}} dx\, dz$$

$$= \int_0^a \int_0^{\sqrt{a^2-z^2}} (x^2 \sqrt{a^2 - z^2} + \tfrac{1}{3}(a^2 - z^2)^{3/2})\, dx\, dz$$

$$= \int_0^a \tfrac{2}{3}(a^2 - z^2)^2\, dz = \tfrac{2}{3}(a^4 z - \tfrac{2}{3}a^2 z^3 + \tfrac{1}{5}z^5)\Big|_0^a = \tfrac{16}{45}a^5$$

Since $M = V = \frac{2}{3}a^3$,
$$I_z = \tfrac{7}{15}Ma^2; \quad I_s = \tfrac{8}{15}Ma^2$$

18.7.5. Find I_z, I_x, I_y for the solid bounded by the plane $z = 1$ and the paraboloid $x^2 + y^2 = z$.

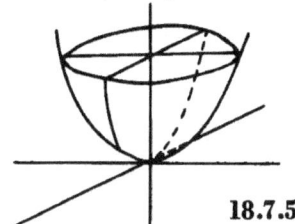

18.7.5.

Solution. It is most convenient here to use cylindrical coordinates, so $z = x^2 + y^2$ becomes $z = r^2$. Our element of volume $r\,\Delta z\,\Delta r\,\Delta\theta$ has a distance r from the z axis and $\sqrt{z^2 + r^2 \sin^2\theta}$ from the x axis.

$$I_x = \int_0^{2\pi}\int_0^1\int_{r^2}^1 (z^2 + r^2 \sin^2\theta)r\,dz\,dr\,d\theta$$

$$= \int_0^{2\pi}\int_0^1 \left(\frac{z^3}{3} + r^2 z \sin^2\theta\right)r\bigg|_{r^2}^1 dr\,d\theta$$

$$= \int_0^{2\pi}\int_0^1 (\tfrac{1}{3}r + r^3 \sin^2\theta - \tfrac{1}{3}r^7 - r^5 \sin^2\theta)\,dr\,d\theta$$

$$= \int_0^{2\pi}\left(\frac{3}{24} + \frac{1}{12}\sin^2\theta\right)d\theta = \frac{\pi}{3}$$

$$I_z = \int_0^{2\pi}\int_0^1\int_{r^2}^1 r^3\,dz\,dr\,d\theta = \int_0^{2\pi}\int_0^1 (1 - r^2)r^3\,dr\,d\theta$$

$$= \int_0^{2\pi}\left(\frac{1}{4}r^4 - \frac{1}{6}r^6\right)\bigg|_0^1 d\theta = \frac{1}{12}\int_0^{2\pi} d\theta = \frac{\pi}{6}$$

18.7.5.

Then $I_y = I_z$ by symmetry.
$$V = \frac{\pi}{2}; \qquad I_z = I_y = \frac{2}{3} M; \qquad I_z = \frac{1}{3} M$$

In spherical coordinates, the element of volume is
$r^2 \sin \phi \, \Delta r \, \Delta \phi \, \Delta \theta; \qquad x = r \sin \phi \cos \theta; \qquad y = r \sin \phi \sin \theta;$
$$z = r \cos \phi$$

18.7.6. Find the centroid of the half of the sphere $x^2 + y^2 + z^2 = a^2$ lying above the xy plane.

Solution. The equation in spherical coordinates is $r = a$; where r varies from 0 to a; ϕ from 0 to $\frac{\pi}{2}$; and θ from 0 to 2π. By symmetry $\bar{x} = \bar{y} = 0$.

$$M_{xy} = \int_0^{2\pi} \int_0^{\pi/2} \int_0^a r \cos \phi \, r^2 \sin \phi \, dr \, d\phi \, d\theta$$

$$= \frac{a^4}{4} \int_0^{2\pi} \int_0^{\pi/2} \cos \phi \sin \phi \, d\phi \, d\theta = \frac{1}{8} a^4 \int_0^{2\pi} \sin^2 \phi \Big|_0^{\pi/2} d\theta$$

$$= \frac{1}{8} a^4 \int_0^{2\pi} d\theta = \frac{a^4}{4} \pi$$

$$\bar{z} = \frac{a^4 \pi}{4} \div \frac{2a^3 \pi}{3} = \frac{3}{8} a$$

18.7.7. Find the moment of inertia of the solid bounded by the upper half of $r = a$ and the cone $\phi = \gamma$ with respect to the z axis.

Solution. Since $x^2 + y^2 = r^2 \sin^2 \phi$,

$$I_z = \int_0^{2\pi} \int_0^{\gamma} \int_0^a r^2 \sin^2 \phi \, r^2 \sin \phi \, dr \, d\phi \, d\theta$$

$$= \frac{a^5}{5} \int_0^{2\pi} \int_0^{\gamma} \sin^3 \phi \, d\phi \, d\theta = \frac{a^5}{5} \left(-\cos \gamma + \frac{1}{3} \cos^3 \gamma + \frac{2}{3} \right) \int_0^{2\pi} d\theta$$

$$= \frac{2\pi a^5}{5} \left(-\cos \gamma + \frac{1}{3} \cos^3 \gamma + \frac{2}{3} \right)$$

18.8. Miscellaneous problems. In the following δ represents a variable density.

18.8.1. Find the mass of a wire bent in the shape of $x^{2/3} + y^{2/3} = a^{2/3}$ if $\delta = k|y|$.

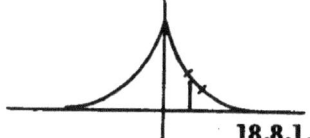

18.8.1.

Solution

$$dM = ky\, ds = ky\sqrt{\frac{x^{2/3}}{y^{2/3}} + 1}\, dy$$

$$M = 4ka^{1/3}\int_0^a y^{2/3}\, dy = \frac{12ka^{1/3}}{5}y^{5/3}\Big|_0^a = \frac{12ka^2}{5}$$

18.8.2. Find the mass of the cardioid $r = a(1 + \cos\theta)$ if $\delta = kr$, that is, if the density is proportional to the distance from the origin.

Solution

$$dM = kr\, dA = kr^2\, dr\, d\theta$$

$$M = k\int_0^{2\pi}\int_0^{a(1+\cos\theta)} r^2\, dr\, d\theta = \frac{k}{3}\int_0^{2\pi} a^3(1 + \cos\theta)^3\, d\theta$$

$$= \frac{5}{3}\pi ka^3$$

18.8.3. Find the centroid of the area bounded by $y^2 = x$ and the line $x = 1$ if the density is proportional to the square of the distance from the origin, $\delta = k(x^2 + y^2)$.

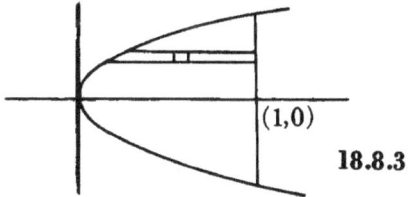

(1,0)

18.8.3.

Solution. For an element of area $\Delta x\, \Delta y$ the moment with respect to the x axis is $k(x^2 + y^2)y\, \Delta x\, \Delta y$, and with respect to the y axis $k(x^2 + y^2)x\, \Delta x\, \Delta y$; x varies from y^2 to 1, and y from -1 to 1.

$$= 2k\left(-\frac{1}{3}\cos^3\theta + \frac{1}{5}\cos^5\theta\right)\Big|_0^{\pi/2} = \frac{4k}{15}$$

$$M_y = k\int_0^{\pi/2}\int_0^{\sin 2\theta} r\cos\theta\, dr\, d\theta = \frac{k}{2}\int_0^{\pi/2}\sin^2 2\theta\cos\theta\, d\theta$$

$$= 2k\int_0^{\pi/2}\cos^3\theta\sin^2\theta\, d\theta = 2k\left(\frac{1}{3}\sin^3\theta - \frac{1}{5}\sin^5\theta\right)\Big|_0^{\pi/2} = \frac{4k}{15}$$

$$M = k\int_0^{\pi/2}\int_0^{\sin 2\theta} dr\, d\theta = k\int_0^{\pi/2}\sin 2\theta\, d\theta$$

$$= -\frac{k}{2}\cos 2\theta\Big|_0^{\pi/2} = k$$

$$\bar{x} = \bar{y} = \frac{4}{15}$$

18.8.5. Find the centroid of the hemisphere of $x^2 + y^2 + z^2 = a^2$ above the xy plane if the density is proportional to the distance above the xy plane.

Solution. Quite evidently $dM = kzr\, dr\, dz\, d\theta$ in cylindrical coordinates. By symmetry $\bar{x} = \bar{y} = 0$.

$$M_{xy} = k\int_0^{2\pi}\int_0^a\int_0^{\sqrt{a^2-z^2}} z^2 r\, dr\, dz\, d\theta = \frac{k}{2}\int_0^{2\pi}\int_0^a z^2(a^2 - z^2)\, dz\, d\theta$$

$$= \frac{k}{2}\int_0^{2\pi}\frac{2}{15}a^5\, d\theta = \frac{2\pi k a^5}{15}$$

$$M = k\int_0^{2\pi}\int_0^a\int_0^{\sqrt{a^2-z^2}} zr\, dr\, dz\, d\theta$$

$$= \frac{k}{2}\int_0^{2\pi}\int_0^a z(a^2 - z^2)\, dz\, d\theta = \frac{k}{2}\int_0^{2\pi}\frac{a^4}{4}\, d\theta$$

$$= \frac{k\pi a^4}{4}; \quad \bar{z} = \frac{8}{15}a$$

If we had chosen to solve the problem by means of spherical coordinates,

$$z = r\cos\phi.$$

$$M = k\int_0^{2\pi}\int_0^{\pi/2}\int_0^a r\cos\phi\, r^2\sin\phi\, dr\, d\phi\, d\theta$$

$$= 2k \left(-\frac{1}{3} \cos^3 \theta + \frac{1}{5} \cos^5 \theta \right) \Big|_0^{\pi/2} = \frac{4k}{15}$$

$$M_y = k \int_0^{\pi/2} \int_0^{\sin 2\theta} r \cos \theta \, dr \, d\theta = \frac{k}{2} \int_0^{\pi/2} \sin^2 2\theta \cos \theta \, d\theta$$

$$= 2k \int_0^{\pi/2} \cos^3 \theta \sin^2 \theta \, d\theta = 2k \left(\frac{1}{3} \sin^3 \theta - \frac{1}{5} \sin^5 \theta \right) \Big|_0^{\pi/2} = \frac{4k}{15}$$

$$M = k \int_0^{\pi/2} \int_0^{\sin 2\theta} dr \, d\theta = k \int_0^{\pi/2} \sin 2\theta \, d\theta$$

$$= -\frac{k}{2} \cos 2\theta \Big|_0^{\pi/2} = k$$

$$\bar{x} = \bar{y} = \frac{4}{15}$$

18.8.5. Find the centroid of the hemisphere of $x^2 + y^2 + z^2 = a^2$ above the xy plane if the density is proportional to the distance above the xy plane.

Solution. Quite evidently $dM = kzr \, dr \, dz \, d\theta$ in cylindrical coordinates. By symmetry $\bar{x} = \bar{y} = 0$.

$$M_{xy} = k \int_0^{2\pi} \int_0^{a} \int_0^{\sqrt{a^2-z^2}} z^2 r \, dr \, dz \, d\theta = \frac{k}{2} \int_0^{2\pi} \int_0^{a} z^2(a^2 - z^2) \, dz \, d\theta$$

$$= \frac{k}{2} \int_0^{2\pi} \frac{2}{15} a^5 \, d\theta = \frac{2\pi k a^5}{15}$$

$$M = k \int_0^{2\pi} \int_0^{a} \int_0^{\sqrt{a^2-z^2}} zr \, dr \, dz \, d\theta$$

$$= \frac{k}{2} \int_0^{2\pi} \int_0^{a} z(a^2 - z^2) \, dz \, d\theta = \frac{k}{2} \int_0^{2\pi} \frac{a^4}{4} \, d\theta$$

$$= \frac{k \pi a^4}{4}; \qquad \bar{z} = \frac{8}{15} a$$

If we had chosen to solve the problem by means of spherical coordinates,

$$z = r \cos \phi.$$

$$M = k \int_0^{2\pi} \int_0^{\pi/2} \int_0^{a} r \cos \phi \, r^2 \sin \phi \, dr \, d\phi \, d\theta$$

$$= \frac{ka^4}{4} \int_0^{2\pi} \int_0^{\pi/2} \cos\phi \sin\phi \, d\phi \, d\theta$$

$$= \frac{ka^4}{8} \int_0^{2\pi} d\theta = \frac{k\pi a^4}{4}$$

$$M_{xy} = k \int_0^{2\pi} \int_0^{\pi/2} \int_0^a r^4 \cos^2\phi \sin\phi \, dr \, d\phi \, d\theta$$

Finally, we shall set the problem up in rectangular coordinates:

$$M = 4k \int_0^a \int_0^{\sqrt{a^2-x^2}} \int_0^{\sqrt{a^2-x^2-y^2}} z \, dz \, dy \, dx;$$

$$M_{xy} = 4k \int_0^a \int_0^{\sqrt{a^2-x^2}} \int_0^{\sqrt{a^2-x^2-y^2}} z^2 \, dz \, dy \, dx$$

Bibiography For Further Reading In Basic Calculus

As I've said several times, the purpose of this series is to revitalize the standard university calculus course to the mathematical level it had been before the American university system decided to monetize it and make it a requirement of far too many students for reasons entirely disjoint from their educational welfare. Calculus should be a revelation for true students seeing it for the first time, regardless of what their future career plans are. Having hopefully stimulated interest in calculus-and mathematics, by extension-it's natural to suggest further reading, particularly for those who wish to study the serious theory of calculus.

This bibliography will focus on sources to supplement our definition of "standard" textbooks in calculus i.e. books that present the main elements of calculus carefully but not completely rigorously, without sophisticated results from analysis. Honors calculus textbooks-of which we have several planned for publication in the near future-*do* give a completely rigorous presentation of calculus. These books will have their own recommended reading lists.

One of the reasons we began this Series was the insane cost of the average calculus textbook today. Hence a guiding principle to this recommendations list will be to suggest inexpensive sources. Fortunately, there are quite a few good ones. Even better, several of these are reprinted classic sources of the same vintage as the books in this series.

The ubiquitous *Schaum's Outlines* series from McGraw Hill are the modern student's preferred shortcut study aids. Many are excellent, but the calculus entries in the series are particularly good. The 2 main entries, (1) and (2) are good enough to function together as an inexpensive calculus textbook. They were both co-written by one of my former professors. Elliott Mendelson, a superior researcher in his day in mathematical logic and a fine teacher. He frequently taught calculus at all levels at my alma mater, CUNY. The first edition of the main book (1) appeared in the 1970's and it hasn't changed in overall style drastically since then. It covers all the major topics of a basic calculus course of one and several variables-algebra and analytic geometry, limits, derivatives, differentials, integration, coordinate systems, vectors, parametric equations, partial derivatives and the differential of several variables, multiple integrals and differential equations-clearly, succinctly and with an enormous number of solved problems in addition to additional exercises for the student to practice. (2) is exactly what the title says it is and supplies about as many solved problems as you need in any given course to act as either examples or as a makeshift solutions manual for your textbook. The books are soft compared to the level most of the books in the OSC series are, there are not many rigorous geometric or analytic problems. But they do provide excellent support for most of the standard subjects in a calculus course. Both are a must have for a basic calculus class.

These days, a wealth of free resources in calculus can be found online. All one needs to do is to go to one's favorite search engine and spend some time looking under "calculus". Many if not all can be found at my website, www.tuloomath.com, where I've collected just about all the major free resources on mathematics available at that time on the internet. In other words, to paraphrase an old detergent commercial, I worked hard so you don't have to. At the basic calculus page, one will find a legion of free lecture notes and textbooks-yes, entire free textbooks in calculus-that are available now. Not only do I have them listed with links (some of which may have to be updated-if the original links are broken and the notes have moved or been deleted, please email me at themathemagician369@gmail.com or bluecollarscholar2018@gmail.com), I've carefully read and reviewed each set of notes at the site so you don't waste your time with them. Of particular interest, in no specific order, for serious calculus students here are the lecture notes of Eleftherios Gkioulekas of the Pan American University, Arthur Mattuck at MIT, Joel Feldman at the University of British Columbia and Ken Kuniyuki of San Diego Mesa College. I'm quite proud of the site-particularly the calculus and analysis pages-and I hope to give the site a much needed updating by the end of 2018. But until then, I hope it remains a major site for math students to find help. The basic calculus page can be found at (3).

(I don't want to begin recommending online material in this bibliography because the purpose of BSC is to make hard copy inexpensive mathematics texts available. It's not that I have anything against online material, don't get me wrong. It's just if we begin listing those sources here, it'll defeat the entire purpose As I said, I have a separate website for that purpose.)

Several quite good and inexpensive calculus textbooks at this level are available in paperback. (4), (5), (6), and (7) are all pitched at about the same level, that is, about the same level as "standard" textbooks in the OSC series-which are a bit more sophisticated than today's books, but still accessible to the average student. They all have very different styles and each has unique characteristics.

(4) is about the same level as Carmicheal, et. al or (1). It was published in 1984 and sadly, it looks it, as the "watering down" of calculus at most American universities was already in full swing. Except for the absence of computer-generated graphs and programs, the book could have been written today. On the plus side, it does have many well-chosen and presented examples and exercises. It's solidly written but pretty run of the mill and pedestrian. Nothing striking about it. Still, with the cost of today's calculus texts, this conceivably could be used as an inexpensive replacement in a standard course.

Morris Kline's (5) is a classic textbook by one of the greatest applied mathematicians and educators of mathematics of the 20th century. It's more sophisticated than the previous texts, about the same level as Phillips. After 25 years of teaching calculus at many different levels to students at New York University, the author wrote this book to serve as a careful calculus textbook that deliberately emphasized applications to the physical sciences without disregarding rigorous proofs of important results. Kline believed that a serious presentation of calculus through its' applications in a detailed and broad presentation is the best way for beginning but serious students to understand it. Only in the last chapter is a brief but lucid discussion of precise proofs of calculus results given. Many applications and geometric explanations are given, but Kline is adamant that these are not proofs, but explanations. A number of these examples in the text are unusual for a first course, such as applications to gravitational and electromagnetic fields. It is masterfully presented and beautifully written by a master teacher and at no time does Kline pretend he's doing a rigorous treatment. It is an applied course in the very best sense-carefully presented arguments based on physics and geometry, while still keeping in mind that a prelude to a rigorous treatment is part of the intent of the course. Indeed, I would recommend all instructors of the subject to read Kline's comments in the prefaces in detail, as they are wonderfully enlightening on teaching calculus to a general audience. The following quote summarizes the method behind his madness:

Rigor has its place in mathematics education. It is a check on the creations and it permits an aesthetic (as well as an anaesthetic) presentation. But it is also to some extent gilt on the lily and an interdiction against the inclusion of functions which rarely occur in practice and which must even be invented with Weierstrassian ingenuity. A rigorous first course in calculus reminds one of the words of Samuel Johnson; "I have found you an argument but I am not obliged to find you an understanding."Even if the rigorous material is understood, its value is limited. As Henri Lebesgue pointed out:"Logic makes us reject certain arguments but it cannot make us believe any argument."

Kline is an absolute jewel and it's a book that should be in every teacher and student of calculus' library at this price. This the book that today's calculus books and courses should take as the model of a general purpose calculus course. A must have.

(6) is another unusual and ambitious book, it combines a relatively careful first course in calculus of one and several variables with a first course in probability and statistics. It's about the same level as (5), although a bit more rigorous. This perspective supplies many unorthodox examples and applications of calculus that aren't usually available to the usual first course in calculus, such as the use of integration in computing expectations and frequency distributions. It's quite well written and clear. An adventurous instructor or student with an interest in probability and statistics, a very popular and lucrative career path today, should be quite intrigued by it.

(7) , by the same author, is a much more standard course in calculus for beginning students that emphasizes geometric intuition, although it does give fairly rigorous definitions, as most books from this period did. It's about the same level as Carmicheal,et. all. With the exception of precision of the definitions, it's a pretty run of the mill calculus text, although the author explains things quite well. Worth a look for the beginner.

No recommendation list for introductory calculus would be complete without the incredible and inexpensive study guide for calculus by Adrian Banner (8). This terrific book gives a detailed, deeply conceptual and visual coverage of all the standard topics and techniques of calculus for beginning students. It can also be a very effective refresher for people who have forgotten calculus. The book is amazingly user friendly and comprehensive. It begins literally from scratch, with basic algebra and geometry and ends with a brief optional discussion of rigorous limits at the end. He presents limits, derivatives, integrals, etc. from all perspectives in a step by step manner and with hundreds of diverse solved examples. Banner writes in a very deliberate, wordy yet warm and conversational style and has a remarkable ability not to miss any steps or observations. This is clearly someone with considerable experience in teaching this subject to students at many levels and backgrounds and it shines through on every page. Indeed, if supplemented with a collection of exercises in calculus, it would make an excellent low priced single variable calculus textbook. This is a book any beginning student of calculus would love their teacher for handing them. My one quibble is that the book only covers single variable calculus. A new edition or sequel that does the same for multivariable calculus would be a godsend for students. An absolute must have for either students or teachers of calculus.

Bibliography

    1) Ayres Jr., Frank, Mendelson, Elliott, *Schaum's Outline of Calculus*, McGraw-Hill Education,6th edition,2012

    2) Mendelson, Elliott, *3000 Solved Problems In Calculus*, McGraw-Hill Education,3rd edition,2013

3) http://www.tuloomath.com/tuloomath-mathematics-site/basic-calculus-calculus-without-theory/ .

4) Gerstling, Judith L, *Technical Calculus with Analytic Geometry* , Dover Books, 1992

5) Kline, Morris, *Calculus: An Intuitive and Physical Approach*, Dover Books, 1998

6) Gemignani, Michael, *Calculus And Statistics*, Dover Books, 2006

7) Gemignani, Michael, *Calculus: A Short Course*, Dover Books, 2004

8) Banner, Adrian, *The Calculus Lifesaver: All the Tools You Need to Excel at Calculus*, Princeton University Press,2007

www.ingramcontent.com/pod-product-compliance
Lightning Source LLC
Chambersburg PA
CBHW071622220526
45469CB00002B/439